Feeding
the
Future

Feeding *the* Future

RESTORING THE PLANET AND HEALING OURSELVES

NICOLE NEGOWETTI

Georgetown University Press / Washington, DC

The publisher is not responsible for third-party websites or their content. URL links were active at time of publication.

Cataloging-in-Publication Data is on file with the Library of Congress.

ISBN 978-1-64712-646-9 (hardcover)
ISBN 978-1-64712-647-6 (paperback)
ISBN 978-1-64712-648-3 (ebook)

♾ This paper meets the requirements of ANSI/NISO Z39.48-1992 (Permanence of Paper).

EU GPSR Authorized Representative
LOGOS EUROPE, 9 rue Nicolas Poussin, 17000, LA ROCHELLE, France
Email: Contact@logoseurope.eu

27 26   9 8 7 6 5 4 3 2   First printing
Printed in the United States of America

Cover design by Brad Norr
Interior design by Westchester Publishing Services

For Alden and Nathan

Contents

Introduction

Over the past decade, I have been tracing the undercurrents of a growing crisis—one that entangles our public health, our ecosystems, and the very narratives we've come to trust about progress and wellbeing. Trained as a lawyer and academic, I once believed that more data, more evidence, and more awareness would be enough to catalyze transformation in our food systems. I now see that the roots of these challenges run deeper—into the ways we learn, parent, market, legislate, and relate. My early entry point into food systems advocacy began with a focused inquiry: How do corporate strategies shape what children eat and what parents are led to believe is nourishing? My first academic publication revealed how food marketing preys on trust and familiarity, using emotional appeal and systemic gaps in regulation to influence consumption in ways that often bypass accountability. This was not just about advertising. It was about power, complicity, and the stories we're sold.

My inquiry deepened when I relocated from the urban corridors of the Northeast to the agricultural landscapes of Indiana, where I began teaching at a law school surrounded by corn and soybean fields. These fields, heavily treated with synthetic inputs, were not feeding local communities but fueling a supply chain designed to sustain industrial livestock operations—distant, enclosed, and often invisible to those not looking. Living beside this system, rather than merely studying it, shifted my understanding. It was no longer an abstract policy issue; it was the air, the soil, the neighbors, the daily dissonance. As I connected with colleagues and community members resisting the spread of factory farms and advocating for localized food sovereignty, it became increasingly clear: Food is not a single issue. It is a site where economic logics, environmental degradation, and social dislocation converge and collide. Immersing myself in the full life cycle of food—its production, marketing, distribution, consumption, and eventual discard—confronted me with a reality more sobering than I had anticipated. What emerged was not just an environmental issue but a planetary pattern of extraction, acceleration, and denial. Within scientific and activist circles, it is no secret that we are in ecological overshoot: consuming the Earth's resources faster than they can regenerate, and producing waste—physical, chemical, emotional—at rates the biosphere can no longer metabolize. The numbers tell a story of urgency: Humanity is using at least 1.7 times more than what the planet can replenish. But behind these figures lies a deeper reckoning. Our current economic systems do not just borrow from the future—they do so in ways that conceal the costs, deflect responsibility, and defer collapse. Like a Ponzi scheme built on planetary credit, we are trading short-term growth for long-term viability while refusing to name what this trajectory demands of us, and who it sacrifices.[1]

The impact of human activities on the environment can be understood through the planetary boundaries framework developed by the Stockholm Resilience Center. This framework identifies the interconnected biophysical and biochemical systems and processes that regulate the planet's state within safe ranges to support human life.[2] Society's actions have pushed six of the nine boundaries, including climate change, biodiversity loss, freshwater, shifts in nutrient cycles (nitrogen and phosphorus), novel entities (human-created chemicals, pollutants, and plastics), and land use, beyond safe limits.[3] These boundary transgressions jeopardize the planet's stability and life-support systems. An international group of scientists issued a sobering *State of the Climate Report* in December 2023: "We are afraid of the uncharted territory that we have now entered. Conditions are going to get very distressing and potentially unmanageable for large regions of the world, with the 2.6°C warming expected over the course of the century." This scenario could lead to the collapse of natural and socioeconomic systems, resulting in unbearable heat, frequent extreme weather events, food and water shortages, rising sea levels, more emerging diseases, and heightened social unrest and geopolitical conflict.[4]

Much of this harm stems from industrial agriculture. Our current methods of food production involve poisoning the land and allowing soil, fertilizers, and pesticides to run off into rivers, lakes, and oceans. This process displaces or destroys vital ecosystems, converting habitats to croplands and pastures, and threatens other species with extinction. Concurrently, it also contributes to a decline in food quality, making people less healthy. Global shifts toward diets high in calories, heavily processed foods, and animal products are exacerbating the burden of obesity and diet-related noncommunicable diseases. These diets pose greater risks to morbidity and mortality than unsafe sex and the combined use of alcohol, illicit drugs, and tobacco.[5] The impact of humans on the Earth has never been so significant or devastating. We have dammed about 60 percent of the world's rivers, cleared nearly half of our temperate and tropical forests, and appropriated about half of the planet's livable surface to feed a population that has increased by nearly 200 percent since 1950.[6]

Unsustainable food production and consumption are major contributors to greenhouse gas emissions, driving up atmospheric temperatures and causing widespread havoc. Climate change in turn impacts agriculture by causing severe droughts, more frequent and intense storms, and reducing the nutrient content of crops due to rising atmospheric carbon dioxide levels. The interconnected relationships among the food system, climate change, and biodiversity loss exert immense pressure on our planet. The stark reality is this: How we grow food, what we eat, and how we discard it pose an existential threat to us and most other species on Earth. As the poet Gary Snyder reflected, "Food is the field in which we daily explore our harming of the world."[7]

The world is significantly off track to meet the Paris Agreement climate targets,[8] commitments to protect biodiversity,[9] and the Sustainable Development Goals,[10] which include ending hunger, food insecurity, and malnutrition in all its forms and realizing the universal right to adequate food for all by 2030. Instead of progress, we have seen a reversal in reducing hunger. In 2023, approximately 733 million people globally suffered from chronic hunger, an increase of about 152 million since 2019.[11] Furthermore, 2.33 billion people, or 28.9 percent of the global population, were moderately or severely food insecure in 2023, lacking regular access to adequate food.[12] Climate change, conflict, economic instability, and growing inequality will continue to threaten food production, distribution, and food security. It's now indisputable that food systems must be fundamentally transformed if we hope to feed ourselves in the future.[13]

Given the magnitude and visibility of the crisis, one might assume that meaningful course correction would follow: that policymakers would enact frameworks to reward health and sustainability, that toxic agricultural practices would be replaced by regenerative ones, that investors would align capital with ecological limits, that corporations would prioritize nourishment over profit, and that individuals would shift their eating patterns accordingly. This logic—rooted in reason, evidence, and incentives—once seemed sufficient to me too. In alignment with that belief, I developed and taught food law and policy courses; published articles, reports, and policy briefs; and contributed to the growing body of work calling out the structural injustices, health harms, and ecological devastation embedded in our food systems.

But over time, it became painfully clear: Knowledge alone does not shift entrenched systems built on extraction and disavowal. Despite the tireless work of researchers, advocates, and community organizers, the industrial food apparatus has continued to consolidate, deepen harm, and resist transformation. Recognizing the limits of academia and policy discourse, I chose to step more directly into frontline advocacy—aligning myself with efforts to dismantle factory farming and to amplify alternatives rooted in justice, care, and interdependence.

SILVER BULLETS

I eventually stepped away from academia and into a more explicitly activist role, becoming the inaugural policy director at the Good Food Institute (GFI), a global think tank animated by the seductive promise of a win–win future: a food system where pleasure, health, and sustainability would no longer be at odds. Their vision was compelling in its simplicity—what if we could replace animal-based products with "alternative proteins" that tasted the same but came without the suffering, the emissions, and the environmental toll? My colleagues were energized and idealistic, with a fervent belief that markets—where moral appeals had faltered—could carry the torch. If consumers could be offered meat without slaughter, indulgence without harm, they argued, then factory farming could be rendered obsolete through innovation, not confrontation.

I entered the role with the conviction that I could help galvanize a broader coalition—uniting public health, environmental, anti-hunger, food justice, and animal protection movements to shift government policy and funding toward a more just and sustainable food future. The premise was alluring:

dietary and systems change without the weight of sacrifice. Yet, not long into my tenure, I began to sense the gap between vision and reality. My work, I discovered, was largely about advancing policies to facilitate the commercialization of cell-cultured meat—biotechnologically produced flesh grown from animal cells without slaughter, marketed under names like "lab-grown," "synthetic," or "cultivated."

To produce cell-based meat, stem cells are taken from a biopsy of a living animal and placed in a culture media with growth-promoting ingredients, antibiotics, and antifungals. As the cells continue to grow, they are placed on a scaffold to help them form traditional meat components, including muscle and fat. The world's first cell-based burger was consumed in London in 2013, and since then, 174 publicly announced companies across the globe have been rushing to get their cell-based inputs or chicken, beef, pork, and seafood products to market.[14] This wasn't just about reducing harm. It was about creating a new market, and with it, a new frontier of techno-optimistic solutionism.

The cell-based meat solution assumes that more technology and products are needed to provide alternatives to the inhumane, unhealthy, and unsustainable animal products currently available. And they don't plan to stop at simply replacing the meat humans already regularly consume. "We can [grow] panda meat if we want to, . . . Or, . . . mammoth or dinosaur meat, the ultimate paleo diet," as one Harvard scientist developing the technology to do so suggested.[15]

A global race was underway. With more than $3 billion in private investment already funneled into food tech ventures,[16] the pressure to deliver returns—and to deliver them quickly—was immense. At GFI, this translated into a focused push: Accelerate the regulatory approval and market launch of cell-based meat products. In some cases, this meant advocating for streamlined pathways in the United States; in others, it involved encouraging entry into countries with more lenient regulatory frameworks. The mandate was clear: Move fast, minimize "burdens," and position these novel products as the transformative fix our food system desperately needs.

Within GFI's ecosystem, cell-based meat wasn't just a technological innovation—it was cast as the panacea for nearly every food-related crisis of our time: environmental collapse, global hunger, diet-related diseases, antibiotic resistance, animal suffering. In this framing, systemic reforms—such as relocalizing food systems, regulating industrial agriculture, or expanding equitable food access—were viewed as peripheral distractions at best and inefficient detours at worst. The preferred narrative was one of elegant simplicity: With the right investment and innovation, we could have our meat and eat it too.

But beneath the optimism, my questions began to gather weight. Were these products truly safe, and who would determine that—especially amid calls for regulatory "efficiency"? What would happen to the farmers, the land, and the animals already enmeshed in this system? Would traditional meat become a boutique commodity for the privileged, while synthetic meat was funneled into low-income communities through the fast-food industry? Who would control the patents and the profit? Would the cellular agriculture boom reproduce or even deepen techno-colonial patterns—extracting resources and labor from the Global South while centralizing ownership and decision-making in the Global North?

And then there was the eerie futurism at the fringes—scientists and entrepreneurs imagining not just ethical steak, but celebrity flesh delicacies, edible avatars of fame and fantasy. The very absurdity of a Beyoncé rump roast or Tom Cruise ribs might seem like satire, but in biotech circles, it was touted as innovation. Cool. Disruptive. Boundless. And yet, I could not help but feel that something vital was being bypassed, not just in our food systems, but in our relationships, our ethics, and our ability to reckon with what harm tastes like.

When I brought these questions forward within the organization, I was met not with curiosity but with caution—and, at times, outright dismissal. The prevailing belief was unwavering: Cell-based meat represented our best and perhaps only chance to avert ecological catastrophe. To publicly acknowledge uncertainties—whether ethical, ecological, or related to food safety—was seen as counterproductive, even dangerous. I was advised to avoid raising concerns that could be construed as skepticism, lest they jeopardize the narrative of progress being marketed to the public and to policymakers.

The unrelenting optimism among my colleagues, scientists, entrepreneurs, and funders created a climate where critique felt like betrayal. My attempts to surface the potential unintended consequences of this technology—consequences that could reproduce the very harms we sought to address—were perceived as discordant, even unwelcome. I began to feel like the organization's wet blanket, an unwelcome reminder that excitement and innovation alone do not guarantee justice or safety. It was in this tension that a deeper realization crystallized. There is no silver bullet. No single innovation—however promising—can resolve the interwoven crises of our food systems.

A few months after joining the organization, we held our first team retreat, where we mapped out the ambitious plan to scale up cell-based meat production with the goal of eliminating animal agriculture by 2050. The vision was clear: Attract capital investments, solve technical challenges, gain the backing of influencers like Bill Gates and nonprofit organizations, bypass burdensome regulations, and convince consumers to embrace these new products. That night, as I sat alone in my hotel room, a realization hit me. In our pursuit to end factory farming, we were repeating the same reductionist and fragmented logic that had created the system we sought to disrupt. The strategy we were advancing was rooted in the same industrial mentality that had given rise to factory farming, focused narrowly on increasing efficiency and maximizing total food production while ignoring externalities and unintended consequences.

The rationale was straightforward. Factory farming exists because people want cheap meat, so we would give them the same experience—meat that looks and tastes like animal products but doesn't require the slaughter of animals. People stop buying meat from factory farms. Those farms shut down. Animals are saved. Multiple problems are solved. However, this equation overlooks the deeper complexity of human dietary choices: the cultural narratives tied to meat consumption, the politics of masculinity and power, the influence of the meat industry, the pervasiveness of meat marketing, and the near-ubiquity of meat in global food systems. Even if a slaughter-free future could be realized, it would still reflect the same industrial mindset that sought to separate farmers from the land, animals from ecosystems, and meat from bodies. Food tech companies could take ownership of animal DNA, further consolidating power and control over what people eat. In that moment, an Albert Einstein quote came to mind and wouldn't leave me: "We cannot solve our problems with the same level of thinking that created them." The more I thought about it, the more I realized that promoting a solution that could potentially deepen the problems we were trying to fix felt dishonest. So, I left my role at GFI.

Around this time, I expanded my research to explore what is called the polycrisis—"the causal entanglement of crises in multiple global systems in ways that significantly degrade humanity's prospects."[17] In other words, from political polarization to economic instability, the rise of authoritarianism, the cascading impacts of climate disasters, it was clear that we are facing an interconnected web of urgent challenges.[18] And, as I dove deeper into these issues, I began to see how the globalized food system was not just a participant in these crises but a major contributor to their worsening effects.

UNLEARNING TO SOLVE PROBLEMS

Responding meaningfully to the crises in our food system requires more than policy tweaks or technical fixes—it begins with a reckoning: with how we diagnose the problem, what frameworks we trust, and what we're willing to see. Over time, I've come to understand that the dysfunctions in our food system are not isolated anomalies; they are symptoms of deeper systemic logics—economic, political, ecological—that are unraveling across the globe. These are not problems to be "solved" in the traditional sense. They are *wicked* problems: layered, shifting, and entangled, where addressing one facet often reveals new tensions or creates unintended consequences. Even the most well-intentioned attempts to eat ethically, grow food sustainably, or support local economies exist within—and are constrained by—an industrial system that seeps into our air, water, soil, and bodies. No one is outside of it. Climate disruption, biodiversity collapse, deforestation, and toxicity do not observe the boundaries of lifestyle, income, or ideology. We are all downstream.

This book shares what I've learned across roles—as an academic, a policy analyst, and an advocate. But perhaps more importantly, it traces the many un-learnings that became necessary: the shedding of my belief that problems as complex and relational as these could be engineered away. What I now seek is not the perfect answer but a more grounded orientation—one that honors complexity, embraces interdependence, and deepens our capacity to act with humility in the face of uncertainty.

When I first embarked on my career in food policy, I did so with the orientation I had been trained in—as a lawyer. This meant approaching the food system through a lens of compartmentalization and linear causality: breaking down complex issues into discrete, seemingly manageable parts—food insecurity here, obesity there, soil degradation over there. Each problem appeared to have its own policy lever, its own technical fix waiting to be implemented. I believed that with the right interventions, each issue could be addressed in isolation and resolved efficiently. The lawyerly mindset I brought into food policy mirrored—and was mirrored by—the fragmented architecture of our regulatory system itself. In the United States, oversight of food is distributed across more than twenty different federal departments and agencies, each with its own narrowly defined mandate. One branch of the US Department of Agriculture (USDA) handles nutrition, another handles meat safety; the Food and Drug Administration (FDA) oversees food safety more broadly, while the EPA monitors agricultural pollution, and the FTC and FCC regulate food advertising and communications. Within the realm of food safety alone, over thirty different federal laws administered by at least fifteen agencies. Each works in isolation, none with a mandate to consider the whole: the interwoven realities of food insecurity, health, environmental degradation, corporate power, marketing, labor, and culture.

This institutional compartmentalization reflects and reinforces the broader tendency to address food system problems as isolated, technical issues—rather than as symptoms of a deeper, entangled crisis. Despite decades of research, innovation, advocacy, and policy proposals, the core challenges remain. The widespread difficulty in implementing transformative solutions points to something deeper than a lack of knowledge or will. It calls for a different kind of seeing. What are we missing? Over time, I came to recognize that the fragmentation is not just structural but cognitive and cultural. It reflects a way of interpreting the world that separates causes from effects, bodies from ecosystems, policies from relationships. As many thinkers and practitioners in agroecology, systems thinking, ecofeminism, sustainable development, and ecological economics have long argued, the dominant narrative treats food system problems as external to us— technological puzzles to be solved with better tools, rather than symptoms of how we've been taught to relate to land, to life, and to one another.

Yes, the crises are embedded in power structures, policy failures, and inequities that disproportionately impact communities historically excluded from decision-making. But they are also rooted in a deeper layer—our collective consciousness. As physicist David Bohm warned, the root of our social and environmental problems lies in a fragmented mode of perception that divides the whole into parts and then forgets that they were ever connected.[19] This is how we end up measuring food system success by calories produced or emissions reduced, while ignoring biodiversity loss, cultural erosion, or the slow unraveling of soil vitality. This is how we come to care for certain people or species while remaining indifferent to others—how harm becomes normalized. Looking back, I can see how my own early attempts to "fix" food system problems replicated this pattern. By applying narrow, linear solutions to complex, living systems, I was often reinforcing the very dynamics I hoped to change. When we isolate one element—hunger, waste, calories—we risk creating false solutions that obscure root causes and spawn new problems. Consider the recurring promise of "efficiency" in food production: Whether through GMOs, cell-based meat, or vertical farming, the emphasis is on yield—on producing more food with fewer resources—while rarely accounting for what is lost or disrupted along the way. These siloed fixes create a comforting illusion of progress. Repurposing food waste to combat hunger, selling "ugly" produce to reduce landfill use, farming insects for protein, or developing pesticide-resistant seeds may appear novel, even virtuous. But without addressing the systemic drivers—extractivism, inequity, disconnection—they often serve as distractions. They make us feel like we're moving forward while the deeper crisis accelerates beneath the surface.

This way of thinking blinds us to the interrelatedness of life. Take the use of insecticides. If insects damage crops, we kill the insects. But in doing so, we may also eliminate the species that controlled fungi. So we apply fungicides. The fungicides then disrupt the mycelial networks that hold soil together and retain water, so we introduce irrigation. That irrigation depletes the aquifer, so we pipe in water from elsewhere. Each "solution" postpones the consequences of the last, layering harm while disconnecting us ever further from the web of life we are entangled in.

SEEING WHOLES

If linear, reductionist thinking has failed to meaningfully address the crises in our food systems, then what kind of approach is truly needed? This question became a guiding thread in my search for deeper insight and relational repair. It led me beyond conventional academic frameworks and into conversations with mentors and guides across a wide array of disciplines—systems science, agroecology, ecological economics, sustainability transitions, biomimicry, Earth systems science, deep ecology, planetary health, and traditional wisdom practices. What I encountered was not a singular answer but a constellation of practices rooted in a different orientation—one that centers care for living systems rather than control over them. I traveled across the United States, meeting farmers, food producers, advocates, nonprofit leaders, philosophers, ecologists, artists, and policymakers, who were not working simply to optimize food production but to restore health to the land, the soil, the waters, and the broader ecologies we are part of. Their work is a form of reweaving—rebuilding relationships with ecosystems through practices that prioritize reciprocity, biodiversity, and long-term regeneration.

But shifting practices alone is not enough. If we want to move in a different direction, we must also examine the deeper logics and mental maps that brought us here—the stories, assumptions, and paradigms that have shaped how we think about food, nature, and progress itself. This book is, in part, an effort to surface those often-invisible frameworks: the beliefs we have inherited, internalized, and rarely questioned, yet which continue to shape the paths we pursue and the futures we foreclose. As systems theorist Donella Meadows insightfully wrote, the deepest leverage point in any system lies not in policy tweaks or feedback loops but in the paradigms—the worldviews—that underlie them. And even more powerful, she suggested, is the ability to *see* paradigms as paradigms: to hold multiple frameworks simultaneously, to loosen our grip on certainty, and to choose, with discernment, how and when to engage them. This is the kind of orientation I believe we need now—not to impose a new doctrine, but to stretch our imaginations, expand our relational sensibilities, and begin living into a different set of questions. Our mindset, comprising worldviews, beliefs, values, and motivations, shapes how we make sense of the world. It influences not just how we interpret problems, but what we even recognize as a problem, and what possibilities we're able, or unable, to perceive.[20] In *Restoring the Kinship Worldview*, Four Arrows and Darcia Narvaez describe two primary orientations: one rooted in Indigenous traditions that regards nature as intelligent and living, with humans as an integral part of the living web; and another, dominant in modern industrial societies, that frames nature as inert and separate from humanity.[21] The contrast is stark: While the kinship-based worldview has sustained communities and ecosystems for millennia, the mechanistic paradigm driving modernity has brought us to the edge of ecological and social collapse.

Restoring human and planetary health demands more than policy reform or technological innovation. It calls for a fundamental shift in how we relate to ourselves, each other, and the world that holds us.[22] This shift is not just conceptual—it is paradigmatic.[23] It is not just a new strategy—it is a new story, one rooted in regeneration.[24] Regeneration offers a pathway that centers the restoration of life, not its extraction. It is inspired by the relational intelligence of living systems, and it values the well-being of the whole—human and more-than-human. It prioritizes resilience, diversity, and interconnection over control, uniformity, and dominance. Rather than treating challenges as isolated problems to be solved, a regenerative approach recognizes that they are entangled and must be met with both inner and outer transformation: shifts in consciousness, as well as shifts in structure.

To think regeneratively is to let go of the fantasy that stability or predictability will return. It means preparing for an uncertain and rapidly changing future—not by fortifying the status quo but by cultivating the adaptive capacities needed to steward change wisely. This includes nurturing local food systems, decentralized economies, and cultural practices that can thrive amid disruption. Regeneration is not a one-size-fits-all solution; it is a practice of discernment, humility, and responsive engagement, rooted in place, context, and relationship.

A regenerative approach calls for a radically different way of thinking about and practicing agriculture. Rather than treating food production as an end in itself, regeneration centers the health of soils, people, more-than-human beings, and entire ecosystems.[25] In this framework, growing food becomes a meaningful *outcome*—a gift—of a system rooted in reciprocity, respect, and re-localization rather than a singular, extractive goal.[26] Though often treated as a novel idea in contemporary discourse, regeneration is anything but new. It has long-established philosophical and practical roots across Indigenous, Eastern, and Western traditions, and it continues to be lived and practiced by many cultures, especially Indigenous communities, for whom the care of land, water, and life is not a strategy but a sacred responsibility.[27] In the Global North, however, the regenerative worldview was largely abandoned in favor of an industrial model of agriculture based on the belief that humans are separate from, and superior to, nature and tasked with conquering it. Regeneration, by contrast, is based on a profound respect for "the interconnected web of biodiversity," recognizing that plants, animals, microbes, and natural cycles co-create the conditions for clean air and water, fertile soil, biodiversity, and an abundance of food and medicine. This holistic orientation reminds us that health is not confined to human bodies alone. It is a continuum—an ecological symphony—from the earth beneath our feet to the food on our plates to the resilience of future generations.[28]

WRITING A NEW STORY

Transforming our food systems demands a fundamental shift in how we see ourselves in relation to the world. This shift calls us to question deeply held assumptions and open ourselves to new ways of thinking that may feel unfamiliar, even uncomfortable. Central to this reorientation is a willingness to examine the myths and metaphors that have shaped our current approaches to food, agriculture, and human flourishing. Perhaps the most enduring and destructive myth is the belief that humans are separate from, and superior to, the rest of nature. In reality, the wellbeing of our planet is inseparable from the health of our ecosystems and communities, just as the health of our bodies depends on the integrity of our cells, organs, and microbiomes. When we begin to recognize this deep interdependence, the root causes of our food crises become more visible, and we can begin to discern the difference between superficial fixes and truly transformative approaches. This shift also invites us to reimagine the meaning of food—not just as fuel or commodity, but as relationship, as ceremony, as kinship. As regenerative educator Carol Sanford emphasized, "transforming the paradigm from which one is thinking is not just desirable or beneficial, it is absolutely necessary at this moment in history."[29] If we are to move toward a food system that heals rather than harms, we must begin with the stories we tell about who we are, how we live, and what we are willing to re-learn in service of the whole.

Unlike many other books about changing the food system, I have deliberately refrained from prescribing a rigid set of steps, processes, technologies, or policies designed to fix the food system. Such an approach would merely reinforce the linear way of thinking that led to our current challenges and perpetuate the illusion that solutions can be neatly mapped in this un-predictable and evolving landscape. Coming to terms with the uncertainty of humanity's future and the path forward can be overwhelming. Yet, what keeps me grounded, and what I share with you in this book, is "active hope," a concept introduced by eco-philosopher Joanna Macy. It offers a way to fo-cus our intentions and actions amid the challenges of a planetary crisis. "Like tai chi or gardening, [active hope] is something we do rather than *have*. It is a process we can apply to any situation, and it involves three key steps. First, we take a clear view of reality; second, we identify what we hope for in terms of the direction we'd like things to move in or the values we'd like to see ex-pressed; and third, we take steps to move ourselves or our situation in that direction."[30]

Mirroring this approach, *Feeding the Future* is structured in two parts. In Part One, I explore the current state of our food system through the lens of three core beliefs that shape the dominant cultural framework of food and agriculture: the notion of human superiority over nature, the prioritization of yield as the agriculture's primary goal, and the reliance on technology as the ultimate solution to our challenges. These mistaken as-sumptions permeate every aspect of our modern civilization, obscuring reality and legitimizing harmful practices within our unsustainable food system. For example, the obsession with productivity—emphasizing plentiful and cheap food—manifests in farm policies, university research agendas, corporate strategies, and agricultural methods. It continues to drive insatiable fossil-fuel consumption along the supply chain, the con-solidation of farms and food corporations, and environmental degradation. These misguided beliefs foster a false sense of hope, lulling many of us into passivity and preventing a clear-eyed seeing of the situation. They also hinder urgent responses needed to address climate change, biodiversity loss, and escalating health crises affecting humans and all life on the planet.

Once these blind spots are exposed, it becomes strikingly clear that our current solutions are not working. No single expert holds the key to solving the issues, and no significant political force is holistically tackling the com-plexity of our predicament. Instead, it will take all of us, each working on ini-tiatives that are responsive to the unique needs of the places we love and inhabit, to address the challenges of our time.

This realization paves the way for Part Two, which will demonstrate how, despite inaction by most national governments and global institutions, a regenerative paradigm shift is not only possible but is already unfolding and gaining momentum. As awareness grows around humanity's precarious future and the fundamental necessity of feeding ourselves, movements, communities, and visionary changemakers are creating alternative economies, strengthening regional food and farming systems, and breathing life back into degraded landscapes. These efforts exemplify new approaches to farming, business, governance, and daily life, grounded in reciprocity and mutually beneficial relationships with the natural world.

As a species, we have always evolved through shared inquiry—adapting not just through competition but through collaboration, reflection, and reimagining our place in the world.[31] Our survival has never depended solely on knowledge but on our willingness to unlearn, to listen differently, and to shift how we relate to each other and to the living systems we're part of. In this moment of overlapping crises and deep uncertainty, our collective capacity to learn—slowly, humbly, and regeneratively—may be one of our most vital resources. This book is offered in that spirit: not as a roadmap, but as a companion in the ongoing work of remembering, imagining, and co-weaving a world where food nourishes not just bodies but relationships—human and more-than-human alike. It is an invitation to wonder about what could still become possible when we root our transformations in care, complexity, and interdependence.

PART ONE

Our Food Stories and Mental Map

1

The Story of Separation

With every breath and bite of food, we share in a web of life in which the energy of plants and other animals becomes part of us. In simple terms, our existence is sustained by billions of years of photosynthesis, a process through which bacteria gradually released oxygen into the atmosphere, making it hospitable for oxygen-dependent organisms like humans, to evolve. We continue to rely on plants and other organisms to harness sunlight, water, and carbon dioxide, converting them into oxygen and chemical energy in the form of sugars such as glucose. The carbohydrates produced by photosynthesis, composed of carbon, oxygen, and hydrogen, are fundamental to life. Plants use glucose to create complex carbohydrates, such as cellulose, which strengthens their cell walls and serves as the fiber we consume in vegetables such as lettuce and broccoli. Plants that produce potatoes, corn, and rice transform glucose into starch, storing energy for future use. When plants convert glucose into sugars like fructose, their fruits become sweet, further illustrating the intricate biochemical processes that connect all living beings.[1]

This process highlights the deep interconnection and entanglement we share with all living beings on this amazing planet. Nature sustains life by nurturing dynamic, interdependent communities. No organism can exist in isolation. Animals rely on plants for energy through photosynthesis, while plants need the carbon dioxide produced by animals and the nitrogen fixed by bacteria at their roots. Plants, animals, and microorganisms work in harmony to regulate the biosphere, creating and sustaining the conditions necessary for life.[2] However, we have been taught to interpret natural processes, like photosynthesis, through a mechanical lens, reducing the complexity of life to mere functions. This perspective is deeply tied to the rise of capitalism, colonialism, the enclosure of the commons in Europe, and the industrial era—fueled by the scientific and technological revolution—all of which profoundly shaped how we perceive and engage with the natural world. In adopting this mindset, we have distanced ourselves from the wonder of the living processes unfolding around us. Anthropologist and systems theorist Gregory Bateson insightfully observed, "The major problems in the world are the result of the difference between how nature works and the way people think." His words challenge us to reconsider not only our social systems but also the very *ways* in which we think. What assumptions do we hold—often unquestioned—that enable and justify the ongoing destruction of the Earth?

Our modern industrial culture often perceives food—and much of reality—as a collection of isolated components rather than an interconnected web of relationships. This prevailing story of separation has profoundly shaped agricultural practices, food policies, and our food environments, often with destructive consequences. As we begin to recognize the flaws in this reductionist approach, it becomes clear that food is not merely a commodity but a dynamic ecological relationship. The misguided notion of separation has led to harmful practices such as the confinement of animals in factory farms, the degradation of North American prairie grasslands, the collapse of ocean ecosystems due to unsustainable fishing and pollution, and widespread food waste by retailers and consumers.

Capitalism, based on transforming natural goods into tradeable commodities and exploiting human labor to accumulate wealth, has created a specific narrative of what nature is. Fundamentally premised on the separation of humanity and nature, capitalism imposes an anthropocentric view, positioning humans at the apex of the species pyramid and viewing nature as a limitless pool of resources for human use. Human exceptionalism suggests that humans are somehow separate from, distinct from, and rightfully holding dominion over nature. This viewpoint assumes that the sociocultural environment is more significant for humans than the ecological context and that social and technological progress can continue indefinitely, solving all social problems,[3] despite the stark reality of planetary boundaries.[4]

These views can be traced back to the Cartesian–Newtonian approach, which views Earth as an inanimate object to be explored, understood, and mastered. This "ontological reductionism"[5] exalts science as the only truthful and useful way of understanding the world and excludes other sources of knowing such as experiential, spiritual, creative, and intuitive ways. It also ignores dynamic relationships. Despite advances in systems thinking,[6] most scientific approaches today still divide and compartmentalize fields of study. Much like the ancient parable of a group of blind men describing an elephant by touching only one part—such as the trunk or tusk—without grasping the whole, we tend to address crises like the warming climate, hunger, poverty, water scarcity, and pollution in isolation. This fragmented approach obscures the deep interconnections between these challenges, making it difficult to develop truly holistic and effective solutions.

This anthropocentric and reductionist view is evident in the model of monoculture industrial agriculture, which partitions nature into units that can be exploited for economic interests and governed by private property rights. According to the *Cambridge Dictionary*, "nature" encompasses "all the animals, plants, rocks, etc. in the world and all the features, forces, and processes that happen or exist independently of people."[7] However, our bodies themselves are ecosystems, with about ten times more bacterial cells than human cells, the majority of which are in the gut. The ratio of microbial to human genes is even more striking, with over 3 million microbial genes compared to 22,000 human genes.[8] Despite this, we have often ignored biology and written ourselves out of the story, failing to grasp that we are nature too. Therefore, all the ways in which we act upon the natural world implicate us individually and collectively.

Black's Law Dictionary defines "animals" as "all living creatures not human."[9] This definition contributes to a prevailing narrative in modern society and its industrial food system, which sees nature as existing outside of us, with humans being superior and thus needing to control, conquer, and utilize nature at will. In contrast, no known Amerindian languages have a term that approximates a Western idea of nature; instead, they observe "a single sphere of life."[10] From an Indigenous perspective, nature is not valued with respect to the material or economic resources it can provide, but rather all its components, including human and nonhuman beings, are regarded as sharing the same life space.[11] We are continuously entangled in an intimate relationship with the more-than-human world.

Viewing the social and natural world as a system that can be neatly predicted and controlled—whether in food, human society, or other species— not only fails to address our challenges but also deepens them. This mechanistic mindset seduces us with a false sense of certainty, reinforcing the belief that identifiable causes lead to predictable effects. In reality, however, such certainty is rare. Ignoring the complexity of living systems in this way blinds us to the unintended consequences of our so-called progress. For example, the genetic modification of plants and animals, such as the approval of genetically engineered salmon for sale and human consumption in the United States, poses serious risks to wild fish populations. These include predation, competition for food and space, transgenic contamination, and the emergence of novel diseases, all of which jeopardize the long-term sustainability of our food supply.[12]

The illusion of human supremacy and separation obscures a fundamental truth: We are nature, and the harm we inflict on ecosystems inevitably harms us in return. This is what Rachel Carson alerted the public to in her book *Silent Spring*, which detailed how DDT and other pesticides had irrevocably harmed animals and had contaminated the world's food supply.[13] Appearing on a CBS documentary about her book shortly before her death from breast cancer in 1964, Carson remarked, "Man's attitude toward nature is today critically important simply because we have now acquired a fateful power to alter and destroy nature. But man is a part of nature, and his war against nature is inevitably a war against himself. [We are] challenged as [hu]mankind has never been challenged before to prove our maturity and our mastery, not of nature, but of ourselves."[14] We haven't heeded the warning. Since 1970, population of birds, amphibians, mammals, fish, and reptiles have declined by 69 percent.[15] Carson's call to action is echoed in a recent commentary on the state of our food system: "Humanity must discern how it will feed nine-billion-plus people without further destroying the environment. If we destroy our environment, we destroy ourselves."[16]

DISENTANGLEMENT

In many ways the story of separation is enshrined in the policies that shape how food is produced, processed, marketed, researched, sold, consumed, and discarded. State, federal, and international laws and regulations—from international trade agreements that determine the price of food imports and exports to local school district wellness policies that establish the types of foods that can be sold at fundraising events—provide the scaffolding for a global food system that fails to connect humans with the planet. We do so at our peril. While the connections between food, health, climate change, and biodiversity loss are increasingly recognized by scientists, food policy experts, and global food and agricultural organizations such as the United Nations Food and Agriculture Organization (FAO), no country has yet developed a comprehensive food system policy that simultaneously advances human nutrition and well-being while safeguarding the environment. Although many nations have individual policies, such as agriculture plans, dietary guidelines, or climate action strategies, these often operate in isolation. They lack integration into a coherent, unified approach that addresses the full complexity of agriculture and food systems. What's more, as we'll explore further, these policies often work at cross-purposes; for example, agriculture subsidies that support the growing of commodity grains undermine efforts to promote nutritious diets through national dietary guidelines.

In the United States and across the globe, food and agricultural policies fail to recognize the interconnections between human and planetary health. A clear example is the ongoing debate over whether sustainability should be factored into national dietary guidelines. Published every five years by the US Department of Health and Human Services (HHS) and the US Department of Agriculture (USDA), the Dietary Guidelines for Americans (DGA) are significant because they form the basis for nutrition standards in nutrition assistance programs across the federal government, including the Special Supplemental Nutrition Program for Women, Infants, and Children (WIC) and the National School Lunch Program.[17] Federal dietary guidance publications are required by law to be consistent with the Dietary Guidelines. As part of this process the Dietary Guidelines Advisory Committee (DGAC)—a group of nationally recognized experts in the field of nutrition, medicine, and public health—reviews the existing guidelines and additional topics for which new scientific evidence is available, culminating in an advisory report. When the Advisory Committee began its review process in 2013, it recommended that sustainability should be a significant component of the 2015 DGA, citing a growing body of high-quality, low-bias peer-reviewed academic studies supporting its inclusion.[18] The DGAC advisory report submitted in February 2015 emphasized the need for "alignment and consistency in dietary guidance that promotes both health and sustainability," describing it as an essential element of food security for current and future generations.[19]

Secretary of Agriculture Tom Vilsack and Secretary of Health and Human Services Sylvia Burwell rejected the Advisory Committee's recommendation, announcing instead that sustainability, defined as "evaluating the environmental impact of a food source," is beyond the scope of the Dietary Guidelines and that the 2015 DGAs were not the "appropriate vehicle for this important policy conversation about sustainability."[20] To make certain that sustainability would not be included in the DGA, the Senate inserted language into the 2016 appropriations bill prohibiting any funds to be used for release or implementation of the 2016 DGA unless the secretaries ensure that the recommendations are "limited in scope to nutritional and dietary information."[21]

Continuing the flawed practice of approaching human nutrition as separate from ecological health, the HHS and USDA once again refused to consider environmental sustainability in the 2020–2025 and 2025–2030 versions of the DGA.[22] This fragmented, reductionist approach to nutrition ignores the bidirectional relationship between food systems and ecological integrity: While environmental degradation threatens the availability and quality of food and water, the entire food system—from production and processing to transportation, consumption, and waste—also significantly contributes to ecosystem disruption and biodiversity loss.

NUTRITIONISM: FOOD IN FRAGMENTS

Food policy not only separates human wellbeing from the health of ecosystems that sustain food production, but it also perpetuates a longstanding legacy of nutritionism. Coined by Professor Gyorgy Scrinis at the University of Melbourne and popularized by Michael Pollan in his best-selling *In Defense of Food*, nutritionism refers to a reductionist view that defines food primarily in terms of its nutritional composition, treating individual nutrients as the key indicators of healthfulness.[23] Nutritionism heavily influences nutrition science, dietary guidelines, and food marketing, as evidenced by the constant stream of often contradictory health claims that promote specific ingredients or superfoods as keys to wellness, or warn that avoiding certain nutrients like fat or sugar is the path to health. Consider the US government's tortured approach to defining "healthy" food for labeling purposes. In 1994 the US Food and Drug Administration (FDA) established a regulatory definition of "healthy" as a "nutrient content claim." Under this rule, a food could be labeled "healthy" only if it meets specific limits on total fat, saturated fat, cholesterol, and sodium and provided at least 10 percent of the daily value (DV) for one or more nutrients such as vitamin A, vitamin C, calcium, iron, protein, and fiber.[24] In 2022 the FDA initiated a process to update the voluntary "healthy" claim, aiming to align it with more current nutrition science and federal dietary guidance. The update also sought to address absurd applications of the original regulation, which permitted a bowl of sugary breakfast cereal to be labeled "healthy" while a bag or bar of nuts could not due to high, albeit natural and nutritious, fat content. The entire framework of evaluating individual foods in isolation reflects a reductionist mindset, rather than a holistic view of nourishment based on overall dietary patterns. A focus on single nutrients paves the way for multinational food companies to use nutritional positioning, influencing policy and public perception about which food should be considered healthy, to bolster their power and influence.[25] This is especially apparent in our society's glorification of protein, the nutrient imbued with special cultural significance as a representation of prosperity, strength, and masculinity.

"Just count calories to lose weight and stay healthy" is another common recommendation that applies an oversimplified and reductionist approach to understanding foods (and food-like substances) and how our bodies respond to them. Most people have been taught that losing weight is a matter of simple math. Cut calories—specifically 3,500 calories—and you'll lose a pound. But as it turns out, experts are learning that this decades-old strategy is actually misguided. "This idea of 'a calorie in and a calorie out' when it comes to weight loss is not only antiquated, it's just wrong," says Dr. Fatima Cody Stanford, an obesity specialist and assistant professor of medicine and pediatrics at Harvard Medical School.[26] Careful calorie calculations do not always yield consistent results because both food and the human body are incredibly complex. How your body processes calories depends on a number of factors, including the type of foods you eat, your metabolism, and the type of organisms living in your gut. Two people can eat the same number of calories derived from different foods yet have very different weights and health outcomes. Responses to foods and nutrients vary due to genetics and other influences, such as the composition of their gut microbiome—the trillions of microbes inhabiting the digestive system—which plays a crucial role in metabolic health and is closely linked with diet.[27] Additionally, we all have different tastes, preferences, and nutrient needs based on life stage, health conditions, and specific goals, such as weight loss or blood sugar regulation. For example, a growing child has different nutritional needs than a pregnant woman or an older adult. Furthermore, the bioavailability of nutrients changes with age, impacting how the body uses them. For instance, some older adults may struggle to absorb vitamin B_{12} as efficiently as younger individuals or produce vitamin D when exposed to sunlight.[28] Those with food allergies or intolerances must make dietary choices that differ from those without such conditions. Other factors, such as meal timing, the sequence in which foods are consumed, physical activity levels, and sleep quality, all play a role in shaping personal responses to food and influencing long-term health.

Modern debates about what should be eaten often overlook the profound historical, environmental, and socioeconomic influences that shape food choices. Across diverse landscapes with varying fertility and climates, people have adapted their agricultural methods to ensure food security. In doing so, distinct local food cultures evolved, alongside genetic adaptations within human populations. There is no universally optimal diet for all humans.[29] A person's dietary needs and responses are shaped by a combination of ancestral heritage, the genetic traits inherited from those ancestors, their unique gut microbiome, and the complex interplay of these and other factors. As a result, a diet that works well for one person may not be equally beneficial for another.

FOOD-LIKE SUBSTANCES

It's indisputable that the standard American diet, aptly abbreviated as SAD, is detrimental to health, given its low intake of fruits and vegetables and reliance on heavily processed foods. This diet reflects the misguided and hubristic belief that the human body operates independently of the natural world as though we no longer rely on living ecosystems to sustain us. We've heard the warnings about processed foods for decades, but defining and analyzing "processed food" remains complex. After all, humans have been modifying food for hundreds of thousands of years by using a wide variety of techniques—milling, shaking, pounding, grinding, fermentating, salting, smoking, drying, freezing, and canning—to preserve and prevent spoilage. But just over a decade ago, Brazilian scientists identified a striking paradox in their national nutrition surveys.[30] Despite declining consumption of oil and sugar, obesity had surged from rarity to the country's leading public health crisis. The data pointed to a rise in industrially processed food such as biscuits, breads, sausages, baked goods, and breakfast cereals as the key dietary shift contributing to this trend. The team of scientists developed the NOVA classification system and introduced the term "ultra-processed" to differentiate traditional foods, whether whole or minimally processed, from products primarily composed of substances extracted from foods, such as oils, fats, sugar, starch, and proteins. These substances undergo chemical modifications before being assembled into ready-to-eat hyperpalatable food and drink items, enhanced with additives for flavor, color, and texture.[31] Their research provided a framework to test the hypothesis that ultra-processed foods themselves contribute to health problems.

The National Institutes of Health (NIH) conducted the first randomized, controlled trial to investigate the sharp rise in obesity in the United States. Twenty volunteers—ten men and ten women—spent four weeks living in an NIH facility, where all their meals were provided for them. They were randomly assigned to consume either minimally processed foods or ultra-processed foods, both with the same amounts of calories, fats, protein, sugar, salt, carbohydrates, and fiber. Researchers initially expected both groups to gain weight at similar rates, given the identical nutrient composition, but they were mistaken. Study participants, free to eat as much or as little as they wanted, consumed more of the ultra-processed meals despite not rating them as tastier than the minimally processed options. Those on the ultra-processed diet ate an additional 500 calories per day, leading to rapid weight gain. When the same people later switched to the minimally processed diet, they lost weight.[32] This is an important finding because it suggests that ultra-processed foods actively drive people to overeat and gain weight, independent of calorie or nutrient content. It raises the possibility that additives and processing methods, not just fat or sugar, contribute to negative health outcomes.

In the United States, ultra-processed foods account for roughly 57 percent of total calorie intake.[33] A 2018 study examining over 230,000 grocery store products found that 71 percent were ultra-processed.[34] As one team of scientists describes it, we can understand the current diet-related health crises as "[a]n evolutionary mismatch between human biology and the modern food environment."[35] As consumption of ultra-processed foods has increased across nearly all segments of the US population over the past two decades, evidence continues to show that our bodies respond differently to artificial ingredients and additives that bear no resemblance to whole, natural foods. A growing body of robust scientific studies has linked the consumption of ultra-processed foods with poor diet quality, increased cardiovascular risk factors, and adverse health outcomes such as obesity and metabolic syndrome, weight gain, stroke, heart attack, cancer, Type 2 diabetes, high blood pressure, fatty liver disease, inflammatory bowel disease, depression, dementia, and early death.[36]

A defining characteristic of ultra-processed foods is the use of additives that preserve, stabilize, flavor, color, and improve functionality.[37] Research has shown that many of the additives are linked to adverse health outcomes. Artificial and nonnutritive sweeteners, once considered to be metabolically inert, have been linked to an increased risk of obesity, Type 2 diabetes,[38] and cardiovascular disease.[39] Higher intakes of artificial food colors have been associated with hyperactivity in children, while nitrates are carcinogenic and may disrupt thyroid function. Certain emulsifiers have been implicated in insulin resistance and weight gain, and flavor enhancers such as monosodium glutamate (MSG) have been connected to various detrimental health effects. Additionally, numerous food additives have been shown to negatively impact gut microbiome health.[40]

The prevalence of these additives in our food supply highlights a disjointed and inadequate regulatory system that fails to consider the cumulative effects of novel substances on our bodies and ecosystems. The FDA, which is responsible for ensuring the safety of most of the US food supply, is not required to evaluate substances, such as spices and preservatives, that are considered "generally recognized as safe" (GRAS) for their intended use.[41] In fact, the FDA only reviews GRAS determinations that companies voluntarily submit through its notification program,[42] meaning that companies can deem a substance safe without the FDA's approval or oversight. Furthermore, the FDA lacks sufficient authority to acquire data on existing chemicals or reassess their safety for human health. Despite its legal obligation,[43] the agency does not routinely account for the cumulative effects of food additives alongside other chemical exposures, such as pesticides and consumer products, that may interact with the same biological mechanisms.[44] As a result, there is *no* consideration of the potential synergistic effects of all the different additives in our foods.

While most additives directly incorporated into processed foods are listed on labels, approximately 12,000 substances from food-packaging materials, including adhesives, dyes, coatings, paper, paperboard, plastic, pesticides, and drugs given to animals also enter the food supply unnoticed. Many chemicals present in food packaging and containers are endocrine disruptors, synthetic compounds that interfere with hormone function.[45] This is particularly concerning, as hormones regulate essential processes such as development, growth, reproduction, metabolism, immunity, and behavior. The risk of endocrine system disruption is especially high during early life, when organ systems are still developing and highly vulnerable to lasting effects.[46] Endocrine disruptors can also impact subsequent generations because they are transferred from a pregnant woman to her developing fetus via the placenta and to infants through breast milk.[47] For example, phthalates—chemicals found in fast foods, food packaging, clear plastic food wrap, and food manufacturing equipment—are linked to increased risk of adverse developmental and reproductive outcomes.[48] Bisphenol A (BPA), a well-documented endocrine disruptor, has been detected in the urine of 93 percent of the US population.[49] We're exposed to it primarily through our diet as BPA leaches into foods from polycarbamate plastics found in water bottles, food containers,[50] and the liners of canned goods.[51] Public concern over BPA's safety has led to its ban in certain products, such as baby bottles and sippy cups. However, this paved the way for BPA analogs such as bisphenol S (BPS), bisphenol F (BPF), bisphenol AF (BPAF), and bisphenol B (BPB)—which share structural similarities with BPA and likely pose the same, or similar, endocrine disrupting risks. Avoiding these chemicals is akin to playing whack-a-mole. They are everywhere—in our foods, ecosystems, and bodies.

A similar issue exists with per- and polyfluoroalkyl substances (PFAS), often referred to as "forever chemicals" due to their resistance to breakdown and the lack of known elimination methods. These widely used compounds are used in coatings and products designed to repel heat, oil, stains, grease, and water—including cookware, fast food containers, microwave popcorn bags, pizza boxes, and candy wrappers. Every PFAS ever produced continues to circulate and accumulate across the planet, contaminating nearly every corner of the Earth. They have been detected in remote environments such as the atmosphere over isolated regions,[52] Arctic and Antarctic seas,[53] and soils of every continent.[54] These highly mobile chemicals have even been found in rain, Arctic snow, and the tissue of wildlife worldwide. Federal studies have identified PFAS in nearly all people tested in the United States,[55] linking them to cancers, immune dysfunction, reduced fertility, obesity, and a wide range of other health concerns.[56]

While carryout food packaging is one of the most direct ways PFAS are transferred to food, it is far from the only route. These chemicals can contaminate drinking water when they enter public drinking water systems and private wells.[57] PFAS also bioaccumulate in fish, shellfish, livestock, dairy, and game animals exposed through contaminated food, soil, or water.[58] Even produce can become contaminated irrigated with polluted water or PFAS-laden compost or biosolids[59] used as "natural" fertilizer.[60] Their widespread presence in the environment and in our bodies is now considered such a critical issue that in 2021 the Biden administration announced a multi-agency plan to prevent PFAS from being released into the air, drinking water, and food and to remediate the impacts.[61] In 2023 the EPA announced its first-ever national standard to address PFAS in drinking water.[62] However, this applies to just two PFAS—perfluorooctanoic acid (PFOA) and perfluorooctane sulfonate (PFOS)—and only in drinking water. Once again, the response to this public health crisis remains fragmented, failing to consider the cumulative effects of these chemicals while implementing patchwork solutions. The EPA typically evaluates individual substances, but PFAS comprise a class of more than 9,000 different chemicals. The full extent of their infiltration into our bodies and ecosystems remains unknown because conventional testing methods can detect only a few dozen.[63]

THERE IS NO SUCH PLACE AS "AWAY"

The industrial food system heavily relies on vast amounts of plastic packaging to preserve and portion convenience foods, such as single-serving snack packs, bottled beverages, frozen meals, and pre-cut vegetables. We throw all this trash "away"—and yet there is no such place. The materials we create on Earth stay here, and they are all rapidly accumulating. The unfettered growth of global plastic production has profoundly polluted our environment to such as extent that scientists have named this era the Plasticene.[64]

By treating nature as an infinite resource, we manufacture vast amounts of materials and products, most of which cannot be reintegrated into natural cycles. Plastic pollution is a critical global environmental challenge, drawing increasing concern[65] as it infiltrates every aspect of life on Earth, from the insides of our bodies to the vast depths of the oceans.[66] Each year, at least 8 million tons of plastics enter the ocean, an amount equivalent to one garbage truck's contents being dumped every minute. If no action is taken, this rate is projected to double by 2030 and quadruple by 2050. We are on a trajectory of filling our oceans with more plastic than fish (by weight).[67] Microplastic contamination has far-reaching consequences. Research shows that zooplankton, a fundamental link in the aquatic food web, consume microplastics, introducing harmful contaminants into marine ecosystems. Moreover, zooplankton play a crucial role in the biological carbon pump, which helps regulate atmospheric carbon dioxide levels.[68] When microplastics replace food in their diets, their ability to consume and export carbon diminishes, potentially accelerating ocean oxygen depletion.[69] This deoxygenation threatens marine biodiversity and contributes to the expansion of algal blooms, further destabilizing oceanic health.[70]

Many environmentally conscious people believe recycling is the solution. However, the harsh reality is that 95 percent of plastic waste cannot be profitably recycled. In 2019, only 9 percent of global plastic waste was recycled,[71] and in the United States, that figure was an alarming 5–6 percent in 2021.[72] The rest ends up in landfills, incinerators, or polluting our waterways—streams, rivers, and oceans alike.[73] As plastic waste degrades, it forms microplastics—tiny particles less than 5 mm in size—that have now become ubiquitous across marine, freshwater, atmospheric, and soil ecosystems. These microscopic pollutants can infiltrate the human body, entering the digestive, respiratory, and circulatory systems. The contamination we have unleashed upon the environment is circulating through our own bodies, serving as an undeniable reminder that the damage we inflict upon the Earth ultimately affects us all.

LIFE KILLERS

The story of separation has also paved the way for extensive ongoing genetic modification of plants and animals for human consumption. Estimates suggest that upward of 75 percent of processed foods in supermarkets contain genetically engineered (GE) ingredients derived from soy, canola, and corn.[74] While farmers have historically selected and bred plants and animals to enhance nutrition, pest resistance, and resilience, major chemical companies such as Monsanto, Dupont, and Bayer have taken genetic engineering much further. By extracting genetic material from one organism and inserting it into the permanent genetic code of another, they have produced potatoes with bacterial genes, "super" pigs carrying human growth genes, fish with cattle growth genes, corn with bacterial genes, and countless other genetically modified species across the plant, animal, and insect kingdoms.

The effects of GE food have been the subject of heated debates for decades. While the National Academy of Sciences[75] and the World Health Organization[76] assert that no evidence links GE foods to human health risks, advocacy organizations such as the Center for Food Safety have raised concerns about potential toxicity, immuno-suppression, allergic reactions, and even cancer associated with genetic modifications in food. What remains undisputed is that GM crops such as soy, corn, and sugar beets are specifically engineered to survive herbicide applications, particularly glyphosate, the key ingredient in Roundup. First introduced by Monsanto in 1974, glyphosate has since become the world's most commonly and intensively used herbicide,[77] applied not only on GE crops but also to home lawns and commercial landscapes. Lesser known is the fact that glyphosate is frequently sprayed on a variety of non-GE crops, including oats, lentils, beans, wheat, chickpeas, potatoes, peas, and millet, just before harvest to act as a desiccant, accelerating drying and ripening. Independent investigations, including lab testing conducted by the Environmental Working Group, have detected traces of glyphosate in popular food products often perceived as healthy, such as oatmeal, breakfast cereals, and hummus.[78] Given that over 200 million pounds of glyphosate are sprayed annually on crops grown in the United States, exposure through food, air, and water is virtually inevitable.[79] In 2022, soybean cultivation spanned 87 million acres nationwide,[80] with 96 percent of those crops genetically modified to tolerate glyphosate and other potent herbicides.[81] Alarmingly, studies indicate that 80 percent of Americans have detectable traces of glyphosate in their urine, raising further questions about its long-term effects on human health and the environment.[82]

The widespread adoption and defense of glyphosate—and the consequences of its prevalence—highlight a fundamental misunderstanding of
human health and its deep connection to the larger ecosystem. The chemical itself is patented as a biocide, a term derived from Greek *bio* and Latin
cide, meaning "life-killer," underscoring its destructive nature. This reflects
not only a fragmented way of thinking but also a lack of awareness about intricate relationships between soils, our microbiomes, and human wellbeing.
Monsanto once promoted glyphosate as a harmless agent, claiming it would
only affect plants because it targets an enzyme absent in humans and other
animals.[83] A commercial from the 1990s even reassured consumers that
"Roundup can be used where kids and pets will play, and breaks down into
natural materials."[84] However, scientific findings later revealed that this enzyme is present in a vast array of microbes, including those essential for
maintaining health in animals, including people.[85] The unintended consequences of disrupting these microbial communities expose the deeper risks
of introducing toxic substances into ecosystems without fully understanding
their long-term impact.

In recent years the potential carcinogenic effects of glyphosate have been
the subject of extensive review and debate among scientific and regulatory
bodies. In 2015 the International Agency for Research on Cancer (IARC), a
division of the World Health Organization, classified glyphosate as a "prob-
able human carcinogen."[86] This determination was based on an analysis of
nearly 1,000 peer-reviewed studies assessing the chemical's carcinogenicity.
Despite these findings, the EPA has dismissed many of these studies and
has repeatedly asserted that glyphosate is "not likely to be carcinogenic to
humans"[87] under typical nonoccupational exposure conditions. However,
since 2016, more than eighty published papers have presented "clear and
compelling evidence" that glyphosate and glyphosate-based herbicide products can damage DNA, potentially leading to cancer.[88]

While past research has primarily examined glyphosate's potential link
to cancer, scientists are now investigating its broader effects—specifically its
impact on the trillions of microorganisms in the human gut that play a crucial role in digestion, metabolism, immune function, and body weight
regulation.[89] Emerging studies from independent researchers suggest that
glyphosate and its metabolites may also interfere with other molecular pathways in animals. These disruptions have been associated with kidney and

liver damage, hormonal imbalances, and deficiencies in essential nutrients,[90] raising further concerns about the chemical's widespread use and its consequences for both human and environmental health.

The effects of glyphosate exposure on children are particularly concerning. Research from the UC Berkeley School of Public Health's Center for Environmental Research and Community Health (CERCH) has linked childhood exposure to glyphosate with liver inflammation and metabolic disorders in early adulthood, increasing the risk of liver cancer, diabetes, and cardiovascular disease later in life. A study of 480 mother–child pairs from the Salinas Valley, California—an agricultural region known as the world's salad bowl—investigated glyphosate exposure during pregnancy and from children at ages five, fourteen, and eighteen. Researchers assessed liver and metabolic health in these individuals at age eighteen, finding that higher levels of glyphosate residue and AMPA (a degradation product of glyphosate) in urine during childhood and adolescence correlated with a greater risk of liver inflammation and metabolic disorders in young adulthood. Additionally the study revealed that agricultural glyphosate use near the children's homes from birth through age five was linked with metabolic disorders by age eighteen. Diet was likely a significant source of exposure, as adolescents who ate more cereal, fruits, vegetables, bread, and carbohydrate-rich foods had higher urinary glyphosate and AMPA concentrations in general.[91] The study results raise serious concerns not just for the children in Salinas but for the broader US population, as glyphosate levels detected in study participants were comparable to those reported nationwide.[92]

Glyphosate has also been associated with another significant health concern affecting many children—autism, which affects one in thirty-six eight-year-olds in the United States.[93] Many children with autism and other brain disorders experience gut-related issues. Stephanie Seneff, a senior researcher at MIT and author of *Toxic Legacy: How the Weedkiller Glyphosate Is Destroying Our Health and the Environment*, has investigated how glyphosate disrupts the gut microbiome, triggering inflammation and a leaky gut barrier.[94] This breakdown allows pathogens and toxic metabolites to circulate throughout the body, contributing to systemic inflammation including inflammation of the brain. Additionally, glyphosate inhibits cytochrome P450 enzymes, disrupting the endocrine system and interfering with the activation of vitamin D—an essential hormone. Vitamin D deficiency has been identified as a contributing factor in autism[95] and remains a growing public health concern in the United States.[96]

Before it became an agricultural staple, glyphosate was originally patented in 1964 by Stauffer Chemical as a metal chelator, used to clean or descale commercial boilers and pipes. Its strong chelating properties enable it to bind and remove essential minerals such as manganese, magnesium, zinc, calcium, and cobalt.[97] This can reduce the uptake of nutrients by plants and subsequently in the people and other animals who consume them.[98] Glyphosate can also indirectly compromise plant health by stunting growth, weakening resistance to disease, and altering soil microflora, which affects nutrient flow and availability.[99] Furthermore, the relationship between pesticides and plant secondary compounds is complex. In the absence of pest pressure, plants no longer activate their own defense mechanisms, which naturally produce compounds beneficial to human health. This diminished biochemical response raises broader concerns about the unintended consequences of pesticide dependence in modern agriculture.

Glyphosate also poses a significant threat to bees. Like humans, bees rely on specialized gut microbiota that support growth and provide protection against pathogens. While most bee gut bacteria contain the enzyme targeted by glyphosate, their susceptibility to the chemical varies, affecting their ability to tolerate exposure. When bees come into contact with glyphosate, their gut microbial communities are disrupted, leaving them more vulnerable to infections from opportunistic pathogens. In experiments involving hundreds of bees, only 12 percent of those exposed to glyphosate survived an infection from *Serratia marcescens*—a bacterium commonly found in beehives and bee guts that can spread throughout a bee's body—compared to 47 percent of bees that were not exposed to the chemical.[100] Additionally, research has shown that glyphosate impairs social thermoregulation in buff-tailed bumble bees, a crucial function for colony growth and overall pollinator health.[101] The disruption of these essential biological processes underscores the broader ecological consequences of glyphosate use.

Glyphosate is the most widely used pesticide, but it is just one of many chemicals permeating conventional farming, our bodies, and ecosystems. The same failure to acknowledge the deep connection between human and environmental health in glyphosate's widespread use is also evident in the prevalence of neonicotinoids, also called neonics. Once again, we are faced with the reality that the harm we inflict on nature inevitably affects us. Neonics are neurotoxic insecticides that permanently bind to insect nerve cells, over-stimulating and destroying them, causing uncontrollable shaking, paralysis, and eventually death. They target nicotinic acetylcholine (nACh) receptors, named for their activation by nicotine. The term "neonicotinoid" itself reflects their nicotine-like properties.[102] While they've been considered safe for humans, nACh receptors are not exclusive to insects; they also exist in human brain cells, playing a key role in learning, memory, mood regulation, sensory processing, and pain perception.[103] This raises particular concerns for children, whose developing brains are more vulnerable. Research has linked prenatal exposure to neonics to birth defects such as heart and brain abnor-malities. About half of the US population is regularly exposed to neonics,[104] with children ages three to five experiencing the highest levels,[105] likely due to consuming conventionally grown, nonorganic foods. Neonics have been as-sociated with neurological issues such as muscle tremors and broader health effects, including lower testosterone levels,[106] disrupted insulin regulation,[107] and altered fat metabolism.[108] These pesticides are persistent contaminants, polluting tap water[109] and resisting removal through standard chlorination treatment. They also infiltrate food supplies. Studies have detected neonic residues in 86 percent of US honey, as well as on widely consumed conven-tionally grown fruits like apples, cherries, and strawberries—favorites among young children. Because neonics are absorbed into the plant itself, they cannot be washed or peeled away, making exposure nearly unavoidable.[110]

The use of pesticides to kill insects not only endangers humans but also jeopardizes the very crops that rely on pollinators—vital contributors to global food production. Staples like almonds, apples, cherries, coffee, blueberries, and watermelons depend upon pollinators for reproduction. In fact, three-quarters of flowering plants and 35 percent of the world's food crops are made possible by pollinators, equating to roughly one out of every three bites we eat.[111] Honey bee populations, among the most crucial pollinators of agricul-tural crops, have declined significantly across Europe and North America.[112] These bees serve as "canaries in the coal mine" for the 4,000-plus species of native bees in the United States, including the rusty-patched bumblebee,[113] which became the first bee in the continental United States to be listed as endangered in 2017. Research reveals that essential fruit crops—apples, blue-berries, and cherries—are now pollinator-limited, meaning that farmers in the United States are experiencing reduced yields due to pollinator decline.[114] Hundreds of studies,[115] including extensive research from Cornell University and the largest industry-funded field study of pesticides, consistently point to neonics as a major factor in global bee and pollinator die-offs.[116] Adding to this crisis, extreme weather and rising temperatures caused by climate change are further straining pollinator populations. Shifting blooming and growing seasons disrupt the delicate balance between plants and the insects they rely on, amplifying the challenges pollinators face in sustaining ecosystems and food production. While we're gaining a better understanding of the damaging impacts of different chemicals used in food production, there is no comprehensive long-term monitoring, evaluation, or regulation of their interactions and cumulative impacts on ecosystems.[117] Agricultural chemical testing is largely focused on acute, short-term exposure and examines substances rather than the complex mixtures found in real-world environments.[118] However, monitoring programs, such as those by the European Food Safety Authority, routinely detect chemical cocktails of more than seven or eight different pesticides in single environmental and food samples.[119] Data from the USDA Pesticide Data Program, which tracks pesticide residues in supermarket produce, further illustrates this issue. Conventionally grown produce often contains multiple pesticide residues—twenty-four different pesticides have been detected on pears, twenty-eight on grapes, and twenty-seven on peaches.[120]

Government policy and dietary guidelines—which, as discussed earlier, shape nutrition programs at schools, for women and children, and at care facilities—also overlook the health effects of additives and other chemicals in food. While fruits and vegetables are a cornerstone of a healthy diet, they are simultaneously one of the primary sources of pesticide exposure. A recent review found that high consumption of fruits and vegetables with low pesticide residues was associated with reduced mortality, whereas this association was absent for those with high pesticide residues. This suggests that pesticide exposure may counteract the health benefits of fruit and vegetable intake.[121]

As explored throughout this chapter, navigating the modern industrial food system involves constant tradeoffs—choosing between consuming more produce while also ingesting pesticide residues, or opting for convenient and flavorful foods that may come with exposure to toxic chemicals. At every decision point, unintended consequences shape our diets and health.

2

The Cult of Productivity

One of the defining traits of reductionist thinking is its fixation on short-term goal achievement—an impulse that has deeply shaped, and distorted, our global food systems. At the heart of this mindset is the story of separation: the belief that humans are detached from, and entitled to dominate, the rest of the living world. Together, these narratives have fueled an approach to food that measures success through a single, narrow metric—*yield*—the quantity of food extracted per acre of land or pulled from bodies of water. This emphasis on yield reflects a deeper belief in endless growth: the assumption that natural resources are infinite, that the planet exists to serve human consumption, and that progress means producing more, faster, and more efficiently, no matter the cost. It is a worldview rooted in productivism, one that sees ecosystems not as relational fields of life but as raw material to be optimized. And it is a system propped up by massive government subsidies and policy frameworks that reward industrial-scale, chemically intensive agriculture over ecological health, community resilience, or long-term sustainability.[1]

Over the past several decades, industrial agriculture has become so normalized that it is now referred to as "conventional," a term that masks how recent and disruptive this model actually is in the broader arc of human history. This rebranding has contributed to a kind of collective complacency. Policymakers, business leaders, philanthropic funders, and even well-meaning advocates and eaters often treat industrial food systems as inevitable, believing that alternatives are either impractical or insufficient to "feed the world." But this belief is not a neutral observation—it's part of a narrative, one that has been carefully constructed and widely disseminated. In this chapter, we begin unpacking the "feeding the world" myth—a story that has justified the consolidation of power in the hands of agri-food monopolies, global institutions, land grant universities, and philanthropic entities that shape the technologies, narratives, and regulations of our current food regime.[2] At the heart of this story is the assumption that the United States and other industrialized nations are generously providing for global humanity. In reality, this framing obscures the extractive dynamics at play: the displacement of small-scale farmers, the degradation of ecosystems, and the global export of highly processed, nutritionally poor foods under the guise of humanitarian aid and progress. It's also worth remembering that the widespread use of synthetic fertilizers, chemical pesticides, and monocultural farming practices is not ancient tradition; it is a post–World War II development. In the grand timeline of human agriculture, these methods are remarkably recent. What we now call "conventional" would have been unrecognizable to most of our ancestors. Recognizing this helps open space for reimagining what is possible—beyond the limits of the industrial imagination.[3]

One of the most critical turning points for conventional agriculture was the so-called Green Revolution, a sweeping initiative that combined modern crop breeding with the expanded use of chemical fertilizers, pesticides, and irrigation to dramatically boost crop yields. Framed as a humanitarian effort to prevent famine and promote development, the Green Revolution began in the mid-twentieth century as a global research and intervention strategy, heavily funded by the Rockefeller and Ford Foundations and backed by the US government. The term "Green Revolution" was strategically coined to contrast with the "Red Revolution" of communism, reflecting Cold War anxieties about political unrest in in Asia, Africa, and Latin America.[4] The project's agricultural ambitions were deeply entangled with geopolitical agendas. In the 1940s, American agronomist Norman Borlaug led a major research effort at the International Maize and Wheat Improvement Centre (CIMMYT) in Mexico, developing shorter, sturdier varieties of wheat that could produce larger heads of grain. Building on rice-breeding techniques developed in China, Japan, and Taiwan, the International Rice Research Institute (IRRI) in the Philippines introduced "dwarf" rice varieties that yielded more grains and were better suited to chemical inputs and irrigation. By 1956, the program moved to India and was later credited with averting famine, improving food availability, and freeing peasants from what was framed as the "drudgery" of traditional farming.[5] Borlaug, celebrated for his innovations, was awarded the Nobel Peace Prize for his contributions to global food production. In the years that followed, high-yielding varieties of other staple crops—sorghum, millet, maize (corn), cassava, and beans—were developed and promoted across the Global South. Over the past decade, however, historians, agroecologists, and social scientists have increasingly questioned the dominant narrative surrounding the Green Revolution. Was it truly a response to urgent food shortages—or, as some now argue, "a solution in search of a problem"?[6]

While the Green Revolution did lead to the widespread adoption of high-yielding crops that may have alleviated short-term human hunger, its longer-term consequences have been profound and far-reaching—undermining the ecological and social foundations upon which future food security depends. The grain varieties developed by Borlaug and others were not standalone innovations; they required an entire infrastructure of support, including chemical fertilizers, synthetic pesticides, expansive irrigation systems, and fossil-fuel-powered machinery like tractors and harvesters.[7] In many ways, industrial agriculture inherited and repurposed the war machine. Chemicals originally developed for use in World War II were rebranded as tools for a new battle—this time, against nature. Explosives like nitrate, ammonium, and phosphorus were turned into synthetic fertilizers, while chemical weapons became pesticides. Companies like DuPont, Dow, and Monsanto led this transformation, promoting an aggressive model of productivity that displaced traditional practices such as crop rotations, intercropping, and other regenerative soil and water stewardship techniques. As Borlaug himself acknowledged during his Nobel Prize acceptance speech, "chemical fertilizer is the fuel that has powered the [Green Revolution's] forward thrust."[8] This fuel-intensive system became the dominant agricultural paradigm across much of the Global North, including the United Kingdom, North America, and Australia, and was rapidly exported to parts of South America, Asia, and Africa.[9] Yet it was a system built on short-sighted logic. The unintended consequences have been severe: widespread soil degradation, erosion, biodiversity loss, hollowed-out rural communities, and a steady decline in the nutritional quality of food. The social costs have mirrored the ecological ones—displacement, consolidation, and growing disconnection between people and the sources of their sustenance.

Though originally framed as a way to support smallholder farmers and combat malnutrition, the Green Revolution's reliance on costly inputs—pesticides, fertilizers, water access—pushed many farmers toward debt, dependency, and eventual displacement.[10] In practice, it incentivized the consolidation of land, making smaller-scale and subsistence farming increasingly untenable.[11] This, in turn, fueled urban migration and accelerated the rupture between communities and the food systems that once nourished them. As high-yield crops took hold, many countries in Asia and Latin America began exporting their agricultural products to the global commodity market. Across the so-called developing world, trade liberalization—the opening up of domestic markets to international competition—was a typical funding condition of the technical assistance, loans, and structural adjustment programs of the World Bank and International Monetary Fund. This has led to specialization of production, in which vast tracts of land were devoted to growing only a few highly productive commodity crops, including wheat, soy, and rice, at the expense of diverse, nutrient-rich food systems.[12] Today, the consequences of this transformation are stark. Of more than 50,000 edible plant species in the world, just three crops—rice, maize, and wheat—make up 60 percent of the world population's energy intake, and they are the staples of diets for over 4 billion people.[13] In total, just fifteen crop plants provide 90 percent of the world's food energy intake, with only a few hundred others cultivated in meaningful quantities.[14] We'll soon explore the implications of this shift toward a global standard diet—and what is lost when diversity, culture, and ecology are sacrificed in the name of uniformity and scale.

SUBSIDIZING PRODUCTION

The industrial food system has not only been shaped by technology and agricultural inputs, it has been sustained and expanded through significant government intervention, particularly in the form of subsidies. In efforts to continually boost yields and shield domestic agriculture from the volatility of global markets, especially under increasingly liberalized trade regimes, governments around the world have poured immense public resources into subsidizing industrial agriculture. Today, more than $850 billion in agricultural subsidies are distributed globally each year, largely drawn from taxpayer and consumer funds.[15]

In the United States, this relationship between government and agriculture has deep historical roots. During the Dust Bowl and Great Depression of the 1930s, federal programs were introduced to stabilize food production and rural livelihoods. Since then, the Farm Bill—reauthorized roughly every five years by Congress—has continued to shape the agricultural landscape through a complex array of financial supports. These supports, while framed as aid for farmers, overwhelmingly reflect a productivist orientation, encouraging commodity crops, machine-managed monocultures, and centralized supply chains. Beneath these programs runs a deeper ideological current: the American myth of limitless abundance and moral obligation to "feed the world."[16] However, much of what is subsidized doesn't directly nourish people. Instead, subsidies are funneled toward crops grown primarily for biofuels, animal feed, and industrial commodities like cotton.[17] Far from incentivizing climate resilience or ecological stewardship, many of these programs actively discourage regenerative practices. A key example is the federal crop insurance program, administered by the US Department of Agriculture's (USDA) Risk Management Agency (RMA), which guarantees farmers a fixed price for commodity crops like corn, soy, and wheat. While farmers pay some of their insurance premiums, taxpayers cover around 62 percent on average. The structure of the program encourages maximum acreage planting regardless of ecological impact. Because farmers are compensated when yields or revenues dip below historical averages, the logic of the system rewards expansion, not adaptation. The result is a kind of "production treadmill": Overproduction drives prices down, forcing farmers to grow even more to make up the difference. In the United States—the world's leading corn producer—roughly 90 million acres of corn are planted each year, concentrated in the Heartland region stretching from the Great Plains through Ohio.[18] Iowa and Illinois alone account for about one-third of the nation's corn crop. Today, despite the ecological costs and market saturation, it remains financially viable to plant more corn. As of this writing, a bushel of shelled corn sells for just $4.74—about eight cents per pound. Yet with government backing, the cycle continues.[19] This is the paradox of industrial agriculture: a system that rewards overproduction, disincentivizes resilience, and externalizes its costs onto ecosystems, communities, and future generations.

The USDA's crop insurance program not only reinforces industrial agriculture—it also disincentivizes the very adaptations needed to respond to climate change. A study by researchers at North Carolina State University found that in counties with crop insurance, corn and soybean crop yield losses from extreme heat were actually higher than in counties without such insurance.[20] This suggests a troubling dynamic: When farmers are assured that their losses will be reimbursed, they are less likely to implement measures that build climate resilience or reduce vulnerability to rising temperatures. As the climate crisis intensifies, bringing more frequent droughts, floods, and heatwaves, the financial burden on taxpayers will only increase unless farming transitions toward regenerative practices, like those explored in Chapter 4. Between 1991 and 2017, rising temperatures triggered $27 billion in USDA crop insurance payouts.[21] Without significant changes, these figures are likely to escalate.

The crop insurance program exemplifies a mechanistic and short-sighted approach to agriculture that views farming primarily as a means to control nature, rather than to work in relationship with it. To qualify for compensation, farmers must demonstrate adherence to a set of regionally defined standards known as Good Farming Practices.[22] Though developed by local experts, these standards historically reflected the norms of industrial, commodity-based agriculture—favoring uniform row cropping, high-input systems, and monoculture. Until a policy revision in December 2023,[23] farmers were sometimes *penalized* for adopting practices widely recognized as ecologically beneficial, such as reducing synthetic fertilizer use, minimizing irrigation, planting cover crops longer than prescribed, using intercropping methods, or deviating from conventional row spacing. Even as some regenerative practices are now permitted within the crop insurance framework, the broader system continues to reward extractive models of agriculture while externalizing their costs—ecological, social, and financial—onto the public and the planet.

Corn farmers in the United States are incentivized to plant more corn not only through crop insurance subsidies but also through energy policy, specifically the Department of Energy's biofuels program and the Renewable Fuel Standard (RFS). Today forty-five percent of US corn grown is used not for food but for ethanol production. Initially, the federal ethanol mandate, which requires a set volume of corn-derived ethanol to be blended into gasoline, was framed as a way to reduce dependence on petroleum and phase out toxic additives like lead. However, this policy has produced a cascade of unintended consequences. The land-use changes necessary to grow bioenergy crops have generated significant emissions, and recent analyses suggest that US ethanol may now have a higher greenhouse gas intensity than gasoline derived from oil.[24] Ironically, rather than accelerating a clean energy transition, the program has made fossil fuels cheaper and more competitive, stalling the momentum for truly renewable energy sources.

At the agricultural level, policies that encourage corn-for-fuel have diverted vast areas of arable land away from food production. The result is a system where only 10 percent of farms in the United States, occupying just 3 percent of cropland, grow fruits and vegetables.[25] As cropland is monopolized by commodity biofuels and animal feed, the availability and affordability of diverse, nutrient-rich foods diminishes. This distortion drives up the cost of food overall, not only for corn-based products but across the board. And because the United States is a major exporter, these effects ripple outward—raising food prices globally. After the implementation of the RFS, the retail price of staples like flour and rice rose by approximately 50 percent. In this way, the ethanol mandate functions as a kind of hidden food tax— one of the most regressive forms of taxation, disproportionately impacting low-income communities and countries dependent on food imports.[26] The impacts are not just economic, but ecological and ethical: A system designed to promote energy independence now undermines both food sovereignty and climate resilience.

Federal agricultural programs that prioritize the overproduction of commodity crops have mostly benefited wealthy landowners, most of whom are white men.[27] This pattern is not incidental; it is rooted in a longer history of dispossession, racialized policy, and systemic exclusion. The hard truth is that the foundation of the US agricultural system rests on stolen land. Laws such as the 1830 Indian Removal Act forcibly removed tens of thousands of Indigenous people from territories they had stewarded for countless generations.[28] The US Army supported the extermination of buffalo—a keystone species and vital food source for many Plains tribes—as a tactic of war and starvation.[29] Following this dispossession, the federal government redistributed over 80 million acres of land through the Homestead Act, offering it to settlers, most of them white, for a nominal fee.[30] This transfer entrenched land ownership as a cornerstone of wealth and power in the emerging agricultural economy. Simultaneously, enslaved Black people generated immense profits for plantation owners and agribusinesses across both the North and South. Even after emancipation, Black Americans faced Jim Crow laws that institutionalized racial subjugation and made access to land, capital, and fair labor nearly impossible. Many formerly enslaved people and their descendants remained trapped in exploitative labor systems like sharecropping or debt peonage—systems that, in many cases, offered little more autonomy or stability than slavery itself.

Those who did manage to acquire land often lost it due to discriminatory lending and support practices by the USDA,[31] as well as acts of racial terror and economic sabotage.[32] Waves of Black families fled rural areas in the South during the Great Migration, escaping violence but also leaving behind land and farming legacies. The impacts of this history are still etched into the structure of today's food system. Low rates of Black land ownership remain a major driver of the racial wealth gap and contribute to persistent disparities in access, influence, and opportunity within US agriculture. According to the most recent US farm census, the number of Black farmers has plummeted—from nearly 1 million in 1910 to 41,807 in 2022.[33] While Black farmers continue to face systemic barriers, large-scale white landowners have historically had the resources to consolidate land, invest in capital-intensive technologies, and weather market shifts.

This trend toward consolidation has been reinforced by policies and market incentives that favor large-scale operations. Since the end of World War II, the number of US farms has dropped dramatically—from nearly 7 million in the 1930s to just 1.9 million farms today,[34] while the average farm size has steadily increased. As smaller farms disappeared, agricultural operations became more dependent on synthetic fertilizers, pesticides, genetically engineered seed, and expensive equipment. Crop and livestock diversity declined, replaced by vast monocultures of genetically modified corn and soybeans. Between 2017 and 2022 alone, the United States lost 20 million acres of farmland, much of it to urban sprawl, infrastructure, and industrial development such as solar arrays. In that same period, the average farm size increased from 441 acres to 463 acres.[35] While large farms—those with sales of $5 million or more—represent less than 1 percent of all farm operations, they now account for 42 percent of all agricultural sales and control two-thirds of the nation's farmland.[36] The concentration of land and power in the hands of a few is not only a threat to economic justice and biodiversity, it also signals a deepening disconnection from the relational practices that once bound land, food, and community together.

Just as government subsidies incentivize the overproduction of commodity crops on land, roughly $22 billion annually in public subsidies are fueling the overexploitation of the oceans. These funds primarily support industrial fishing fleets by artificially lowering the costs of fuel and vessel construction, enabling them to travel, fish longer, and harvest more.[37] Without these subsidies, deep-sea fishing on the high seas—defined as waters beyond 200 nautical miles from any shore—would be unprofitable.[38] This mirrors the extractive logic of land-based industrial agriculture: a productivist paradigm that prioritizes volume and profit over sustainability, equity, or ecological integrity. High seas fishing disproportionately benefits wealthy nations with the capacity to operate far-ranging industrial fleets, while sidelining the food sovereignty of coastal and small-scale fishers in the Global South.[39] Some of the most intensive fishing occurs just off the coasts of Africa, South America, and South Asia—regions where local communities rely heavily on fish as a primary source of nutrition. These extractive practices are, in effect, removing food from the plates of those who can least afford to lose it. According to the FAO, West African fisheries are now severely overexploited, with coastal fish populations declining by 50 percent in the past thirty years. This means that millions of people who depend on fish as a vital protein source face deepening food insecurity—and, for many, the erasure of cultural and economic lifeways rooted in small-scale fishing.[40]

Meanwhile, industrial fishing technologies have been described as among the most ecologically destructive activities on earth.[41] Trawlers scrape the ocean floor through bottom trawling, devastating fragile marine habitats. Enormous quantities of non-target species, known as "bycatch," are discarded, dead or dying. And the fossil fuel consumption required for these operations produces exorbitant carbon emissions, intensifying the climate crisis they also suffer from.[42] Globally, this has reached alarming levels: The FAO estimates that 35.4 percent of fish stocks are now being exploited beyond sustainable thresholds.[43]

As fish populations are steadily depleting, so are their habitats. The ocean is not a series of isolated zones—it is a deeply interconnected system, where high seas and coastal ecosystems influence one another in complex and often invisible ways. Mismanagement of biodiversity in international waters doesn't stay "out there"; it reverberates through coastal regions, undermining the availability of key fish species and disrupting local food webs. Apex predators like sharks and tuna play an essential role in maintaining marine ecosystem balance. Their presence helps regulate species populations, ensuring the stability of entire oceanic food chains. When these predators are removed, especially through industrial overfishing, the consequences cascade across ecosystems, weakening the resilience of the seas and threatening the livelihoods of coastal communities.[44]

Recognizing the urgent need for global coordination, the United Nations adopted the High Seas Treaty in 2023, aimed at protecting marine biodiversity in the half of the planet comprised of international waters. But for the treaty to become legally binding, it must be ratified by at least sixty countries.[45] As of March 2024, eighty-eight nations have signed the High Seas Treaty, signaling their intent to ratify, but only two countries, Chile and Palau, have officially completed the process.[46] The gap between symbolic support and enforceable protection remains wide.

Across both land and sea, the pattern repeats: Industrial extraction, whether through agriculture or fishing, is subsidized by governments and externalized onto ecosystems, communities, and future generations. Meanwhile, those most responsible for the degradation are often insulated from accountability. Until this underlying logic is addressed—where profit is protected and life is expendable—efforts at reform will struggle to meet the scale and urgency of the crisis.

HOMOGENIZATION OF DIETS AND ECOSYSTEMS

As Lebohang Liepollo Pheko, decolonial Afrikan feminist and senior research fellow at the think tank Trade Collective, has pointed out, global markets are underpinned by a colonist logic—one that assumes people in the Global North are entitled to eat whatever they desire, whenever they desire it, with little regard for ecological rhythms or seasonality.[47] The world's wealthiest nations continue to dictate which foods are most profitable, determining global commodity prices and influencing where and how resources are extracted. In this way, consumption patterns in the United States, Europe, and increasingly China do not just reflect preferences—they actively shape how land and ocean ecosystems are used, managed, and often degraded in service of distant appetites. Consider the case of the avocado, a fruit deeply rooted in Mexican culinary traditions and now marketed globally as a "superfood." What was once a regional staple has become a high-value commodity in international markets. Mexico, the world's leading producer and exporter, supplies roughly 80 percent of the avocados consumed in the United States, generating an export industry worth over $3 billion annually. In total, Mexico grows 40 percent of the world's avocados, with 80 percent of its national production concentrated in the state of Michoacán.[48] While domestic consumption of avocado in Mexico has remained steady, reflecting the fruit's cultural and nutritional importance, rising global prices have made it increasingly difficult for local populations to afford. Export markets, particularly the United States, are prioritized by producers, especially those with access to export certifications.[49] In recent years, Mexican avocados have also entered other markets in Europe, Canada, and Asia, with more than $2 billion worth exported outside the United States over the past five years. This surge in international demand is not accidental. It has been actively shaped by aggressive marketing campaigns that frame avocados as both healthy and "sustainable." US consumption has tripled since 2000, driven in part by industry-funded promotions like the Super Bowl ad featuring Eve in the Garden of Eden, accompanied by the slogan: "Avocados make everything better."[50]

But beneath the seductive marketing lies a more troubling reality. The export-driven monoculture model incentivizes widespread ecological harm, particularly deforestation in Michoacán and Jalisco, the two states responsible for all Mexican avocados shipped to the United States. Forests are being cleared to make way for orchards, destabilizing ecosystems and displacing biodiversity. Avocado farming is also extraordinarily water-intensive. To meet growing demand, some producers illegally extract water from rivers, springs, and aquifers, diverting it from communities already facing scarcity. The environmental toll is stark: depleted water sources, increased landslide and flood risks, and growing vulnerability for rural populations.[51]

As global demand for avocados has surged, profits for growers and exporters in Mexico have risen sharply. Financial incentive has fueled both the expansion and intensification of avocado cultivation, transforming diverse landscapes into monocultural orchards with far-reaching environmental, social, and political consequences. In Michoacán, where much of this production is concentrated, native oak–pine forests are being cleared to make way for avocado plantations. This shift disrupts entire ecosystems, leading to biodiversity loss, diminished wildlife habitat, compromised water and soil regulation, and reduced carbon sequestration. One of the most visible casualties of this transformation is the Monarch Butterfly Reserve, an internationally renowned sanctuary in Michoacán that serves as the southern migration site for millions of monarch butterflies each winter.[52] As avocado plantations encroach on forested areas, the delicate overwintering habitat of the monarchs is increasingly at risk, threatening not only a vital ecological phenomenon but also a deep cultural connection for many communities.[53] The economic benefits of avocado production are also unevenly distributed. Profits are concentrated in the hands of a few large and powerful producers, leaving limited gains for local farmers and communities. This disparity often fuels land grabs, exacerbates pollution, and has been linked to the displacement and intimidation of Indigenous peoples and other local residents. In some cases, violence has accompanied the scramble for control over profitable avocado-producing regions.[54] The story of the avocado is a microcosm of a much larger pattern—one where global appetites, driven by wealth and marketed ideals, reshape landscapes and livelihoods far beyond the consumers' view.

THE PARADOX OF PRODUCTIVISM

"Efficiency!" is the rallying cry of industrial food system proponents. Large-scale, highly mechanized farming and globalized trade are heralded as the pinnacle of resource allocation—streamlining production to feed a growing world. The USDA, for instance, celebrates agricultural productivity for making food more abundant and cheaper, even amid population growth.[55] Yet at the heart of this achievement lies a troubling contradiction: As the efficiency of production has increased, the efficiency of the global food system, in terms of delivering nutritious food sustainably, is declining.[56] Modern food systems depend on vast, opaque supply chains that disconnect consumers from the conditions under which their food is grown, processed, shipped, and discarded. From an energy perspective, this system is anything but efficient. It is profoundly fossil-fuel dependent. Across the four stages of the value chain—input, production, processing and packaging, and consumption and waste—80 percent of the energy comes from fossil fuels. The majority of that energy use occurs not on farms but in processing (42 percent) and consumption and waste (38 percent). Just 20 percent of energy use occurs during input and production, with fertilizer manufacturing expected to drive even greater fossil fuel dependency through 2050.[57] This energy-intensive infrastructure has made the global food system a major driver of climate change, responsible for over one-third of global greenhouse gas emissions.[58]

Yet the dominant narrative, upheld by agribusiness, international development agencies, government bodies, and philanthropic institutions, continues to emphasize the noble goal of "feeding the world." This framing fixates on the delivery of calories, often in the form of cheap, processed foods, while sidestepping essential questions: *How* is the food grown? *Where* and *by whom?* Who has access to land, water, capital, and decision-making power?[59] The reality is stark. In 2020 the global food supply provided an average 2,947 calories per person per day,[60] well above the 2,000-calorie benchmark for basic energy needs.[61] In North America that number jumped to 3,548.5 calories per person.[62] Yet, despite this surplus, between 690 and 783 million people faced hunger in 2022,[63] and an estimated 2.4 billion people—nearly 30 percent of the world's population—did not have access to nutritious, safe, and sufficient food throughout the year.[64] The lingering effects of the Covid pandemic, rising food prices, inflation, and the cost of a healthy diet have further compounded the challenges, especially for low-income populations. Malnutrition persists in multiple forms: 148 million children are stunted, 45 million are wasting, and 37 million are overweight.[65] In the United States alone, 12.8 percent of households—17 million families— faced food insecurity in 2022,[66] uncertain whether they could afford enough food to meet basic needs.

Taken together, these data make it clear that hunger is not the result of a global shortage of food. Attempts to address food insecurity through narrow, reductionist strategies—such as simply producing more—fail to grasp the deeper dynamics at play. The persistence of hunger is not a supply issue; it is a justice issue. At the heart of the industrial "cheap food" model lies a system that relies on, and reproduces, conditions of poverty and inequality. The drive for ever-lower food prices is built on the backs of those who grow, harvest, process, and serve food—many of whom work in precarious, hazardous conditions for inadequate wages. The food system does not merely reflect economic disparities; it actively sustains them.[67]

CORPORATE CONSOLIDATION

Consolidation is not limited to land ownership; it pervades every stage of the industrial food system, from production and processing to distribution and retail. Despite being framed as a strategy to enhance efficiency and reduce costs, the consolidation of corporate power has, paradoxically, made food both less affordable and less accessible.[68] Since the 1980s, a wave of mergers and acquisitions has left just a few multinational firms in control of nearly every step of the food chain,[69] especially in key sectors such as seeds, meatpacking, and food retail.[70] Among the most dominant are Walmart in retail, Nestlé in processing, and Bayer in seeds and chemicals. Across many common grocery categories, a striking concentration of market share exists:[71] Just three companies control 79 percent of the dry pasta market, 93 percent of soda sales, and 73 percent of breakfast cereals in the United States.[72] The illusion of choice on supermarket shelves conceals a deeper uniformity—one shaped by a few powerful entities with outsized influence over what food is produced, how it is grown, who grows it, and who gets to eat it.[73] This consolidation of control has far-reaching consequences not only for consumers, who face higher prices and fewer real options, but also for farmers, farmworkers, and rural communities whose livelihoods depend on agriculture. As corporate power grows, so too does its leverage to dictate the terms of production. In the seed and agrichemical sectors, monopolies have driven up the cost of inputs like seeds, pesticides, and fertilizers.[74] Meanwhile, increased concentration among meatpackers, grain traders, and food retailers has pushed down the prices paid to producers—squeezing margins for farmers and workers while boosting profits for intermediaries.[75]

An investigation by the Biden administration in 2021 highlighted the extent of this imbalance, revealing that four conglomerates overwhelmingly control meat supply chains. Over the past fifty years, companies like Cargill, JBS, Tyson, and Marfrig have steadily absorbed smaller operations, creating near-total control in key markets. Today, the top four beef packers control 82 percent of the market; the top four poultry processors control 54 percent;

and the top four hog processors hold 66 percent.[76] When dominant middle-men control so much of the supply chain, they can extract value from both ends, paying farmers less while charging consumers more.[77] As journalist Christopher Leonard documents in *The Meat Racket: The Secret Takeover of America's Food Business*, this system increasingly mirrors the exploitative dynamics of sharecropping: Farmers bear the risk and labor, while corporations reap the rewards.[78]

Poultry giants like Tyson, Pilgrim's Pride, and Perdue operate through a model of vertical integration, meaning they control every link in the supply chain, from feed mills and hatcheries to processing plants and distribution. In this highly consolidated system, nearly all broiler chickens in the United States (99.5 percent by production value in 2020) are raised under strict production contracts, which set terms for how the chickens are raised, which inputs the farmer and the company provide, and how the farmer is paid.[79] Farmers have little negotiating power with these corporations—known in the industry as "integrators"—because they depend on them not only for the animals but also for access to the processing infrastructure that turns live birds into salable meat. Critically, poultry farmers do not own the chickens they raise. They also do not control the feed, medications, or other essential inputs, all of which are provided by the integrator. While the corporations retain ownership of the birds and control over production standards, they shift most of the financial risk onto the farmers.[80]

Farmers are responsible for building and maintaining the chicken houses, supplying labor, and covering utilities. This requires substantial upfront investment. The average broiler operation consists of four chicken houses valued at approximately $1 million, often financed through loans.[81] Once a farmer takes on this debt, exiting the contract system becomes almost impossible. This dynamic creates deep financial vulnerability. Farmers are left with few options if their contracts are terminated or if terms are changed unilaterally by the integrator. The imbalance of power means farmers often have to comply with new demands without recourse, even when those changes increase their costs or reduce their income.[82]

One of the more insidious tools used by integrators to externalize risk is the so-called "tournament" payment system. In this model, integrators deliver chicks to several farmers at the same time, with the birds destined for the same processing facility. Once processed, the performance of each flock is assessed relative to the others, measuring factors such as growth rate and feed efficiency. Farmers whose flocks perform better than average receive a bonus above the base pay rate; those who perform below average receive a deduction. In 2020, the median payment was 6.79 cents per live weight. But actual fees ranged from 4.29 cents for the bottom 10 percent of growers to 9.64 cents for the top 10 percent.[83] Crucially, many of the variables that influence flock performance, such as the quality and health of the chicks, the consistency and nutritional content of the feed, and the timing of pick-up, are entirely controlled by the integrator.[84] This rigged system creates the illusion of meritocracy while functionally redistributing financial pressure from corporations to farmers. It pits growers against one another in a zero-sum game, where bonuses for a few are funded by penalties for others—a clear example of how power consolidates profit by offloading risk.

As consolidation in the food system accelerates, most farmers find themselves with few, if any, options when it comes to selling their products. With limited buyers and virtually no negotiating power, the share of every food dollar that goes to farmers has steadily declined. In the 1980s, farmers received 37 cents for every dollar; today, they receive less than 15 cents.[85] This economic squeeze leaves many with only one path to survival: scale up, produce more, and hope to stay afloat.

Despite claims that consolidation brings efficiency, it has not translated into consistently lower prices for consumers. In fact, in highly concentrated markets, food prices tend to rise. From 2020 to 2024, the cost of feeding a family of four grew at 2.5 times the rate of inflation, while corporate food sector profits surged, growing five times faster than inflation from 2020 to 2022, often hitting record highs.[86] In this equation, both farmers and consumers lose, while concentrated corporate power wins.

That same concentration extends beyond economic dynamics into political and informational control. A handful of powerful companies now hold disproportionate influence over what food is produced, how it's produced, and who gets to access it. These firms shape public policy, fund industry-friendly research, and influence media narratives.[87] In 2023, agribusiness spent $177.25 million on lobbying efforts,[88] and during the 2020 election cycle alone, food and agriculture sectors—including meat and dairy processors, sugar, tobacco, crop producers, food manufacturers, and retail chains—spent nearly $200 million on political contributions, lobbying, and campaign influence.[89]

The corporate dominance has also weakened local and regional food infrastructures, making nations like the United States heavily dependent on long, fragile global supply chains. These chains not only obscure the relationships between producers and consumers, they also create profound vulnerability. A single breakdown in the chain can trigger systemic failure. The reliance on a few "chokepoints," like the Panama Canal or the Strait of Malacca, for international food trade has heightened exposure to disruption.[90] These structural weaknesses were thrown into sharp relief during the Covid-19 pandemic. With restaurants, hotels, and schools shuttered, demand for certain food products evaporated, leaving farmers with no buyers. As a result, tens of millions of pounds of fresh food were destroyed, and millions of gallons of milk were dumped daily, even as food banks struggled to meet skyrocketing demand from newly unemployed families. Meatpacking plants, which were already consolidated and tightly scheduled, shut down as workers fell ill. This led to a loss of up to 45 percent[91] of slaughtering capacity in some sectors. The response was brutal: Millions of animals were culled—"depopulated"—on-site through suffocation, drowning, or shooting, their lives and deaths rendered invisible by the very systems designed to maximize output.[92]

Globally, farmers who had been funneled into monoculture cash crops like coffee, cotton, or export wheat faced similar devastation. When borders closed and markets collapsed, they were left with no way to sell their crops, many of which rotted in the fields, while food shortages rippled across the world, hitting the most economically vulnerable communities hardest.[93] This is not just a supply chain issue; it is a reflection of a food system designed around efficiency for the few at the expense of resilience, justice, and care for the many.

FOOD "WASTE"

The Covid pandemic laid bare the fragility of the industrial food system—revealing, with painful clarity, its inability to connect abundant food with human need. Crops were left to rot in fields, milk was dumped in ditches, and meat was "depopulated" by the millions, even as hunger surged. But this was not an anomaly. The staggering wastefulness on display was not a temporary crisis response; it is a built-in feature of the system. Food loss and waste occur at every stage of the global supply chain, but the majority happens at the consumer-facing end: in grocery stores, restaurants, and other food service establishments. Waste begins at the point of production, where low market prices and high harvest costs often make it financially irrational for farmers to harvest all that they grow. Cosmetic standards eliminate fruits and vegetables deemed too "imperfect" for shelves. Labor shortages add further constraints. And while gleaning programs and farm-to-food-bank partnerships recover some of this produce, most of it is tilled back into the soil—untouched, un-eaten. Retailers, meanwhile, have come to see waste not as a flaw but as part of the business model. Overstocked displays, oversized "family packs," shelves of pre-made meals available until closing, rigid adherence to "sell-by" dates unrelated to food safety, seasonal over-purchasing, and the disposal of dam-aged or unpopular goods all reflect a culture where surplus is synonymous with abundance, and loss is absorbed as a cost of doing business.[94]

But the consequences of this systemic waste are staggering. The United Nations Environment Programme (UNEP) estimates that more than one-third of all food produced globally—worth over $1 trillion—is wasted each year. This waste consumes 28 percent of the planet's agricultural land; squan-ders labor, energy, and water; and contributes significantly to climate change.[95] Marine food systems are no exception: The FAO reports that 8 percent of fish caught in global marine fisheries, approximately 78.3 mil-lion tons annually, are discarded, a testament to the same extractive mindset that plagues land-based systems.[96]

When food is wasted, so are the resources embedded in it. Imagine if all the uneaten food in the United States were grown on a single "waste farm." That farm would span 80 million acres—three-quarters the size of California—and require all the water used annually by California and Idaho combined. It would generate enough food to fill a 40-ton truck every 15 sec-onds, only for much of that food to be refrigerated, shelved, and ultimately thrown away. Instead of feeding people, it would be hauled to landfills, where it would decompose and release potent greenhouse gases into the atmosphere.[97]

In the United States, food waste is the most common material in both landfills and incinerators, making up 24 and 22 percent of municipal solid waste, respectively.[98] Its environmental toll is massive: equivalent to green-house gas emissions of more than 42 coal-fired power plants, enough water and energy to supply over 50 million homes, and an agricultural footprint equal to the combined landmass of California and New York. All this for food that never nourishes a single body.[99]

Even the term "waste" reflects the industrialized food system's dysfunction—a system that generates its own crises and then repackages them as solvable. In nature, there is no waste. What one species discards becomes nourishment for another. In circular, regenerative systems, surplus becomes soil, excess becomes energy, and everything has a place in the web of life. But within industrialized models, overproduction and discarding are normalized, even strategic. Many advocates frame food waste as a straightforward, solvable issue. Project Drawdown ranks reducing food waste as one of the most impactful climate solutions.[100] Media outlets like *Vox* have dubbed it the world's "dumbest problem"—not because of its consequences, but because of how allegedly easy it would be to fix.[101]

Meanwhile, the food industry, policymakers, and some charitable organizations have positioned surplus food as a ready-made solution to hunger. From this angle, food that would otherwise be discarded can simply be "rescued" and redirected to feed those in need. But this framing masks deeper questions. Why are we producing so much food that isn't eaten in the first place? And why does hunger persist despite such abundance? As Andy Fisher, author of *Big Hunger: The Unholy Alliance Between Corporate America and Anti-Hunger Groups*,[102] observes: "If the food industry is addicted to overproduction, then the emergency food system is its enabler."[103] Take, for example, the shelves of food pantries overflowing with bakery goods such as breads, pastries, and sugary snacks. These are byproducts of a system where cheap inputs like sugar, flour, and labor make it profitable for retailers to overstock, knowing the excess can be offloaded as a charitable contribution. Efforts to rescue and redistribute surplus food, while well-intentioned, often divert attention from the structural roots of waste: the commodification and devaluation of food, aesthetic standards that exclude "imperfect" produce, and policy frameworks that incentivize overproduction without regard for ecological boundaries or social wellbeing. These conditions are not accidents—they are designed features of a system that externalizes harm while maximizing throughput.

And this waste is not just inefficient; it's dangerous. We are squandering not only food but also the finite resources it takes to produce it: land, water, labor, energy, and biodiversity. Although the Green Revolution dramatically increased yields[104]—global cereal production rose 175 percent between 1961 and 2014—those gains are no longer holding. Wheat yields climbed from 1.1 tons to 3.4 tons per hectare,[105] and livestock productivity improved through intensified breeding and feeding regimens.[106] However, in recent decades, yield growth in major crops such as corn, wheat, and rice has begun to stagnate or even decline in regions from Kansas to northwest Europe to Hokkaido, Japan.[107] A global meta-analysis of yield trends from 1961 to 2008 found that in 24–39 percent of the areas studied, productivity either plateaued or collapsed after initial gains.[108] The industrial system's once-touted efficiency is faltering under its own weight. Environmental degradation, including soil erosion, nutrient depletion, salinization, and biodiversity loss, is stripping ecosystems of the very functions needed to sustain future yields.[109]

Climate change is already reshaping the conditions under which agriculture can function, and its impacts are only deepening. In 2023, Earth's average surface temperature reached the highest level on record, rising 2.1 degrees Fahrenheit (1.2 degrees Celsius) above the baseline set by NASA for the period 1951–1980.[110] From June through December of that year, every single month broke global heat records. At the time of writing, July 2023 was marked as the hottest month ever recorded.[111] By the time this chapter was finalized, that record had been surpassed by July 2024.[112] Over the past thirty years, rising temperatures have extended the average length of the growing season by more than two weeks across nearly every US state except Alabama and Georgia, compared to the early twentieth century.[113] While a longer growing season might appear beneficial at first glance, the broader impacts of elevated temperatures paint a more sobering picture. Heat stress, increased evaporation, and reduced soil moisture all compromise crop health and productivity.[114] In fact, scientific studies consistently show that warming temperatures have already begun to reduce yields for major commodity crops around the world. These rising temperatures also intensify the water burden on agriculture. As soils dry and plants require more water to thrive, the demand for rainfall or irrigation increases—at a time when water availability is becoming increasingly precarious. In the western United States, 20 percent of the region is facing extreme drought conditions, while another 40 percent is grappling with severe drought. Just 5 percent of region is free from any drought designation.[115] Drier conditions not only threaten agricultural viability but also heighten the spread and severity of wildfires by reducing moisture in both soil and vegetation. These patterns are not confined to North America. The climate crisis is global, driving an upsurge in extreme weather events, including droughts, tropical cyclones, floods, and violent storms, that are becoming more frequent, more intense, and less predictable.[116] Over the past fifty years, the number of weather-related disasters has increased fivefold.[117] In the United States alone, by August 2024 there had already been nineteen climate-related disasters exceeding $1 billion in damages. The previous annual record—twenty-eight events—was set just a year earlier, in 2023.[118] Climate change is also disrupting ecological relationships. Longer growing seasons provide conditions for additional generations of pests, pathogens, and invasive species, many of which would have previously been controlled by winter die-offs. These biological shifts create cascading vulnerabilities across food systems, compounding the stressors already posed by heat, drought, and disaster.[119]

THE TRUE COST OF "CHEAP" FOOD

The industrial food system's fixation on producing ever-cheaper calories drives a vicious cycle of degradation and harm. The pursuit of low-cost, high-volume food demands intensive, large-scale agricultural operations— systems that erode soil, pollute air and waterways, and hollow out rural economies. To maintain profitability, these systems continue to intensify and consolidate, increasing greenhouse gas emissions through fossil fuel dependency, synthetic inputs, and the expansion of livestock operations and cropland. Climate change further intensifies these pressures. As yields become more volatile and harvests less reliable, land is pushed harder and stretched thinner. The result is a cascade of ecological and social consequences: accelerating land degradation, mounting food waste, deepening health crises linked to diet-related diseases, and rising social and environmental costs, all treated as externalities by dominant economic models.

This self-perpetuating cycle is sustained not just by subsidies and incentives, but by a policy architecture rooted in what's known as "agricultural exceptionalism."[120] Emerging during the era of slavery and sharecropping, this doctrine reflects a longstanding effort, initially led by Southern plantation owners, to shield agriculture from the labor protections afforded to other sectors. Its legacy persists: Farmworkers remain excluded from foundational labor laws like the National Labor Relations Act, which protects workers' rights to organize, and the Fair Labor Standards Act, which governs minimum wage, overtime, and child labor protections. Agriculture is similarly exempt from nearly every major environmental regulation enacted since the 1970s. The Animal Welfare Act does not apply to animals raised for food, leaving billions of chickens, pigs, and cows—most of them confined in factory farms—without federal protection.[121] At the state level, livestock are often excluded from anticruelty laws, and "ag-gag" legislation criminalizes whistleblowers who expose abuse or contamination. As of now, six states have active ag-gag laws, while courts have struck down five others as unconstitutional.[122] Meanwhile, every US state maintains a "right-to-farm" law, shielding agricultural operations from lawsuits over pollution, odor, and other harms that diminish residents' quality of life.[123] These legal protections, alongside policy omissions, uphold entrenched patterns of exploitation and ecological harm. They obstruct the emergence of more just and regenerative alternatives, all while undermining public health, farmer autonomy, worker dignity, animal welfare, and long-term agricultural viability.

To confront these hidden costs, a growing number of researchers, advocates, and institutions are turning to True Cost Accounting (TCA), a systems-based approach that reveals the full spectrum of ecological, social, and economic impacts embedded in food production. Launched by the UNEP in 2012, the Economics of Ecosystems and Biodiversity for Agriculture and Food Programme (TEEBAgriFood) brought together 150 scientists from thirty-three countries to develop a comprehensive framework for understanding the total value and cost of food systems.[124] The 2018 TEEBAgriFood Framework offers a tool for governments, investors, and producers to assess food systems through a whole-systems lens. It accounts for every stage of the supply chain—from farms and processors to retailers, consumers, and the ecosystems that sustain them.[125] In one application, Mexico's National Commission for the Knowledge and Use of Biodiversity (CONABIO) used the framework to evaluate maize production across North America. Their analysis revealed that the genetic uniformity of US and Canadian corn made these crops highly susceptible to pests and disease, leading to yield losses exceeding $27 billion between 2012 and 2015.[126] This data was used to advocate for the economic and ecological value of preserving agrobiodiversity.[127]

Building on the foundational work of TEEBAgriFood, the Rockefeller Foundation undertook the first comprehensive national-level TCA analysis of the US food system. The initial figure was striking: Producing, processing, distributing, and retailing the food purchased in the United States amounts to approximately $1.1 trillion. But this number tells only part of the story. It excludes the immense hidden costs that the industrial food system externalizes onto people, communities, ecosystems, and future generations. These hidden costs include the soaring burden of diet-related diseases, the health impacts of food production-related pollution, biodiversity loss, and the climate consequences of greenhouse gas emissions from agriculture and packaging. When these broader impacts are fully accounted for—spanning human health, ecological degradation, economic inequality, animal welfare, and systemic vulnerability—the *true cost of food* in the United States is estimated to be at least three times the current market expenditure: $3.2 trillion annually. Perhaps most revealing is that the greatest portion of these unaccounted-for costs—around $1.1 trillion—comes from health impacts. This includes direct medical care for diet-related illnesses, lost productivity due to chronic disease, the toll of food insecurity, and injuries and illness from unsafe or exploitative labor conditions in food production. In other words, the amount Americans currently pay for food at the register is matched—nearly doubled—by what we pay in medical bills, worker harm, and lost livelihoods due to the very systems that produce that food.[128]

These costs are not evenly distributed. They disproportionately fall on communities of color, who are more likely to live in food-insecure households, suffer from chronic diet-related conditions, and work in the lowest-paid, most hazardous food system jobs. Obesity is 1.2 times more prevalent among Black Americans than the national average. Diagnosed diabetes is 1.7 times higher among Latinx communities and 1.5 times higher among Black communities compared to white Americans. Meanwhile, Black households face food insecurity at more than twice the rate of white households.[129]

Environmental costs are similarly immense. The US food system contributes significantly to greenhouse gas emissions, plastic pollution, and ecosystem collapse—costs estimated at $900 billion per year.[130] Yet even this staggering number cannot capture the irreversible losses underway. The clearing of wild lands for farming, development, and resource extraction has catalyzed what many scientists describe as the planet's sixth mass extinction.[131] In just the last fifty years, it's believed that half of all animal species on Earth have been lost—a disappearance so profound it defies economic valuation.[132] Together, these findings point to a systemic truth: The market price of food is an illusion. What we pay at the checkout counter hides a complex web of harm that is absorbed not just by ecosystems and future generations but by those most economically and socially marginalized in the present.

SQUANDERING SOIL AND WATER

We are only beginning to fully comprehend the devastating impact industrial agriculture has not just on life above ground but on the vast and largely unseen web of life beneath our feet. Soil, the most overlooked and undervalued ecosystem on Earth, is a living, breathing matrix of interdependent beings and relationships. It is home to 90 percent of the world's fungi, 85 percent of plant species, 50 percent of all known bacteria, and even 3 percent of global mammal life. A 2023 study found that an astonishing 59 percent of our planet's species reside within the soil microbiome, making it the most biodiverse habitat on Earth.[133] Soil is not merely a medium for growing crops; it is the foundation of terrestrial life. It supports nearly every ecological function that makes life possible: filtering and storing water, cycling nutrients, regulating climate and flood patterns, sequestering carbon, and producing 95 percent of the food we eat.[134] And yet, in modern life, fewer people than ever understand this truth, let alone feel it. As human labor has increasingly moved away from the land, our physical and relational contact with soil has diminished, creating a rift in our embodied knowledge of what sustains us.[135]

The more disconnected we become, the more invisible soil becomes, even as our dependence on it deepens. While we now know more about soil's composition and processes than ever before, the cultural and practical awareness that soil is *alive* and *sacred* has largely been lost. But emerging science is helping to reawaken this understanding. A growing body of research shows how soil health directly impacts the vitality of plants and animals, and by extension, human health. The rhizosphere—the root zone of plants—is a critical site of exchange, where microbial communities mediate nutrient uptake, influence plant resilience, and even shape crop nutrition.[136] These microbes depend on soil organic matter, which acts as their food source. In turn, healthy microbial activity transforms raw minerals into bioavailable nutrients and produces metabolites that benefit both plant and human health. Soils have a remarkable ability to store, transform, and cycle the eighteen nutrients essential to plant life—fifteen of which can only be supplied by soil that is alive and intact.[137] Biodiversity is the heartbeat of soil resilience, just as it is the foundation of human resilience. Healthy soils require a complex weave of organic matter, moisture, minerals, and vibrant microbial communities. Human bodies require the same: diversity, nourishment, and freedom from chemical residues that harm the delicate balance of our internal ecosystems. The parallels between the gut microbiome and the soil microbiome are more than metaphor; they are relational truths.[138]

The industrial food system is rapidly undermining one of Earth's most vital and complex life-support systems: the soil ecosystem. Today, one-third of the world's soils are *already* degraded, and this number continues to rise. The causes are interconnected and systemic—ranging from wind and water erosion, tillage-induced degradation, and the depletion of soil organic carbon to salinization, chemical pollution, compaction, acidification, and the rapid decline of soil biodiversity.[139] These conditions are driven by extractive agricultural practices: widespread deforestation to expand croplands, routine tilling with heavy machinery, overgrazing by livestock, and the persistent use of synthetic fertilizers, herbicides, and pesticides. Such methods disrupt microbial life, reduce organic matter, and create sterile growing environments that rely increasingly on chemical inputs to remain productive.[140] The result is a downward spiral—soil erosion and climate disruption demand even more inputs, which further degrade microbial diversity and diminish the land's regenerative capacity.

The consequences of this degradation are far-reaching. Soil loss diminishes agricultural productivity, increases the risk of floods and landslides, pollutes waterways, and in severe cases, displaces communities altogether.[141] But the damage doesn't stop there. As awareness grows about what's in our food, concern is also rising about *what's missing* from it. Industrial farming, focused on speed and scale, often disrupts the plant–microbe relationships that are essential for nutrient-rich crops. Without their microbial allies, conventionally grown plants may fail to synthesize or absorb key micronutrients and phytochemicals that support both plant resilience and human health. This has contributed to a global crisis of "hidden hunger," a form of malnutrition where people consume enough, or even too many, calories but lack essential vitamins and minerals such as iron, zinc, iodine, and vitamin A. This condition now affects approximately 2 billion people worldwide,[142] including those who are simultaneously overweight and undernourished.[143]

While the Green Revolution helped stave off acute hunger by increasing calorie availability, it left behind a food system built to maximize yield, not nourishment. Decisions about how crops are grown—what varieties are planted, when they're harvested, and how long they're stored or shipped—have profound impacts on food quality. The industrial model's emphasis on uniformity, shelf life, and transportability has indeed lowered food prices, but it has also stripped our food of flavor and vitality.[144] Many people can taste the difference—between a tomato grown in a backyard garden and one shipped across continents, or between a freshly picked peach and one bred for durability, not sweetness. Scientific analysis confirms this experience. Studies show that nutrient content in fruits and vegetables has declined significantly over the last fifty years.[145] Since 1940, magnesium levels in vegetables have dropped 24 percent; spinach has lost 53 percent of its potassium and 60 percent of its iron; cherries, peaches, and strawberries show a 27 percent decline in zinc; and carrots have lost up to 75 percent of their nutrient density.[146] This is not just a story of poor soil—it is a story of distorted priorities. As agricultural systems have chased higher yields, nutrient density has suffered.[147] The wheat developed by Norman Borlaug, for instance, produced more grain per plant, but not more nutrition. As Professor Steve McGrath of Rothamsted Research in the UK explains, while the nutritional content per kernel remained constant, the starch content doubled or tripled, resulting in a dilution of nutrients when processed into flour. In other words, we are growing more but getting less.[148]

Modern high-yield crop breeding has not only prioritized productivity at the expense of biodiversity, it has also led to a significant loss of critical secondary metabolites that benefit both plant and human health.[149] Take flavonoids, for example: This diverse group of plant compounds possesses powerful anti-inflammatory, antiviral, and anticancer properties. In many industrial cultivars, these compounds have been bred out or diminished to enhance taste or shelf appeal. One striking case is the domestication of Brassicaceae vegetables like cabbage and cauliflower, where levels of glucosinolates—a class of compounds responsible for their bitter flavor—have been deliberately reduced. Yet these very compounds are not only vital for plant resistance against pests and pathogens but are also thought to function as potent prebiotic and anticancer agents in the human diet.[150]

This nutritional erosion is exacerbated by another emerging threat: the climate crisis. Industrial agriculture already depletes the living soil systems essential for growing nutrient-rich food, but rising atmospheric carbon dioxide (CO_2) levels are now compounding the problem. Since the dawn of the Industrial Revolution, atmospheric CO_2 has increased by 50 percent from roughly 280 parts per million (ppm) to 425 ppm as of early 2024. Scientists estimate we could reach 550 ppm within the next 50 years, double the concentration that existed when American agriculture began mechanizing with tractors.[151] This dramatic, human-driven rise in atmospheric CO_2—faster than any natural increase seen in the last 20,000 years—has profound implications for plant metabolism and the nutritional quality of food.[152]

While elevated CO_2 can accelerate plant growth,[153] particularly for crops that use the C_3 photosynthetic pathway (such as wheat and rice),[154] it also dilutes the nutritional content of those crops.[155] Studies consistently show that increased CO_2 levels lead to reductions in protein concentrations and essential minerals, including iron, zinc, calcium, magnesium, and phosphorus.[156] In one study, grains of wheat, rice, barley, and tubers like potatoes grown under elevated CO_2 showed 10–15 percent less protein.[157] A 2018 study of rice varieties also found troubling declines in B vitamins (B_1, B_2, B_5, and B_9), even as vitamin E levels increased.[158] Additional research has confirmed these declines in nutrient density across staple crops like wheat, corn, rice, and soy.[159] Forecasts predict reductions in iron and zinc content of up to 17 percent by mid-century.[160] These declines are not limited to human food systems—plants like goldenrod, a crucial late-season food source for North American bees, also lose nutritional quality under elevated CO_2, affecting pollinator health and ecosystem integrity.[161] What's happening here is not just a series of isolated biochemical shifts. It is a fundamental breakdown of the relational integrity between plants, soil, atmosphere, and human health.

Alongside soil degradation, the industrial food system is also accelerating a water crisis, driven by productivist policies that prioritize yield and export over long-term ecological viability. Today, 2.4 billion people live in water-stressed regions,[162] many of them are smallholder farmers who already face daily challenges accessing safe drinking water, nutritious food, and essential services like hygiene and sanitation. The burdens of water insecurity fall most heavily on women, Indigenous communities, migrants, and displaced peoples. Globally, per capita freshwater availability has declined by 20 percent over the last two decades. This is the result of decades of uncoordinated water management, overextraction of groundwater, pollution, and the intensifying impacts of climate change. Recent research reveals that groundwater reserves—our hidden lifeline—are shrinking at alarming rates,[163] with nearly one-third of aquifers worldwide now experiencing accelerating depletion. Agriculture is the primarily driver of this trend, consuming 72 percent of all global freshwater withdrawals—more than any other sector.[164]

In the United States, this crisis is most acute in California and other Southwestern states, where agriculture relies heavily on dwindling water supplies from the Colorado River. After more than twenty years of sustained drought, the region faces severe shortages.[165] In California alone, approximately twenty groundwater basins are considered "critically overdrafted,"[166] a designation indicating that more water is being pumped than can be naturally replenished. This has far-reaching implications. The Southwest currently supplies more than half of the fruits, nuts, and vegetables consumed in the United States.[167] Consolidation, efficiency-driven logic, and a climate that allows for year-round production have concentrated 90 percent of the nation's leafy greens production, including lettuce, spinach, and cabbage, in California and Arizona.[168] California farmers also grow nearly all of the country's almonds, artichokes, celery, garlic, grapes for raisins, kiwi fruit, honeydew melons, nectarines, olives, pistachios, plums, and walnuts—crops that require significant water inputs.[169] Compounding the strain is California's industrial dairy industry, which uses staggering volumes of water to feed and cool cows, flush manure from barns, and irrigate thirsty feed crops like alfalfa.[170] According to the advocacy organization Food & Water Watch, California's dairy operations consume roughly 142 million gallons of water per day, enough to meet the daily recommended water use for every resident of both San Jose and San Diego.[171] But the problem is not just overuse; it's contamination. Industrial dairies often dispose of manure by spraying it onto cropland. This practice leaches nitrates into the soil, eventually contaminating the groundwater below. In California's Central Valley, home to the largest concentration of factory farms in the state, up to 40 percent of domestic wells show nitrate levels that exceed federal safety standards.[172] Nitrate contamination is linked to a host of health issues, including cancer, miscarriages, birth defects, and a life-threatening condition in infants known as blue baby syndrome.

The southwestern United States has become a strategic target for corporations seeking access to water—a pursuit enabled by legal structures that heavily favor agricultural interests.[173] In California, an arcane and deeply inequitable water rights system allows companies to acquire land bundled with entitlements to large amounts of water, often at artificially low costs and with minimal oversight. These water entitlements, some of which date back over a century, function as de facto subsidies, even amid deepening drought and ecological collapse. A striking example lies in the Palo Verde Valley, where historic appropriative water rights give local users first priority to Colorado River water, even during times of scarcity. In Blythe, California, a desert community adjacent to the lower Colorado River, an 1877 water rights claim guarantees "unquantified water rights for beneficial use." This means that landowners within the district can use as much water as they deem necessary, with virtually no constraints, regardless of the broader ecological crisis or competing needs.[174] Water travels through a vast infrastructure of dams and canals to reach this arid region, reinforcing an extractive logic that dislocates water from its ecological and communal context.

Global agribusiness firms are capitalizing on these legal loopholes. For example, Fondomonte Farms, a subsidiary of the Saudi food giant Almarai, has purchased 15,000 acres in Blythe, along with massive storage facilities, to grow alfalfa for export to Saudi Arabia, where it is used to feed dairy cattle.[175] Almarai, one of the world's largest producers of dairy products and packaged baked goods, relocated its forage operations to the United States after the Saudi government, recognizing its own ecological limits, banned domestic cultivation of wheat and livestock feed in 2016 due to severe water scarcity.[176] This practice—growing water-intensive crops like alfalfa in drought-stricken regions and exporting them overseas—is a glaring example of ecological outsourcing. Water, a vital public commons, is effectively being privatized and extracted for corporate gain, while local ecosystems and communities shoulder the consequences.

These water-intensive agricultural operations are especially alarming in the context of worsening climate volatility. The US Global Change Research Program has warned that climate change will intensify drought, heatwaves, and the reduction of winter chill hours—all of which directly undermine crop viability and livestock health. At the same time, growing competition for water among agriculture, energy, and municipal uses will heighten the risk of food insecurity.[177] The crisis unfolding in the US Southwest is not simply about resource mismanagement; it is about the systemic privileging of profit over planetary boundaries.

FACTORY FARMING NATION

In the twentieth century, as industrial capitalism reshaped the contours of economic life, its logic was also applied to agriculture—particularly to the raising of animals. Inspired by factory production models, a "productivist" approach took root, treating pigs, chickens, and cows as units of output rather than sentient beings embedded in complex ecologies. This mechanized model facilitated a dramatic increase in the number of animals raised for food, driving down the market price of animal protein. But this apparent efficiency has come at tremendous hidden costs that are not reflected in the price tag of meat, milk, or eggs. These costs include the accelerated emergence of zoonotic viruses, the proliferation of antibiotic-resistant bacteria, and widespread, institutionalized cruelty inflicted on animals confined in factory farms. Globally, more than 100 billion land animals are killed each year for meat and other animal products—hundreds of millions every single day.[178] There are now more farmed animals on Earth than wild ones by a wide margin. Farmed poultry alone accounts for 70 percent of all birds on the planet, with wild birds making up just 30 percent. For mammals, the disparity is even more severe: 60 percent are livestock, primarily cows and pigs, 36 percent are humans, and only 4 percent are wild animals.[179]

In the United States, though data is fragmented and not regularly updated, estimates suggest that over 10 billion land animals are raised for food each year.[180] The vast majority—around 99 percent—are confined in concentrated animal feeding operations (CAFOs), where thousands of animals are packed into crowded, unsanitary facilities that deprive them of even the most basic forms of agency and comfort.[181] In such conditions, animals are often unable to walk freely, lie down on natural surfaces, or engage in instinctual behaviors like rooting (for pigs) or dust-bathing (for chickens). These conditions not only cause deep physical and psychological distress but also raise profound ethical questions—questions that demand more than technical fixes and touch on the nature of our relationship with other sentient life. Because factory-farmed animals are subject to chronic stress and frustration of their natural instincts, their immune systems are often compromised. To prevent widespread illness, they are routinely fed antibiotics, not for healing, but as a prophylactic measure to keep them alive in environments that would otherwise kill them. Yet these measures do little to prevent the spread of viral diseases, which have become increasingly common and catastrophic. When outbreaks occur, they often lead to mass culling of animals—millions at a time—not because all are infected, but to preempt potential spread.

This is not an unfortunate byproduct of an otherwise functioning system. It is a predictable outcome of a global food regime that disrespects ecological boundaries, violates animal integrity, and disturbs viral equilibria. The industrial model places human, animal, and planetary health at constant risk. Consider recent examples: In 2019, African swine fever, a highly contagious and fatal disease, swept through China, the world's largest producer and consumer of pork. Millions of pigs were exterminated in an attempt to contain the outbreak.[182] In the United States, the 2022–2023 spread of avian influenza (H5N1) became the deadliest on record.[183] Hundreds of poultry operations across nearly every state were affected,[184] resulting in the culling of over 50 million chickens,[185] turkeys, and other birds, regardless of whether they were infected,[186] due to the virus's rapid spread and high mortality rate.[187] The economic toll was staggering: $757 million in federal spending and over $1 billion in industry losses. As of April 2024, another outbreak of avian flu is unfolding in the United States—this time with even more alarming implications, as the virus has been detected not only in poultry but also in dairy cattle and in at least one farmworker.[188]

The expansion of industrial livestock production, driven by escalating demand for meat and dairy, has become a significant contributor to environmental degradation, climate disruption, and intensified competition for essential resources like land and water. This model, rooted in productivist logic, has been extended to aquaculture, which now supplies half of global seafood consumption. Aquaculture practices often rely on feeding farmed fish with wild-caught fish, soybeans, and grains, mirroring the intensive animal farming systems on land and perpetuating similar ecological and ethical concerns. Shrimp farming, for example, predominantly conducted in saline coastal waters, exemplifies the environmental toll of such practices. It has led to the destruction of 38 percent of the world's mangrove forests—ecosystems that play a crucial role in coastal protection, carbon sequestration, and biodiversity support. The accumulation of waste, antibiotics, and pesticides in shrimp ponds, coupled with the loss of mangroves that naturally filter these pollutants, often renders these aquaculture sites ecologically unsustainable.[189]

The proliferation of factory farming, both terrestrial and aquatic, is facilitated by the overproduction of subsidized crops like corn and soybeans. These crops, primarily used as animal feed, contribute to increased greenhouse gas emissions through their cultivation, harvesting, transportation, and processing. This system is notably inefficient in converting plant-based calories into animal-based nutrition. For instance, in the United States, 40 percent of all corn and 97 percent of soybean meal are allocated to livestock feed,[190] yet meat, dairy, and farmed fish collectively provide only 17 percent of global caloric intake and 38 percent of protein supply.[191] Livestock production occupies a disproportionate amount of agricultural land— accounting for 80 percent when combining grazing areas and cropland used for feed—while delivering a relatively small share of human nutritional needs.[192] This imbalance underscores the concept of "opportunity food loss,"[193] where the resources dedicated to animal agriculture could potentially yield greater nutritional benefits if redirected toward plant-based food production.[194]

Since the last US agricultural census in 2017, there has been a consolidation of farmland and a decline in small-scale farming operations. The 2022 census reports a 34 percent reduction in dairy farms, 9 percent in hog farms, and 7 percent in cattle farms, with overall livestock numbers remaining relatively stable. This trend indicates a shift toward larger, more concentrated CAFOs, which are associated with various environmental and social harms,

including water and air pollution, compromised animal welfare, labor exploitation, and exacerbated climate impacts. CAFOs in the United States produce approximately 940 billion pounds of manure annually, double the
amount of sewage generated by the entire human population.[195] Unlike
human waste, which undergoes treatment, the majority of this animal waste
is applied untreated to land, leading to nutrient runoff that contaminates waterways, contributes to eutrophication, and fosters harmful algal blooms.[196]

A notable example is the recurring algal blooms in Lake Erie. Each year,
the cumulative impacts of industrial agriculture are made disturbingly visible in western Lake Erie, where a surge of microscopic organisms paints a
toxic green slick across the water's surface—so vast and vibrant it's visible
from space.[197] At its peak, the algal bloom coated roughly 620 square miles
of the lake, an area more than twice the size of Chicago, according to satel-
lite imagery.[198] This annual phenomenon reveals a cruel irony. Even as global
communities grapple with escalating drought and water insecurity, residents
living along the shores of one of the largest freshwater systems on Earth,
upon which more than 11 million people depend for drinking water, are
sometimes forced to rely on bottled water. Despite billions of dollars invested
in water quality monitoring, upgrades to drinking water infrastructure, and
sewage system overhauls, the primary source of Lake Erie's algal eruptions—
agricultural runoff—remains largely unchecked. The problem originates
upstream, where the seasonal melt of snowpack and spring rains cascade
across the rural landscapes of northwestern Ohio, carrying an immense burden of nutrients. These waters sweep manure from nearly 80 concentrated
hog, cattle, dairy, and poultry operations, along with synthetic fertilizers
from industrial crop fields, into the Maumee River. This 137-mile artery col-
lects agricultural waste from across Ohio, Indiana, and Michigan before dis-
charging it into the shallow western basin of Lake Erie. Compounding the
issue, the city of Detroit regularly discharges wastewater and sewage into
the Detroit River, which also empties into the lake. This nutrient-rich cock-
tail, particularly high in phosphorus, triggers explosive algal growth, leading
to hypoxic conditions that suffocate aquatic life and threaten public health.[199]

Despite mounting evidence and repeated warnings, the regulatory frameworks responsible for addressing these nutrient inflows remain insufficient.
According to the joint "State of the Great Lakes" assessment by the governments of Canada and the United States, the ecological health of the Lake Erie

basin is officially rated as "poor"—a status that, tragically, has not improved.[200] What we witness in Lake Erie is not an anomaly but a symptom of a deeper pattern, an industrial food system fixated on throughput, scale, and short-term gains, regardless of long-term costs. This logic of relentless output doesn't just shape current practices; it also distorts our imagination of what counts as a "solution." Instead of confronting the root causes of ecological degradation, many so-called innovations double down on the same extractive paradigms.

Take, for example, the recent construction of a 26-story pig farming facility in China, designed to house 650,000 animals and process 1.2 million pigs annually. Marketed as a marvel of efficiency, this vertical factory farm replicates, and magnifies, the harms of industrial animal agriculture: from severe animal suffering to heightened public health risks due to the dense concentration of disease-prone livestock. Despite clear evidence that such high-density facilities accelerate the spread and mutation of infectious pathogens, they continue to be championed by investors and officials as models of productivity.[201] This trajectory reflects a broader systemic pattern where human activities, especially those framed as noble efforts to "feed the world," exert relentless pressure on the Earth's life-support systems: the atmosphere, geosphere, hydrosphere, and biosphere. By prioritizing scale over stewardship and yield over relational integrity, we have destabilized the very ecologies that make food, and life, possible.

3

The Progress Trap

A decade ago, I attended an event at a vertical vegetable-growing facility tucked into an industrial office park in northwest Indiana. Marketed as cutting-edge and celebrated in local media, the "farm" was a climate-controlled warehouse outfitted with LED lights and hydroponic trays—no soil in sight. That evening's main event was the unveiling of a mural commissioned to represent the future of agriculture. The painting traced a linear progression: from subsistence farmers with mules, to the introduction of mechanized plows, to scenes of industrial-scale agriculture powered by fossil fuels. The final panel, heralded as the apex of innovation, depicted the very facility we were standing in—glowing under artificial light, sanitized and soil-less, as if the ecological mess outside had finally been solved. As the crowd erupted in applause, I felt my heart sink.

Missing from the mural, and from the narrative it reinforced, was any recognition of the vast and diverse traditions of Indigenous and regenerative foodways, which have sustained life through reciprocity and ecological intimacy. In their place was a vision of progress narrowly defined by mechanization, control, and the substitution of living systems with engineered environments. This moment crystallized for me the seduction of a dominant worldview, what has come to be known as "solutionism." It's a perspective that assumes every ecological and social problem can, and should, be solved through technological ingenuity. Embodied by figures like Norman Borlaug, who helped usher in the Green Revolution, this techno-optimist mindset posits that with enough innovation and investment, human cleverness will engineer us out of the crises we've engineered ourselves into.[1]

Since that visit, I've collaborated with many well-intentioned thinkers, entrepreneurs, and policymakers who earnestly believe that technology holds the key to transforming our food system. I understand the hope it offers, especially amid urgent crises. But as I've delved deeper into the structural roots of our interlocking emergencies, I've come to see why we must interrogate this reflex. If we are to respond meaningfully to the polycrisis, we cannot simply reach for new tools—we must reckon with the very logics and values that have brought us to this brink.

AGRICULTURE 4.0

The indoor hydroponic farm I visited a decade ago is no longer a futuristic novelty; it has become emblematic of a growing trend. Vertical farming, while often reliant on costly artificial lighting and energy-intensive HVAC systems for climate control, continues to expand, especially in urban centers where it promises a steady supply of locally grown produce.[2] These technologies are part of what's being heralded as Agriculture 4.0, the so-called fourth agricultural revolution. This emerging paradigm is defined by the convergence of digitalization, artificial intelligence, robotics, nanotechnology, synthetic biology, and molecular technologies. Its advocates promise not just to optimize agricultural production but to fundamentally reengineer it— minimizing uncertainty, removing human labor, and positioning AI as the new farm manager.[3]

The question of how to feed a growing population has motivated technological innovation across food systems, stretching back to the first agricultural revolution. Since the Neolithic era, human ingenuity has transformed food through technologies like fermentation, breadmaking, cheesemaking, and irrigation—tools that have shaped not only our diets but also key indicators of societal wellbeing, including longevity, disease control, and poverty reduction.[4] The history of agricultural revolutions is often narrated in stages: Agriculture 1.0 marked the shift from foraging to settled cultivation; Agriculture 2.0 brought mechanized innovations like Jethro Tull's seed drill during the British Agricultural Revolution; and Agriculture 3.0 emerged with the Green Revolution's chemical and genetic technologies aimed at maximizing yield.[5] But Agriculture 4.0 signals a departure from previous eras. Where earlier revolutions framed nature as the problem, something to control, weed out, or overcome, this new wave of techno-agriculture is responding to crises largely of our own making. Today's challenges—ranging from microplastic pollution and soil collapse to metabolic disease and oceanic dead zones—are byproducts of the very technologies once hailed as solutions.

SOLUTIONISM AND SEPARATION

The allure of ag-tech solutions is powerful. It offers the comfort of continuity, the possibility of sustaining meat-centric diets through alternative proteins, optimizing nutrition through algorithmically curated meals, and sidestepping environmental collapse with high-tech fixes. These narratives reassure us that we can emerge unscathed from our centuries-long fossil-fueled binge without fundamentally altering our consumption habits or the extractive worldview that underpins them. This techno-optimism rests on a familiar logic: Localize food by building vertical farms in urban skyscrapers; reduce fossil fuel dependence through a booming biofuel industry; increase nitrogen efficiency with precision agriculture; replicate meat in industrial bioreactors; remove excess atmospheric carbon with machines. Each proposal offers a shiny, surface-level intervention, but few ask us to reckon with how we got here.

While some of these innovations may offer partial or transitional benefits, they often sidestep the deeper truth: that our ecological and food system crises are not merely technical in nature but profoundly relational. These crises arise from a breakdown in our connection to land, place, each other, and the more-than-human world. If, as Einstein warned, we cannot solve problems with the same level of thinking that created them, then we must be vigilant. Many of the solutions we are now celebrating may actually deepen the very disconnection—the story of separation—that got us into this mess. At the core of ecomodernist philosophy lies the belief that human development can be decoupled from environmental harm through technological advancement. This perspective is explicitly laid out in the *Ecomodernist Manifesto*, published in 2015 by the Breakthrough Institute, which proclaims: "[W]e affirm one long-standing environmental ideal, that humanity must shrink its impacts on the environment to make more room for nature, while we reject another, that human societies must harmonize with nature to avoid economic and ecological collapse."[6] Like earlier frameworks such as ecological modernization theory, developed in Europe during the 1980s and 1990s, ecomodernists advocate for "land sparing": the intensification of activities like agriculture, energy production, and urban development on smaller land footprints, in order to leave more of the Earth untouched. This ideology is anchored in a particular vision of human flourishing,[7] one that equates the "good life" with material abundance, personal autonomy, high consumption diets (especially of meat and dairy), and liberation from the physical labor of food production.

Rather than seeking to restore or sustain reciprocity with living ecosystems, ecomodernism promotes substitution—replacing ecological functions with engineered solutions. Nutrient cycling, for instance, can be handled through synthetic fertilizers; trees' role in filtering air can be mimicked by industrial purifiers; freshwater scarcity can be offset by desalination;[8] and livestock-free protein production can replace traditional animal agriculture.[9] Proponents argue that this shift toward high-efficiency, high-tech systems will allow us to meet human demands while minimizing environmental impact. Journalist and environmentalist George Monbiot champions this vision in his book *Regenesis: Feeding the World Without Devouring the Planet*, where he praises precision fermentation as a breakthrough technology that can create animal-free protein and fats identical to those derived from cows, pigs, sheep, and fish. "It is time to develop a new and revolutionary cuisine," Monbiot writes, "based on farmfree food. . . . We can now contemplate the end of most farming, the most destructive force ever to have been unleashed by humans."[10]

The ecomodernist proposal to concentrate human activity, such as intensified agriculture and dense urban settlement, into discrete zones in order to "spare" wilderness from further human intrusion echoes the old story of separation. It upholds the idea that nature is something external, something to be partitioned off from human life. As ecomodernist Ruth DeFries bluntly put it, "there's no intrinsic value to nature for most people and that's okay."[11] French philosopher and anthropologist Bruno Latour critiques this worldview, describing the *Ecomodernist Manifesto* as "written entirely as if humans were still alone on stage, the only being who out of its own free will is in charge of apportioning space, land, money, and value to the old Mother Nature."[12] This philosophical stance positions nature and agriculture as mutually exclusive, and by extension, casts farming communities, especially peasants, Indigenous peoples, and small-scale producers, as inherently harmful to ecological integrity.[13] This logic elevates urbanization and automation as forms of liberation from what is portrayed as the drudgery of working with the land, while obscuring the relational richness and ancestral wisdom embedded in land-based life.[14]

In Chapter 4, we will explore in greater detail why this land-sparing logic is not only flawed but dangerous. It erases the longstanding, reciprocal relationships that humans have held within ecosystems across the globe. Far from being passive or destructive inhabitants, many societies have co-shaped richly biodiverse environments for millennia. Archaeological, anthropological, and ecological research shows that over 12,000 years ago, nearly three-quarters of the Earth's land was stewarded by human communities. Today, lands under Indigenous governance remain among the most biodiverse in the world.[15] From the rangelands of sub-Saharan Africa, where pastoralists have used fire, rotational grazing, and herd migration strategies for 5,000 years,[16] to the Amazon rainforest, where Indigenous land care has shaped forest structure, biodiversity, and soil fertility, these examples reveal a truth ecomodernism tends to obscure: What is often labeled as "pristine wilderness" is, in fact, the outcome of enduring and respectful human–nature relationships.[17] These practices are not remnants of the past but living models of sustainability, rooted in connection, not separation.

Dismissing the relational interdependence between people and land echoes the Green Revolution's reductionist paradigm, which largely overlooked the ecological foundations of agriculture, particularly the crucial role of soil health, its symbiotic connection with root systems, and the broader ecological and social functions that agricultural landscapes provide. This narrow framing ignores the multi-functionality of farming systems: their capacity to support clean water, foster biodiversity, sustain rural livelihoods, and sequester greenhouse gases.[18]

Ecomodernism extends this reductionist impulse by rejecting the need to curb material consumption. In doing so, it reinforces the industrial food system's foundational myths—the story of separation, the dogma of efficiency, and the lure of techno-optimism. This is the seductive pull of solutionism: a narrative that reassures us we can engineer our way out of existential crises without altering our behaviors, relationships, or values. Its message, often echoed by advocacy groups, policymakers, and tech entrepreneurs, is deceptively simple: "We've got this. A smarter policy or newer gadget will secure a more abundant, just, and sustainable future—no need for real sacrifice."[19] As Bruno Latour sharply observed, this mindset resembles the fantasy that an e-cigarette might cure a chain smoker's addiction, not by changing the habit but by upgrading the delivery system. Ecomodernism, in this sense, is just another way of insisting we can have our cake and eat it too.[20]

HUMAN HUBRIS

The ecomodernist posits that human development can be decoupled from environmental impacts through technological innovation. This perspective manifests in agri-food technologies that aim to produce food independently of natural ecosystems. Vertical farms eliminate the need for sunlight, utilizing artificial lighting systems; hydroponics allows for soil-less cultivation; and projects like Harvard's RoboBees endeavor to replicate the pollination role of bees through robotic means.[21] The development of RoboBees at Harvard's Wyss Institute aims to mimic the intricate behaviors of bees, including sustained flight, stability in varying conditions, and complex decision-making within hives.[22] While these robotic pollinators represent a significant engineering achievement, they underscore a broader tendency to favor technological fixes over addressing the root causes of pollinator decline, such as habitat loss and pesticide use. Similarly, scientists in New Zealand are exploring the cultivation of fruit tissue from plant cells, aiming to produce consumable fruit without traditional agricultural processes. This initiative seeks to address issues like food waste and the challenges posed by urbanization and climate change.[23]

The belief that humans can endlessly enhance nature with minimal consequences is also foundational to dominant techno-optimist narratives, particularly those promoting genetic engineering as a solution to global hunger. This view, rooted in a reductionist paradigm, justifies the deliberate reconfiguration of organisms by altering their genetic codes—framing such interventions as benevolent progress.[24] One prominent example is Golden Rice, genetically modified to produce beta-carotene, a precursor to vitamin A.[25] While approved for cultivation in the Philippines in 2021,[26] its implementation was halted by the Philippines Supreme Court in 2023.[27] Proponents frame it as a humanitarian intervention to address malnutrition,[28] yet critics caution against treating complex nutritional and systemic inequities with single-nutrient fixes. They argue that biofortification oversimplifies the multifaceted realities of food insecurity and erases the relational, ecological, and political contexts that shape malnutrition.[29] Moreover, technical limitations, including low and unstable beta-carotene levels and the need for plastic-based storage to slow degradation, compound the issue, introducing new ecological harms under the guise of "sustainability."[30]

Genetically engineered crops are also marketed as tools for "sustainable agriculture," with claims that they reduce pesticide use and improve resilience. A key example involves Bt crops, plants engineered to produce insecticidal proteins derived from the bacterium *Bacillus thuringiensis* (Bt). In its natural form, Bt has long been used as a biological pesticide in organic agriculture, where it is applied externally and breaks down quickly in the environment. However, the logic shifts profoundly when Bt genes are embedded directly into plant DNA, turning every cell of the plant into a pesticide-producing factory. Traits like herbicide resistance and Bt toxin production have spurred cycles of resistance, leading to "superweeds," secondary pest outbreaks, and increasing chemical dependency. Glyphosate resistance is now widespread, affecting numerous US crops and accelerating a return to intensive mechanical tillage and escalating herbicide use.[31] This illustrates a pattern familiar in modernity: the promise of progress giving way to new, often more intractable forms of harm. Rather than addressing the root causes of agricultural vulnerability—monocultures, extractive farming models, disrupted ecosystems—these technologies risk reinforcing a treadmill of dependency that undermines ecological integrity and community self-determination.[32]

When farmers first adopt synthetic pesticides, they often encounter immediate, visible benefits: rapid pest suppression, increased yields, and a sense of control. What tends to remain obscured are the long-term ecological disruptions and dependencies that such interventions set into motion. Insecticide sprays, for instance, do not discriminate. While targeting insect pests, they also eliminate beneficial predators and pollinators that play vital roles in regulating agroecosystems. Over time, this imbalance fosters conditions for pests to adapt, developing resistance that renders previous chemical interventions ineffective. This dynamic unfolded vividly in India during the 1990s, when cotton farmers faced a cascade of resistant pests following widespread insecticide use.[33] As traditional agroecological practices and local pest-management knowledge eroded, displaced by faith in technological solutions, farmers found themselves increasingly vulnerable. With natural checks on pest populations dismantled and few relational alternatives available, two dominant responses emerged: intensification and substitution. The first response was to escalate chemical use. In the 1970s, Indian cotton fields were typically sprayed two or three times per season. By the early 2000s, that number had risen to over thirty applications in some regions. Insecticide costs soared, consuming over half of total input expenses in certain areas, driving many farmers deeper into cycles of debt and dependency. The second response was to seek new chemical solutions—a reactive desperation that fed an ever-churning market for next-generation pesticides.[34] Industry capitalized on this treadmill by embedding multiple resistance traits into genetically engineered seeds and formulating ever-stronger chemical compounds. The result has been not a reduction in pesticide use, as once promised, but an alarming increase, including greater reliance on highly toxic substances such as neonicotinoids, glufosinate, 2,4-D, and dicamba.[35] These compounds are now implicated in ecological degradation, soil microbiome disruption, and impacts on human and nonhuman health.

What emerges is not just a technological failure but a systemic pattern: the displacement of ecological memory, the sidelining of traditional practices, and the substitution of complexity with control. Scientists now find themselves locked in an escalating contest with nature, racing to develop novel chemicals in response to evolving resistance. Yet many acknowledge the uncomfortable truth: Evolutionary processes are likely to outpace chemical innovation.[36]

Biotechnologists are now advancing a new suite of genetic engineering techniques under the banner of gene editing, heralding it as a transformative solution for agriculture's mounting challenges. These technologies are promoted as precision interventions capable of forestalling pesticide resistance, enhancing the flavor and nutritional profile of food, extending the shelf life of perishable produce, and buffering crops and livestock against the impacts of climate volatility, including shifting precipitation, temperature extremes, and disease pressures.[37] Echoing the grandiose claims of the Green Revolution technologies, gene editing is being promoted as a central strategy to eradicate hunger, malnutrition, and food insecurity.[38] By altering

specific genes, proponents suggest, crops can be made resistant to particular pathogens, thereby reducing the need for fungicides and other chemical
treatments.[39] In livestock, gene editing is being applied to increase disease
resistance and purportedly improve animal welfare in the face of intensifying climate-related stress.[40] Examples include engineering animals less susceptible to swine flu[41] or enhancing the stress tolerance of certain African
cattle and chicken breeds.[42]

It is within this context of amplified techno-optimism that the 2020
Nobel Prize in Chemistry was awarded to Emmanuelle Charpentier and
Jennifer Doudna for their development of one of the most celebrated gene-
editing tools: clustered regularly interspaced short palindromic repeats
(CRISPR-Cas).[43] This recognition helped solidify CRISPR's status as a flag-
ship innovation of the "precision agriculture" movement, which promises to
engineer crops and animals not only for efficiency and resilience but for sensory appeal and market desirability. Researchers are already applying
CRISPR to a wide range of crops: non browning white button mushrooms,
seedless tomatoes, herbicide-resistant canola, extra-starchy potatoes, cacao
engineered to resist fungal and viral diseases, strawberries designed to be
sweeter and longer-lasting.[44] In some countries, including the United States
and Japan, gene-edited soy oil with reduced trans and saturated fats is on
store shelves,[45] as is a tomato infused with gamma-aminobutyric acid (GABA),
marketed for its purported health benefits like stress relief and blood pressure regulation.[46]

CRISPR's appeal stems partly from how it differs from earlier genetic
engineering methods. Rather than introducing foreign genes from unrelated
organisms, CRISPR works by editing existing genes within the plant or animal's own genome, often drawing from genetic diversity already present
within a species' broader family. Using bacterial-derived molecular "scissors,"
scientists can cut, disable, or rewrite specific DNA sequences with remarkable
precision, sidestepping some of the regulatory and public resistance faced by
traditional transgenic techniques.[47] Proponents of CRISPR frame it as an extension, or acceleration, of natural evolutionary processes or conventional
breeding, minimizing perceived risk and distancing the technology from the
controversies surrounding first-generation genetically modified organisms
(GMOs). They argue that the refined accuracy of CRISPR reduces off-target
effects and preserves desired traits. However, critics point out that even small

genetic alterations can result in cascading and unpredictable impacts, particularly in complex ecological systems where gene expression is not isolated from environmental, microbial, and interspecies interactions.[48] From this perspective, gene editing reproduces a familiar reductionist pattern: treating life as a modular system of traits to be optimized rather than a relational and adaptive web. By focusing narrowly on single-gene interventions, rather than on biodiversity, soil health, or collective food sovereignty, CRISPR technologies risk amplifying ecological homogeneity. Engineered traits could enable crops to outcompete wild relatives, reducing genetic diversity and destabilizing local ecosystems. These outcomes are not theoretical; they echo long-standing concerns about industrial agriculture's role in degrading the commons. The same logic is extending to animals, where gene editing is increasingly applied to reduce the symptoms of suffering rather than the structures that create it. One disturbing example involves the engineering of blind chickens for industrial egg production—a move intended to reduce stress responses in hyper-crowded environments. Such interventions do not address the systemic violence of factory farming; rather, they normalize and deepen it, removing the biological "indicators" of distress without dismantling the conditions that cause it. The result is not welfare, but the optimization of suffering.[49] In this light, gene editing reflects not just a new technology but a deepening entrenchment in the modernist fantasy of control, one that avoids the harder relational questions about how we live with, consume, and relate to more-than-human life. Rather than transforming the system that produces hunger, degradation, and violence, CRISPR risks perfecting it.

HACKING PHOTOSYNTHESIS

Not content with modifying individual traits of plants and animals, many scientists have turned their attention to the ultimate optimization project: reengineering the very process that sustains most life on Earth: photosynthesis. While often presented in textbooks as a simple, elegant equation—plants absorbing carbon dioxide and water, transforming them with sunlight into sugars for growth, and releasing oxygen—photosynthesis is, in fact, one of biology's most intricate and delicate processes. It involves over 170 biochemical steps, operates through two interdependent cycles, and includes three different carbon-fixing pathways.[50] First documented by scientists in the eighteenth century, photosynthesis has long been considered a marvel of evolutionary refinement—complex, but already optimized by millions of years of adaptation. More recently, however, this evolutionary complexity is being reframed not as a finely tuned system but as an inefficient machine in need of repair. As scientific understanding deepens, certain features of photosynthesis are being recast as design flaws, targets for technological intervention.[51]

One of the most prominent efforts in this direction is the RIPE (Realizing Increased Photosynthetic Efficiency) project at the University of Illinois.[52] Under the familiar narrative of feeding a growing population—a rhetorical device that often masks deeper structural issues in food distribution—researchers are attempting to "fix" photorespiration, a process seen as a drag on productivity. Photorespiration occurs when RuBisCo, a key enzyme that typically catalyzes the conversion of atmospheric carbon dioxide into sugars, mistakenly binds with oxygen instead. This error produces toxic byproducts that the plant must then detoxify through a labyrinthine process involving multiple steps across four cellular compartments. This metabolic detour is like a road trip from Maine to Florida via California: energy-intensive and inefficient. In response, the RIPE team has designed a streamlined East Coast route by genetically engineering tobacco plants to perform this detoxification within a single cellular compartment,[53] thus simplifying the chemical pathway and boosting efficiency.[54] Encouraged by these results, the RIPE project is pushing forward with more ambitious alterations,[55] including the insertion of synthetic carbon-concentrating pumps into chloroplast membranes. The goal is to increase the local concentration of carbon dioxide around RuBisCo, which researchers claim could yield up to a 60 percent increase in crop output.[56]

The RIPE project's mission statement crystallizes the dominant productivity-driven narrative: "to end hunger worldwide by improving the complex process of photosynthesis to increase crop production. By equipping farmers with higher-yielding crops, we can ensure that everyone has enough food to lead a healthy, productive life."[57] Similarly, a team of Brazilian researchers recently argued that: "As traditional plant breeding is most likely reaching a plateau, there is a timely need to accelerate improvements in photosynthetic efficiency by means of novel tools and biotechnological solutions."[58] This framing reflects a powerful continuity: the persistent belief that technological intensification is the only viable path forward, especially under the specter of population growth and climate disruption. But this enthusiasm often bypasses a more grounded reality, namely, that nutrient availability, water access, and ecological limits play a far greater role in determining crop yields than the efficiency of photosynthesis alone. Despite decades of effort, there remains scant evidence linking enhanced photosynthetic capacity to meaningful increases in grain production for human consumption.[59] Some media outlets describe these efforts as "hacking" photosynthesis, with a cautious footnote: It's still too early to know what unintended consequences might arise.[60] But the very framing of photosynthesis as something to be hacked—optimized, accelerated, made more productive—exposes a deeper cultural orientation.

In the face of ecological collapse, instead of reckoning with unsustainable consumption, extractive agricultural systems, or unjust food distribution, we invest our hope in molecular interventions that keep the broader structure intact. This is not simply about scientific curiosity or technical challenge. As difficult as it may be to modify one of life's most complex biochemical processes, such pursuits are, in many ways, more palatable than confronting the harder questions: Who benefits from these technologies? Who bears their risks? And what does it cost to continue doubling down on the industrial logic of "feeding the world" when the actual crisis lies not in food scarcity but in relational breakdown between people, land, and systems of care? Choosing to edit plants rather than address the political and ecological dimensions of food insecurity reveals a profound disinvestment in transformation. It reaffirms a familiar story: that nature must be conquered, not listened to; that human ingenuity, rather than relational accountability, will carry us through collapse.

MAKING BAD WORSE

The ecomodernist narrative reassures us that the very tools and logics that contributed to the breakdown of food systems can now be redeployed to repair them. It suggests that with enough technological refinement, we can innovate our way to a more sustainable future, without fundamentally altering the structures that created the crises in the first place. This narrative conveniently sidesteps the fact that many of the problems biotech companies and agricultural researchers claim to be solving, such as pesticide resistance and soil degradation, are direct outcomes of industrial monocultures and extractive farming models. What remains unexamined are the systemic forces that entrench poverty, accelerate climate instability, and fuel ecological unraveling. By framing the core challenge narrowly as "feeding a growing population" and measuring success primarily through yield and efficiency metrics, the dominant paradigm of industrial agriculture is not questioned; it is further legitimized. This framing perpetuates a path toward greater corporate consolidation and control over food production, distribution, and consumption, while accelerating trends toward digitization, surveillance, and algorithmic governance under the guise of "sustainability."[61] At the heart of this worldview is the unspoken belief that technology is socially neutral: that innovations developed in one context can be transplanted anywhere without friction, resistance, or unintended consequences. In this view, tools are presumed to be universally beneficial and free of cultural, political, or historical entanglements.[62]

This technocentric approach is not confined to the realm of genetic engineering. It reflects a broader political and epistemological orientation—one that prioritizes scale, efficiency, and standardization. Within this framework, "solutions" must work at the level of large-scale industrial farms, often to the exclusion of smallholder, diversified, or land-based food systems. Innovation pathways are thus shaped by the demands of capital-intensive operations, leaving little space for agroecological approaches rooted in local knowledge, relational accountability, or interdependence with land and community.

One prominent example of this logic is the rise of precision agriculture. Positioned as a corrective to the degradation caused by past practices—overuse of fertilizers, pesticides, water, and land—precision agriculture introduces a new suite of technologies: satellite-guided tractors, aerial drones, soil sensors, and GPS-mapped irrigation systems.[63] These tools promise to increase yields while reducing input costs, aligning with the familiar promise of doing more with less. Farmers can now level fields using lasers, optimize planting patterns with Global Positioning System (GPS) algorithms, and use drones to map moisture, deploy seed pods, or apply exact doses of chemicals at optimal times. Thermal imaging is used to monitor livestock, while 3D mapping analyzes runoff and drainage potential.

The integration of robotics and artificial intelligence into agricultural labor is increasingly being framed as a cost-effective response to labor shortages and a liberatory innovation that frees farmers from the burdens of repetitive, physically demanding tasks. At first glance, this appears pragmatic—relieving human toil and enhancing productivity. For example, researchers at Cambridge University have developed a prototype known as Vegebot, a robotic harvester capable of delicately picking lettuces, a task once considered too complex and fragile for mechanization.[64] By combining computer vision with real-time processing, Vegebot scans the field, assesses ripeness, and executes a calculated harvest.[65] Meanwhile, companies like Nexus Robotics have designed autonomous weed-removal systems, such as R2Weed2, that use machine learning to distinguish crops from weeds and simultaneously perform soil analysis.[66]

These technologies extend the promise of precision agriculture: to extract maximum value from each plant, reduce waste, and mitigate environmental impacts compared to older, more polluting farming methods. As one 2015 article in *Foreign Affairs* optimistically put it, precision farming could help feed a projected 9.6 billion people by 2050, revolutionize land and livestock management, and ultimately "eliminate the drudgery that has characterized agriculture since its invention."[67]

Yet behind this narrative lies a deeper question: What is being optimized, and for whom? Precision farming may reduce agrochemical exposure and offer marginal gains in efficiency, but it also accelerates the mechanization and digitalization of rural life, often at the expense of smallholders, laborers, and the socio-ecological fabric of agricultural communities.[68] Rural livelihoods depend not only on access to land but also on access to labor opportunities in agricultural sectors. When jobs are replaced by machines, and knowledge by algorithms, we risk deepening patterns of displacement, urban migration, and unemployment. This, in turn, heightens social precarity, particularly in areas already struggling with weak infrastructure and minimal support for land-based economies.[69] The turn toward "smart farming" technologies also carries significant economic thresholds. High up-front costs make these systems largely inaccessible to small- and medium-scale farms, ironically, the very producers most impacted by labor shortages and climate volatility.[70] While automation may reduce certain operational costs, such as labor or pesticide use, it can increase others—particularly energy consumption and maintenance. The question then becomes not only whether these technologies work, but *who* can afford them, and what relational costs they carry.

Perhaps most concerning is how digital agriculture reconfigures power. As decision-making becomes increasingly mediated by software, sensors, and proprietary algorithms, farmers risk becoming data laborers in a system they do not control. Corporations that manage farm-level data, gathered through precision tools, gain unprecedented access to intimate details about farming operations, often on a field-by-field, crop-by-crop basis. In the United States, it's estimated that data aggregators already track information from over a third of farmland.[71] Confidential investigations have shown that this data is not always kept within agricultural networks; some is sold to hedge funds and real estate investors seeking to acquire farmland.[72] In such cases, the farmer is no longer just a food producer but a data source feeding financial speculation—disadvantaged in land markets now influenced by predictive analytics and private capital.

Centering precision farming as the primary solution to food system challenges reinforces a narrative that legitimizes chemical-dependent, industrial farming and deepens our reliance on monocultures.[73] Much like genetically engineered seed systems, the focus of corporate research and development in precision agriculture is narrowly trained on major commodity crops—those most profitable within the existing global trade infrastructure.[74] The data streams that feed commercial agritech platforms are generated by expensive sensing technologies, often embedded in machinery outfitted with telematic systems that merge GPS navigation, sensors, and onboard informatics. These tools are typically accessible only to large-scale industrial operations, not to smaller or diversified farms. To make use of the insights produced by these platforms, farmers must already possess machinery capable of "variable-rate" input applications, for example, targeted deployment of pesticides, fertilizers, or irrigation. Without such infrastructure, the location-specific data is essentially unusable. This architecture mirrors earlier patterns in industrial agriculture: Just as monocropping was necessary to maximize the use of large-scale planting and harvesting equipment, today's data-driven farming maps assume a uniform layout—straight rows, bare soil in between, and minimal variation. The technology prescribes a specific form of land use, reinforcing a standard of efficiency that is blind to ecological nuance and cultural difference.[75]

Moreover, automation technologies are inherently optimized for homo-geneity. Robotic systems work best in controlled, repetitive environments—conditions most easily achieved through monoculture. As a result, emerging agri-robots are designed to serve specific grain, vegetable, or fruit crops that can be made to conform to mechanized patterns. This not only narrows the range of crops prioritized for development but also pressures farmers to shift their practices, either by consolidating land into larger, uniform tracts or by abandoning diverse crop systems that do not fit the robotic mold. This tech-nological lock-in exacerbates ongoing trends of farmland consolidation and corporate concentration. Small and mid-sized farms, particularly those en-gaged in agroecological practices, are increasingly pushed to the margins: economically, politically, and ecologically. The implications extend far beyond economics. As the diversity of what we grow shrinks, so too does the resil-ience of the ecosystems we rely on and the nutritional biodiversity of our di-ets. What is at stake is not merely agricultural productivity but the capacity of food systems to sustain life in dynamic, responsive, and place-based ways.

GREENING FACTORY FARMS

A critical component of any meaningful food system transition in response to climate change and the intersecting crises of human and planetary health is a significant reduction in the consumption of meat and other animal-based products, particularly in industrialized Western nations.[76] In the United States alone, the average person consumes approximately 225 pounds of poultry and livestock products annually, including around 110 pounds of pork and beef.[77] Globally, meat consumption has been steadily increasing over re-cent decades.[78] To meet this demand, it's estimated that between 3.4 and 6.5 billion animals are slaughtered every day for food—an industrial scale of kill-ing with profound ecological, ethical, and public health implications.[79] As discussed earlier, industrial animal agriculture is one of the most destructive drivers of greenhouse gas emissions, land degradation, biodiversity loss, water contamination, pesticide overuse, and antibiotic resistance. Among the scientifically recognized strategies to reduce one's environmental impact, avoiding factory-farmed meat and dairy remains one of the most effective.[80]

And yet, even as the consequences become more widely known, many remain resistant to significantly reducing their meat consumption.[81] In

Western societies especially, high levels of meat intake are sustained not only through structural access and dietary habit but also through psychological strategies of dissociation, such as concealing the animal origins of food[82] or engaging what scholars call "strategic ignorance."[83] This internal dissonance, between caring about animal suffering and continuing to consume meat, is known as the "meat paradox."[84]

Into this tension steps the technological fix. Instead of confronting the deeper ethical and ecological questions—such as why we continue to funnel massive amounts of land, water, and energy into growing corn and soy to feed animals, or whether the cultural obsession with protein is scientifically warranted—the dominant response has been to make factory farming more "efficient." The preferred approach of most governments, including the United States, has not been to question the scale or structure of livestock production but to retrofit it with emissions-reducing technologies. As then–Secretary of Agriculture (and former dairy lobbyist) Tom Vilsack stated in 2021, "I do not think we have to reduce the amount of meat or livestock produced in the U.S." to meet climate goals.[85] He reiterated the familiar productivist refrain: To feed a global population of 9 billion, we need all kinds of protein: plant, animal, and fish. In order to meet international climate commitments,[86] the Biden administration chose to invest in technical solutions like vaccines, feed additives, and methane capture systems rather than shift consumption patterns or dismantle industrial meat production.[87]

One such solution is the use of anaerobic digesters, large machines that trap methane emitted from manure lagoons on factory farms and convert it into so-called biogas for energy production.[88] Though marketed as a renewable energy source, these digesters are costly to construct and operate.[89] Their proliferation has only been made possible through significant public subsidies: programs like California's $200 million Dairy Digester Research and Development Program,[90] the USDA's Environmental Quality Incentives Program (EQIP), and the Rural Energy for America Program (REAP) have poured funds into this infrastructure. The EPA's AgSTAR Program also provides free technical support,[91] and federal and state renewable energy mandates such as the federal Renewable Fuel Standard (RFS) create markets for this energy, further entrenching the factory farm model by rebranding its waste as a resource.[92]

Though marketed as a dual-benefit strategy for reducing greenhouse gas emissions and generating renewable energy, the expansion of biogas infrastructure is generating perverse incentives that ultimately deepen the harms it purports to address. Subsidies intended to reward methane capture are, in practice, encouraging the expansion of herd sizes, the proliferation of factory farms, and a corresponding increase in overall emissions. These financial incentives, layered on top of existing government support, create new revenue streams from manure, disproportionately benefiting large-scale operations and widening the gap between industrial megafarms and smaller, more diversified producers. In fact, some industrial dairy operations are now profiting more from waste than from milk. *Hoard's Dairyman* reports that manure supplied to digesters can generate greater returns than dairy production itself. As Steve Sheheady of Bar 20 Dairy remarked: "We used to joke about how funny it would be if we could make more money off the poop than the milk. And now we're essentially here."[93] When animal waste becomes an asset, it creates an economic incentive to produce more of it, rewarding scale and reinforcing consolidation.[94] According to the EPA, a farm is considered a viable candidate for digester installation if it maintains at least 500 cows or 2,000 hogs and has a liquid or semi-solid manure system— conditions typically met only by large CAFOs.[95] The push to expand digesters thus aligns with the continued industrialization of livestock production. And with that comes a host of externalized harms: air and water pollution, odor, pathogen exposure, and disproportionate health impacts on frontline communities.[96]

In places like North Carolina, these harms are not hypothetical; they are lived realities. For decades, researchers have documented the consequences of hog CAFO concentration in low-income, predominantly Black and Latinx communities.[97] One study found that residents living near these operations suffer significantly higher rates of all-cause mortality, infant mortality, and deaths from diseases such as anemia, kidney disease, tuberculosis, and septicemia. Hospitalizations, emergency room visits, and low-birth-weight infants are also more common.[98] Despite this extensive evidence, state legislation has supported projects to connect CAFOs to methane capture and gas pipeline infrastructure. One such initiative plans to retrofit 19 existing hog farms with massive lagoon-based digesters, processing methane into so-called renewable natural gas. For many environmental justice and community groups, this signals not a shift toward sustainability but the entrenchment of a toxic system under a green veneer.[99]

Framed as an eco-friendly innovation, digesters function as a costly distraction, what some call a "false solution," that blocks necessary public dialogue about the fundamental incompatibility between industrial animal agriculture and long-term climate, ecological, and public health goals. Rather than confronting the root causes of agricultural emissions—excessive meat production, unsustainable herd sizes, and extractive waste management systems—digesters allow business-as-usual to continue, merely rebranded in the language of renewable energy.

ALT PROTEIN AND THE PROTEIN WAR

Another emerging "solution" to the harms of industrial animal agriculture is being offered by a growing alliance of start-up food companies, techno-optimist think tanks, philanthro-capitalists, and advocacy organizations: a technologically driven protein transition.[100] In September 2019, the futurist think tank RethinkX proclaimed, "We are on the cusp of the deepest, fastest, most consequential disruption in food and agricultural production since the first domestication of plants and animals ten thousand years ago."[101] At the heart of this projected upheaval are alternative, or "alt," proteins—framed by ecomodernists as a key pathway to feed the world while decoupling food production from ecological destruction.

While vegetarian and vegan diets have existed across cultures for millennia, and meat substitutes like tofu, tempeh, and seitan have long been staples in many traditions, recent developments in molecular food science have expanded the capacity to simulate meat more precisely. Researchers and companies now have the biochemical tools to deconstruct and replicate the specific interactions of amino acids, proteins, lipids, carbohydrates, and salts that give meat its distinctive flavor, texture, and mouthfeel. Capitalizing on these innovations, companies like Beyond Meat[102] and Impossible Foods[103] are producing plant-based alternatives that mimic the sensory profile of meat, milk, eggs, and seafood. To optimize their production processes, companies are increasingly turning to artificial intelligence. For instance, Kraft Heinz has partnered with NotCo, a food tech start-up powered by AI, to reverse-engineer animal products using a proprietary algorithm named Guiseppe.

This technology enables them to recreate familiar animal-based foods, molecule by molecule, using plant-based ingredients.[104]

Another rapidly expanding frontier is cellular agriculture.[105] Since the debut of the first lab-grown burger in 2013 by biochemist Mark Post, over 150 companies have entered the race to commercialize "cultivated" meat, dairy, eggs, and seafood, produced by growing animal cells in controlled environments without raising or slaughtering animals.[106] Cultivated meat has now reached limited restaurant menus and, as of 2023, has received regulatory approval for retail sale in the United States.[107] Alongside this, precision fermentation is being used to produce animal-identical proteins via genetically engineered microbes like yeast or bacteria.[108] These are now found in several consumer products: Impossible Foods' burgers include bioengineered heme, a molecule that gives meat its bloody taste when raw and meaty flavor when cooked,[109] Perfect Day's Brave Robot offers dairy-like ice cream made from animal-free milk proteins,[110] and Every is marketing an egg white created through microbial fermentation.[111]

The scope of innovation continues to expand. Companies have replicated a growing array of animal proteins—from fish and chicken to butter and even mammoth, with some developers celebrating this as evidence that "we can produce animal proteins without ever touching one."[112] One company, Solar Foods, claims to be producing protein powder from air-metabolizing microbes and suggests their process heralds a new era: "For the first time in history, humankind can produce food without burdening our home planet."[113] The promises bundled into this alt-protein revolution are undeniably seductive. Consumers are invited to imagine a world where animal suffering is eliminated, environmental degradation reversed, and hunger solved—all through market innovation. As one promotional narrative boasts: "Imagine producing the entire world's protein on an area of land the size of Greater London. Imagine rewilding 3/4s of today's farmland. Imagine eating guilt-free meat, milk, and cheese without ever having killed an animal. Imagine providing abundant food to the world's poorest."[114]

Impossible Foods founder and former CEO Pat Brown once stated, "You don't solve the problem [of climate change, environmental degradation, etc.] by asking people to change their diets."[115] This statement encapsulates the techno-optimist ethos—rooted in the mid-twentieth-century thinking of nuclear physicist Alvin M. Weinberg—which proposes that complex social and environmental problems are best addressed through "quick technological fixes" rather than the messier, slower work of changing human behavior, values, or cultural systems.[116] The underlying assumption is that meat consumption is too deeply embedded in cultural norms, especially in the United States, to be meaningfully challenged. Therefore, the "pragmatic" solution is to create seamless substitutes: products that mimic meat so closely in taste, texture, and price that consumers can make the switch without any ethical or behavioral shift. This framing renders ethics optional, and in many ways, irrelevant. When plant-based alternatives become indistinguishable from animal products in flavor and convenience, consumption is reduced to an automated market transaction. Philosopher of technology Langdon Winner has called this stance "technological orthodoxy," a worldview in which technological development is perceived as separate from, and even superior to, moral or political deliberation.[117] Within this logic, the new generation of alt protein products is not just a business opportunity but a supposedly apolitical way to "save the planet" without requiring consumers to sacrifice taste, habit, or affordability. Fueled by this seductive narrative, and the enduring trope of "feeding the world," alt protein advocates have succeeded in mobilizing substantial public investment. Governments worldwide are now funding partnerships and infrastructure projects exceeding $1 billion,[118] with many positioning alternative proteins as essential components of national food security strategies.[119] WePlanet, an ecomodernist network of environmental organizations,[120] is urging the European Union to back a $25 billion "moonshot" investment to accelerate a rapid protein transition.[121] Their Open Letter to the EU President declares that "sustainable protein production shows the potential to displace the most harmful forms of animal agriculture at a speed and scale previously unimaginable."[122]

Yet these sweeping proclamations often ignore key nuances, particularly the diversity of global food systems and cultural relationships to land and animals. Pastoralist traditions, agroecological practices, and community-based foodways are largely absent from the alt protein narrative.[123] Instead, the conversation is cast in binary terms: Industrial meat is bad, so industrial alternatives must be good. This false opposition ignores the ecological and social variation among different animal and crop production systems. Most alt protein proponents are "production agnostic," reluctant to distinguish between conventional, regenerative, or biodiverse methods of cultivation and processing.[124] In many cases, they rely on the same extractive logics that undergird the very systems they claim to replace. Embedded in much of the alt protein discourse is a view of agriculture itself as inherently harmful to nature. As Solar Foods has proclaimed: "Protein production is a massively disproportionate squanderer of the Earth's resources. It's time to enter the era of sustainable food production to liberate our planet from the burdens of agriculture."[125] This narrative, while attention-grabbing, flattens the profound complexity of land-based relationships and treats food not as nourishment or culture but as an engineering problem to be solved with synthetic biology.

A hallmark of this framing is nutritionism: the reduction of food to its molecular components, especially protein. Rather than asking how to nourish communities in just and ecologically grounded ways, the focus is on optimizing protein delivery systems. Beyond Meat, for example, has trademarked the slogan "The Future of Protein" and envisions transforming the meat aisle into a generalized "Protein Section."[126] This rhetoric is now being adopted by industrial giants like Tyson, Hormel, and Cargill, which have rebranded themselves as "protein companies," selling both animal and plant-derived products. In this model, food is no longer embedded in place, memory, or relationship; it becomes a neutral vehicle for nutrients.

While factory farming is indisputably destructive, the alt protein industry may ultimately reproduce many of the same paradigms that justified it: industrialization, corporate consolidation, worker exploitation, the commodification of life (even down to animal cell lines), and input-intensive monocultures reliant on chemical-laden soy, pea, and wheat crops. The question is not just what we replace meat with, but whether we are reimagining the conditions under which food is grown, shared, and valued—or simply repackaging the same logic under a new label.

THE CURSE OF EFFICIENCY

Beyond the unresolved technical debate over whether alternative proteins produced via fermentation and cellular agriculture are genuinely more efficient or sustainable than conventional options,[127] a broader systemic blind spot persists: the rebound effect, also known in environmental economics as Jevons's paradox.[128] First articulated by British economist William Stanley Jevons in 1865, the paradox notes that increased efficiency in resource use often leads to *more* consumption of that resource, not less. Jevons observed that as coal-burning technologies became more efficient during the industrial era, coal consumption didn't decline; instead, it surged, driven by expanded industrial activity and lower operating costs. This pattern has held across generations. In a comprehensive study, researchers at the Massachusetts Institute of Technology found that nearly every material and energy source used in industrial production has seen increased overall consumption despite becoming more efficiently utilized. The only exceptions tend to be substances banned or heavily regulated due to their toxicity—like asbestos and mercury—not because efficiency curbed demand.[129] The prevailing assumption in many environmental and technological discourses is that improving efficiency, of land, fuel, water, or synthetic biology, naturally leads to conservation. But in practice, efficiency gains often reduce costs and increase profitability, driving expansion and intensifying extraction. Unless bounded by strong regulatory or ecological limits, these gains can accelerate resource depletion.[130]

This phenomenon has been observed repeatedly across sectors. Between 1975 and 2010, refrigerator efficiency in the United States improved by 10 percent, yet the number of refrigerators in use rose by 20 percent. Over the same period, airplanes became 40 percent more fuel-efficient per mile, while total aviation fuel consumption increased by 150 percent. Efficiency, without structural constraints, fuels scale—not restraint.[131] In agriculture, this paradox is particularly potent. Advanced irrigation technologies, intended to reduce water use, have in many regions led to higher total water extraction as farmers expand cultivated areas.[132] Oil palm is another example. It is far more land-efficient to grow than crops like sunflower or canola oil, requiring five to eight times less land to produce the same volume of oil.[133] Yet rather than sparing land, this "efficiency" has fueled a dramatic expansion in palm plantations, with land use for palm oil quadrupling since 1980, primarily in Southeast Asia and South America, at the cost of tropical forests and Indigenous territories.[134] Despite this, some plant scientists continue to call for genetically engineering more "efficient" palm varieties as a solution to deforestation, perpetuating the cycle.[135]

Jevons's paradox has also undermined assumptions about the Green Revolution's land-sparing benefits. While crop yields per acre increased threefold during that era, land conversion for agriculture in regions like Africa, Asia, Oceania, and Latin America has also expanded. The promised conservation did not materialize.[136] In fact, global fossil fuel inputs for agriculture, especially through machinery and synthetic fertilizers, have grown by 137 percent per hectare over the past fifty years.[137]

When viewed through the lens of capitalism, Jevons's paradox exposes a core contradiction in the belief that environmental degradation can be solved through technological innovation alone.[138] In an economic system oriented around perpetual growth, profit maximization, and accumulation, any gain in efficiency is more likely to fuel expansion than conservation. Rather than reducing overall resource use or environmental impact, efficiency tends to lower costs and raise output, thereby accelerating extraction and ecological harm. This dynamic directly challenges the ecomodernist faith in technological intensification, especially in agriculture, as a pathway toward sustainability. A clear example is the industrial production treadmill of genetically engineered soy. Since 2002, soybean yields in the United States have increased by roughly 30 percent, primarily through the adoption of genetically modified varieties and intensified input use. Yet instead of leading to land sparing, soybean acreage has grown by 18 percent, increasing about 1 percent a year. These yield gains are underpinned by heavier reliance on fertilizers, fungicides, insecticides, and precision agriculture technologies—all of which raise per-acre production costs. But the logic of expansion persists: Increased yields offset rising input costs, making expanded cultivation profitable and, ultimately, encouraging more land conversion.[139]

Jevons's logic doesn't just apply to efficiency-enhancing technologies. It also manifests in what might be called the "displacement paradox," the failure of supposedly "sustainable" substitutes to reduce reliance on the very systems they aim to replace. For instance, decades of global energy data show that as non-fossil fuel sources (like solar and wind) have expanded, overall energy consumption has grown, while fossil fuel use has declined only marginally. In other words, renewables have supplemented, not supplanted, carbon-intensive energy sources.[140]

This same trend appears in the global food system. Chicken, often touted as a more sustainable protein due to its relatively lower greenhouse gas emissions per pound compared to beef or pork, has seen its consumption skyrocket.[141] Over the last sixty years, per capita chicken consumption rose from 6.3 to 37.4 pounds—a nearly 600 percent increase.[142] But this hasn't come at the expense of other meats. Instead, it has added to the total volume of animal products consumed. Industrial chicken farms, like other factory farms, still contribute substantially to emissions, rely on monocultured feed crops like corn and soy, and generate pollution through manure and fossil fuel use.[143] A similar pattern emerges with aquaculture. Despite being hailed as a more efficient alternative to wild-caught seafood, the rise in farmed fish production has not led to a reduction in wild capture rates. Instead, global seafood consumption has increased overall.[144]

Alt proteins are following this same trajectory. Although these products are marketed as tools to displace meat consumption, the evidence suggests they are being layered onto existing consumption patterns rather than replacing them. Sales of plant-based meats have plateaued or declined in key markets such as the United States[145] and Australia.[146] Where growth has occurred, it has often paralleled rising meat sales,[147] with the possible exception of Germany.[148] Moreover, even where plant-based meat sales have grown, they have often coincided with increasing meat consumption rather than replacing it. One study even found that plant-based meat and beef function as price complements, not substitutes: Consumers are simply buying both.[149]

Major food conglomerates appear to be banking on this outcome. Cargill's investment strategy exemplifies this "both/and" model. With a growing portfolio of plant-based and cell-cultured products, the company has no intention of scaling back traditional meat. As Florian Schattenmann, chief technology officer of Cargill, put it, "That's why this is an 'and' story for us. We expect, over the next 25 to 30 years, 70 percent more protein consumption [among] an increasing world population. That protein has to come from somewhere. We want to be that diversified protein supplier."[150] Tyson Foods expressed a similar stance in its 2019 alt-protein launch, with then-CEO Noel White stating bluntly: "We're creating new products for the growing number of people open to flexible diets that include both meat and plant-based protein. For us, this is about 'and'—not 'or.'"[151] Far from dismantling the logic of industrial protein, alternative proteins are being positioned as another high-growth market within it. These dynamics underscore a deeper problem: So long as growth remains the measure of success, and industrial logic the baseline, new technologies are unlikely to displace harmful systems. Instead, they risk extending and repackaging them under the banner of sustainability.

AVOIDING THE SIREN SONG OF SOLUTIONISM

Unless we rigorously interrogate the potential for perverse incentives, unintended consequences, and tradeoffs, the allure of cutting-edge technologies—and the seductive narratives of surpassing nature's limits—risks overshadowing other, often more grounded and transformative, pathways to food security and sustainability. As a group of critical food scholars aptly asked in the context of protein innovation: "Are there roads not being taken because the map has been drawn by those who already know where they want to go?"[152] When technology is driven by preset outcomes rooted in industrial paradigms, it often narrows the field of imagination, sidelining plural, relational, and justice-centered approaches. The techno-optimist worldview also tends to treat social impacts as secondary or peripheral. A 2022 literature review on emerging agricultural technologies found that most studies focused on productivity gains to feed a growing population. About half referenced potential environmental benefits. Yet fewer than half addressed social dimensions, and when they did, the focus was typically limited to consumer concerns like traceability, dietary change, or public acceptance.[153] Deeper issues of justice, equity, labor, and ethics were largely absent, as were the relational entanglements that food systems both depend upon and disrupt. As we'll explore in Chapter 6, technological innovation can indeed play a role in shaping regenerative food futures. But the interconnected ecological and social breakdowns we now face cannot be treated as engineering problems alone.[154]

Technology is never neutral; it reflects the intentions, worldviews, and blind spots of its creators. When designed to serve narrow metrics like efficiency, profitability, or productivity without attending to ethical, cultural, and systemic concerns, it not only falls short—it often generates new layers of harm. There is a tangible risk, already playing out, that the hype surrounding food tech "solutions" will divert resources and attention away from already existing, community-rooted, and ecologically sound strategies that could catalyze real transformation. These are the very strategies we will turn to in the next chapter. The deeper danger is not merely technological overreach but the complacency it fosters: a collective sigh of relief that "they've fixed it," just when we most need to abandon the comfort of fixes and step into the discomfort of reimagining the social, economic, and political architectures—and underlying cosmologies—that have shaped the industrial food system.[155]

PART TWO

Restoring Our Relationship to Food

4

Farming with Nature

Coming to terms with the immense and intertwined harms wrought by the industrial agricultural system—its contribution to climate disruption, biodiversity collapse, and the erosion of human and planetary health—can be overwhelming. The scale and complexity of the crisis often provoke a sense of paralysis or despair. These are not just policy failures or technical glitches; they are symptoms of a worldview that has normalized separation, extraction, and control.

To genuinely transform food systems, we must first face a sobering reality: We are not going to "save the world." There will be no sweeping top-down revolution that dismantles global industrial agriculture or mandates a universal return to localized, organic systems. Nor is there a technological fix waiting in the wings to solve the cascading crises we face. As Buckminster Fuller reminds us, "You never change things by fighting the existing reality. To change something, build a new model that makes the existing model obsolete." The work ahead does not lie in global overhauls but in grounded, place-based acts of repair. While we cannot engineer a singular solution, we *can* regenerate communities, restore relationships with land, and build resilient local food systems that reflect values of reciprocity, care, and interdependence. This requires moving beyond the old narratives of growth and mastery, and cultivating a new orientation—one that reconnects food with the web of life.

There are many worthwhile and necessary interventions underway: campaigns to ban harmful pesticides, policies to rein in factory farming, efforts to reward regenerative practices, and calls to ensure living wages for food and farm workers. All of these matter. But in this chapter, and throughout the rest of this book, we turn our attention to a deeper, systemic shift: a reweaving of food into a new operating system, not just for agriculture but for how we live together on a rapidly changing planet. The shift has been described as moving from *ego* to *eco*. As Otto Scharmer of the Presencing Institute puts it, the dominant societal model is still anchored in "me first," while an eco-systemic model centers the wellbeing of all life.[1] It echoes the African wisdom of ubuntu: "I am because we are." What follows are stories of people across the world who are reclaiming degraded lands, reimagining food not as a commodity but as a sacred relationship, and rebuilding cultural narratives that nourish both ecosystems and communities. Within the framework of regeneration lies a powerful possibility: to cultivate food systems that are not only resilient and adaptive in the face of climate uncertainty but also rooted in equity, dignity, and mutual care. At the heart of this vision is a simple but profound recognition: that the Earth, our food, and our bodies are not separate—they are interdependent, living systems, co-shaping one another with every breath and bite.

BEYOND SUSTAINABILITY

Our current systems of food production and consumption have pushed well beyond the ecological thresholds of a livable planet. Corrective action is no longer optional; it is urgent. Yet what's needed is not just new tools or better policies but a fundamental shift in mindset. The concept of *regeneration* speaks to this deeper reorientation. Regeneration is not only about how we grow food—it's about how we live as parents, citizens, educators, entrepreneurs, and members of an Earth community.[2] As Regeneration International, a global network launched in 2015 to mobilize farmers on all six arable continents, puts it: "Regenerative systems improve the environment, soil, plants, animal welfare, health, and communities."[3] Rooted in the Latin *regenerare*— "to create again"—the regenerative movement invites us to imagine a future in which systems not only avoid harm but actively heal what has been damaged.[4]

After decades of degradation to soils, waterways, and biodiversity, and bodies, simply slowing the rate of destruction is insufficient. Becoming "less bad" is no longer a viable aspiration. We must move beyond harm reduction to cultivate systems that are net-positive, systems that repair, nourish, and regenerate life. Importantly, "regeneration" is not just a trendier substitute for the overused term "sustainability." While sustainability, as defined by the 1987 United Nations Brundtland Commission, aims to meet present needs without compromising future generations, its mainstream interpretation has often been diluted.[5] The UN's Sustainable Development Goals (SDGs) established in 2015 to address poverty, inequality, climate change, environmental degradation, peace, and justice offer a useful global framework.[6] Yet despite these well-intentioned efforts, the crises have deepened. Climate instability, resource exhaustion, hunger, and systemic inequity persist, often worsened by the very paradigms sustainability efforts have failed to disrupt. Government responses, philanthropic investments, and corporate sustainability initiatives often double down on the same industrial logic that created these crises. Emissions are managed through "net-zero" targets, soil and water are extended as "resources" for longer use, and harm is framed as acceptable if efficiently mitigated. These are incremental responses to systems already in collapse.[7]

Sustainability, while valuable for its recognition of intergenerational stewardship, often remains trapped in a resource-use paradigm: nature as a tool, humans as its managers.[8] What's missing is a confrontation with root causes: capitalism, colonization, commodification, and the foundational narrative of separation from Earth, from one another, and from more-than-human kin. Incrementalism and technocratic solutions not only fall short, they narrow our imagination, making it harder to envision radical alternatives.[9] Regeneration invites us to expand that imagination.[10]

Though the term is gaining traction, its roots run deep. Philosophies of regeneration are embedded in Indigenous, Eastern, and Western cosmologies and continue to be lived in practice in many communities around the world. For example, the Andean concept of *sumak kawsay*—translated from Kichwa as "the fullness of life"—underpins *Buen Vivir*, a cultural and spiritual principle enshrined in the constitutions of Ecuador and Bolivia. It centers collective wellbeing, ecological harmony, and relational interdependence over individual gain. It offers an embodied alternative to the Western individualist notion of "wellbeing," which often ignores systemic and ecological context.[11] Similarly, the Rarámuri (Tarahumara) concept of *iwígara*, described by ethnobotanist Enrique Salmón, articulates a worldview where all beings share a common breath. Rocks, bugs, fish, humans—we are all kin, equally enmeshed in the web of life.[12] This resonates with the Māori understanding in Aotearoa (New Zealand), where health and identity are inseparable from community and ancestral land. These are not metaphorical relationships; they are epistemologies rooted in the living world.[13]

Regeneration calls for a profound shift, not just in farming practices but in our systems of governance, economics, and social organization. It asks that we re-pattern our lives and structures to generate reinforcing cycles of wellbeing, not only for humans but for the entire Earth community.[14] Central to this is a living systems worldview: the ability to see life not as a collection of parts to be managed but as interwoven patterns of relationship, rhythm, and renewal.[15] As transformations scholar Sandra Waddock observes, "breaking living systems into their parts quite literally destroys them."[16] When we disassemble a living whole, be it an animal's body, a forest ecosystem, or a cultural foodway, we cannot simply reassemble it without altering or losing its integrity. Something essential is lost in the process. This insight speaks to the core of ecological thinking: Life cannot be understood through fragmentation.[17]

Just as ecology illuminates the interconnections between species and their environments, transforming the food system demands that we approach it not as a collection of isolated parts but as a complex, interdependent whole. Too often, efforts to address food system issues operate in silos, focusing narrowly on soil, or labor, or emissions, without recognizing the interwoven nature of these elements. This piecemeal approach fragments a system that is inherently whole, leading to interventions that may address symptoms while deepening root causes. A more generative approach requires us to think relationally—to attend to connections, patterns, feedback loops, and dynamic interdependencies. This is the essence of what Anishinaabe systems thinker Melanie Goodchild refers to as *relational systems thinking*, a perspective grounded in Indigenous knowledge traditions that see the world as alive with reciprocal relationships, not static components.[18] Modern sciences, including ecology, physics, botany, biology, and systems theory, converge on this insight: The world functions in nonlinear, dynamic, and deeply interconnected ways.[19] Health is not a separate domain from land use. Economics cannot be disentangled from soil degradation. Policy decisions ripple across ecosystems in ways that often escape siloed analysis.

This interdependence is at the heart of the emerging field and movement known as "planetary health," which frames the wellbeing of human societies as inseparable from the health of Earth's natural systems. Led by networks like the Planetary Health Alliance, institutions across academia, government, and civil society are collaborating to study and communicate the cascading impacts of ecological disruption on human health and collective flourishing.[20] Similarly, the One Health framework recognizes that human, animal, plant, and environment health are not separate silos but dimensions of one interconnected reality.[21] Industrial agriculture and deforestation, for instance, not only deplete ecosystems, they destabilize habitats, increase the risk of zoonotic spillover, and fuel the spread of diseases like avian influenza among both farmed and wild species.[22] One Health affirms the need for holistic, systems-based approaches to policy and practice, capable of addressing the ripple effects of each decision across land, water, air, and all forms of life.[23]

To address the interlinked crises of biodiversity collapse, climate disruption, and food insecurity, we must situate production, consumption, and waste within the living ecologies they affect and depend upon. Consider industrial agriculture: Its impacts extend far beyond the boundaries of a single field. Through nitrogen runoff, pesticide drift, groundwater extraction, and habitat loss, it alters ecosystems at regional and global scales. To respond meaningfully, we must surface the often invisible ties between ecosystem vitality and human wellbeing. This means making visible how soil degradation affects not only plant health but human nutrition and immune function. It means understanding how food system injustices are entangled with poverty, displacement, and inequality. And it means restoring awareness of the intimate relationships between the food we eat, the systems that produce it, and the bodies we inhabit.[24] For example, policies that support soil health might gain broader traction if people understood the symbiosis between the soil microbiome and our own. Healthy soil microbial communities nourish plants by enhancing nutrient uptake, disease resistance, and climate adaptation. Likewise, the human microbiome regulates our metabolism, hormone cycles, circadian rhythms, and ability to process nutrients.[25] These are not separate systems; they are reflections of a shared, living continuity.

A REGENERATIVE STORY OF AGRICULTURE

In light of the damage caused by industrial food systems, it's fair to ask: Can we even imagine producing food in ways that restore ecosystems and human

health? We've been conditioned to believe that agriculture is inherently destructive to nature—a belief that has justified ongoing degradation and extraction. This assumption is central to the ecomodernist narrative and has
been echoed by scholars like Jared Diamond, who famously described the
advent of agriculture as "a catastrophe from which we have never recov-
ered."[26] Indeed, many historical civilizations have collapsed under the
weight of unsustainable farming practices and widespread soil erosion.
Geologist David Montgomery, in *Dirt: The Erosion of Civilizations*, traces
how early societies often emerged on rich, naturally formed soils. As populations grew and hillsides were cultivated, erosion set in, depleting the fertility that had once supported them. In search of new lands, these civilizations
expanded into more fragile ecosystems, triggering rapid degradation and
ultimately collapse, followed by long periods of ecological recovery in their
absence.[27] These historical cycles offer a clear warning: Industrial agriculture
is not immune from the same fate. If we do not heed the lessons of past failures, we risk repeating them on a global scale. Yet history also offers models
of resilience.

Across cultures and centuries, many traditional and Indigenous food systems have sustained life in regenerative ways, often adapting to shifts in
climate and population without depleting their ecological base. This isn't to
romanticize the past, but to acknowledge that there are cultural and ecological experiments that have worked—systems where food was grown in harmony with local ecosystems and human needs. These systems were rooted
in closed nutrient cycles, localized production and consumption, and deep
ecological knowledge. Examples include forest gardens, Mediterranean sil-
vopastoralism, Asian rice–fish systems, chinampas in Mesoamerica, Hopi
dryland farming, pastoralist traditions across Africa, and many others.[28]
Traditional diets emerging from such systems were diverse, minimally processed, and often protective against modern chronic diseases.[29] In India, for
instance, more than 1,400 species of wild greens, seeds, and tubers form the
basis of high-fiber, diversity-rich diets that support both human and ecological health.[30] These ways of eating and cultivating food reflect a deep relationality between people, plants, animals, and landscapes that is often
erased by industrial models of nutrition and monoculture.[31]

What we now call "regeneration" in Western discourse is a concept long
lived and practiced by Indigenous peoples around the world. Numbering

over 476 million across more than ninety countries, Indigenous communities steward lands that are among the most biodiverse on Earth.[32] These
communities carry profound biocultural knowledge, passed down through
generations, about growing nutrient-dense food in regenerative ecosystems—
knowledge rooted in land-based worldviews that see all beings as alive, animated, and valuable beyond material worth.[33] This is not incidental. Studies
show that over 80 percent of global biological diversity is found in Indigenous-
managed territories, and 92 percent of Indigenous and local community
lands act as net carbon sinks.[34] These landscapes are biodiverse *because of*
Indigenous stewardship, not despite human presence. Their relationships
with forests, mountains, waterways, and wildlife are not extractive but reciprocal, based in spiritual connection, communal governance, and intergenerational responsibility.

Understanding this is critical, especially in light of a painful paradox:
Efforts to "protect" nature through conservation have often led to the displacement, criminalization, or erasure of Indigenous peoples, undermining the
very systems that have made those places ecologically rich.[35] The expansion
of protected areas, when imposed without Indigenous leadership and consent, has contributed to the destruction of cultural landscapes and regenerative lifeways. Lands often labeled as "pristine" or "wild" are, in fact, deeply
storied places shaped by long histories of human care and cultural
relationship.[36]

Contrary to popular portrayals of the Amazon as a pristine, untouched
wilderness, mounting evidence reveals that this remarkable ecosystem has
been profoundly shaped by human hands, specifically, by generations of
Indigenous stewardship. The Amazon basin, home to an unparalleled diversity of life, including 9 percent of the world's mammals, 14 percent of birds,
8 percent of amphibians, 13 percent of freshwater fish species, and
22 percent of vascular plants, is increasingly understood as a co-created
landscape. In some areas, scientists estimate that up to 90 percent of the species have yet to be formally identified, testifying to its ecological richness
and ongoing mystery.[37]

Prior to European colonization, which brought violence, exploitation, and
epidemics that decimated Indigenous populations,[38] between 6 million and
10 million Indigenous people lived across the Amazon.[39] Their presence was
far from passive. Over countless generations,[40] they engaged in intentional

ecosystem management, including the creation of *terra preta*, or Amazonian dark earth. This rich, black soil, formed by incorporating charcoal from cooking fires, broken pottery, compost, bones, and manure, continues to support agriculture at hundreds of archaeological sites across the basin.[41] Today, communities like the Kuikuro of the upper Xingu River in Brazil still practice these ancestral techniques. Known as *eegepe* in the Kuikuro language, this soil is formed through daily practices: discarding organic and fire waste into piles, then planting fruit trees and crops in the enriched ground after several years. The result is soil high in nitrogen, phosphorus, calcium, and carbon—far more fertile than the naturally poor soils typical of the region. This wasn't an accidental byproduct of habitation; it was a deliberate, long-term design for ecological and agricultural abundance.[42] These soil-building practices also contributed to the spread of beneficial species and enhanced biodiversity throughout the forest, demonstrating that human cultivation, when relational and regenerative, can be a driver of ecological flourishing.

Similar findings have emerged from research in West Africa. In Liberia, Sierra Leone, Guinea, and Ghana, scientists have documented widespread use of high-carbon, nutrient-rich African dark earths still being created and maintained through traditional practices. When asked about these fertile soils, Loma farmers in northwest Liberia responded with profound clarity: "God made the soil, but we made it fertile."[43] This reflects a shared regional understanding: Human activity, when rooted in care, can enhance rather than degrade the land. These ancient practices are being revisited today through the modern application of biochar—a substance produced by pyrolysis, a process that chars organic matter while retaining much of its carbon content.[44] When incorporated into depleted or acidic soils, biochar creates hospitable conditions for microbial life, improves nutrient retention, and increases water-holding capacity.[45] Depending on how it's made, biochar can sequester carbon for centuries, offering a potent tool for both soil restoration and climate mitigation.

But terra preta and biochar are just one example among many. Indigenous regenerative ecosystem design (IRED), a term coined by Diné, Tsétsêhéstâhese, and European scholar-activist Dr. Lyla June Johnston, describes the intentional, place-based systems through which Indigenous nations have cultivated food and biodiversity over millennia.[46] In her doctoral work, Johnston details how practices such as selective harvesting, nutrient cycling, species facilitation, habitat creation, and genetic diversification were not only aimed at sustaining human life but were integral to supporting vibrant, more-than-human ecologies across Turtle Island (the Americas). Far from practicing static subsistence, Indigenous peoples shaped whole landscapes for resilience and regeneration. Fire was used not just for clearing land but for encouraging nutrient cycling and promoting habitat diversity. Competing plant species were removed to favor crop allies; nutrients were returned to the soil through intentional composting and animal integration; and landscapes were managed to support both cultivated plants and desired wildlife populations.[47]

Intercropping, often now reclaimed as a "new" sustainable innovation, has deep roots in Indigenous agricultural systems. The Haudenosaunee (Iroquois) and many other Native nations developed the Three Sisters method: corn, beans, and squash planted together in a mutually supportive system. Corn provides a trellis for beans, beans enrich the soil with nitrogen, and squash shades the ground to retain moisture and suppress weeds. Agroforestry and silvopasture, where trees are integrated into crop and grazing systems, are also longstanding Indigenous practices, guided by a kincentric ethic that sees all beings as relatives, not resources.[48] This worldview is encapsulated in a powerful insight from Indigenous scholars and elders: "Indigenous societies did not plant gardens—they managed whole forests. They did not run oyster farms—they managed whole estuaries."[49]

Johnston's research into IRED reveals that the vitality of these systems is rooted not merely in technique but in worldview. The success of Indigenous regenerative practices is inseparable from the values that animate them—principles such as "relationality, reciprocity, respect, reverence, restraint, regenerative practice, responsibility to homeland, kinship with life,"[50] and ongoing regenerative care. From this foundation arise several guiding truths:

- *"non-humans are equal to or Elder to humans,*
- *non-human lifeforms are our relatives,*
- *all lifeforms have an ecological and spiritual role (including humans),*
- *humans have a sacred covenant to protect and care for their respective homelands,*
- *humans have a responsibility to create a home for future generations, and*
- *creation is sacred."*[51]

These principles challenge the very architecture of the industrial food system, where life is reduced to inputs, resources, and yields. Johnston's work points to a profound insight: The values we embed in our food systems, explicitly or implicitly, will determine whether they support life or hasten its unraveling.[52] Without a grounding in relational ethics, even the most well-intentioned sustainability or regeneration efforts can be easily co-opted—rebranded without re-rooting, tweaked without transformation. And this co-optation is already underway, as we'll explore more fully in Chapter 5. Governments and agribusinesses increasingly adopt the language of regeneration while maintaining extractive practices, substituting optics for ethics. As Johnston writes, "Until our societies are rooted in reciprocity, reverence, responsibility, and respect for life, we may have a new face, but our heart will remain the same."[53]

In this pivotal moment, we are being called to ask: What is our rightful role as humans within the web of life? This is not a technical question; it is a cosmological one. It marks a critical divergence between the ecomodernist vision, which casts humans as managers or outsiders to nature, and regenerative paradigms, which locate us as participants, co-creators, and relational beings within Earth systems. IRED reframes humanity not as an ecological mistake to be minimized but as a potentially generative force—when guided by values aligned with life. Rather than imagining ourselves as separate from the natural world, we are invited to see that we belong to it, and that we can become a keystone species: organisms whose presence enhances the diversity, resilience, and vitality of entire ecosystems.[54] The creation of *terra preta* and the diverse Indigenous foodways of Turtle Island are not simply examples of historical ingenuity; they are reminders of what is possible when culture, governance, and spirituality are aligned with ecological integrity. These systems not only reflect the regenerative capacity of human activity but also underscore the central role of Indigenous leadership, knowledge, and sovereignty in shaping viable, life-affirming futures.

FOODWAYS FOR THE FUTURE

Johnston offers a compelling hypothesis: If we can shift from an extractive orientation toward nature to one grounded in respect, reciprocity, and deep relationality, then our societies will begin to flourish in ways that mirror

Indigenous regenerative systems.[55] Rather than positioning humans as con-
querors of nature, regeneration invites us to become partners in co-creating
abundance within our local lifesheds—the specific ecological contexts that
nourish and shape us.[56] In light of the repeated failures of industrial agriculture and the limited transformative capacity of technologies rooted in extractive paradigms, a central question emerges: How can we co-create a
food future that is nourishing for all life? The answer is complex, but the direction is clear. We need an agricultural model grounded in the relational
principles of kincentric ecology and buen vivir: one that supports dignified
livelihoods, maximizes biodiversity, and restores the long-term health and fertility of agroecosystems.[57]

Fortunately, such models already exist in many forms. Agroecology, bio-
dynamics, permaculture, organic farming, agroforestry, silvopasture, and
other regenerative approaches have long drawn from Indigenous and traditional ecological knowledge systems. These methods enrich soil, replenish
water systems, foster biodiversity, and generate food without reliance on synthetic fertilizers or toxic pesticides. Data increasingly shows that diversified
farming systems not only hold their own against industrial agriculture in
terms of productivity but are far more resilient to climate disruption,
something we urgently need in an era of ecological volatility. What sets regenerative agriculture apart is not just its practices, but its paradigm. It
represents a fundamental departure from the industrial model, which reduces agriculture to a narrow metric: yield. Even many so-called alternative
food systems remain tethered to reductionist labels—organic, grass-fed, non-
GMO, fair-trade—none of which guarantee that a farm or food system contributes to ecosystem restoration, biodiversity, or community wellbeing. Truly
regenerative food systems are holistic, relational, and place-based. They hold
as central the imperative to grow nourishing food *and* restore the land. They
see farming not as a technical operation to be managed but as a living relationship with soil, water, plants, animals, and people: a form of stewardship
deeply embedded in ecological rhythms and cycles.[58]

The term "regenerative agriculture," first rooted in the teachings of
Dr. George Washington Carver and later popularized by Robert Rodale in the
1980s,[59] is not a fixed list of techniques. It is a living system of thought and
action, grounded in agroecology—the integrative study of the entire food system in its ecological, social, and economic dimensions.[60] As Aldo Leopold

reminds us, to live ecologically is to act in ways that "preserve the integrity, stability, and beauty of the biotic community."[61] Regenerative agriculture takes this seriously by treating farms not as factories but as vibrant, adaptive ecosystems that nourish both human and more-than-human life.[62] Rather than merely "doing less harm," regenerative agriculture aims to *heal* the land.[63] It employs organic, chemical-free practices[64] such as cover cropping, crop rotation, composting, and managed animal integration, to enhance soil structure, boost nutrient density, increase farm incomes, and sequester carbon.[65] These practices are guided by the "Law of Return": What is taken from the soil must be returned to it.[66]

Central to this model is the re-centering of biodiversity, not just as a goal but as a methodology. Regenerative farming works with ecological complexity rather than against it, mimicking the resilience and abundance of natural ecosystems. And it understands that the agroecosystem is not limited to the field; it includes farmworkers, animals, community members, and all of us who participate in the food web. Regeneration is also relational. It requires rebuilding the connection between those who grow food and those who eat it, cultivating mutual responsibility and transparency.[67] As Jeff Moyer, former CEO of the Rodale Institute and cofounder of the Regenerative Organic Certification, explains:[68]

> When Bob Rodale talked about regenerative agriculture, . . . He would say regeneration really goes beyond agriculture. If you can regenerate a farm's soil, you can regenerate a farmer; if you regenerate a farmer, you can regenerate a farm community; if you regenerate a farm community, you can regenerate urban areas. He saw it as a series of concentric rings that keep growing out from this very concept of first focusing on soil health.[69]

This is the heart of regeneration: a ripple effect that starts from the ground up, anchored in soil and expanding toward systemic transformation. It is not merely about food; it is about reweaving life.

A powerful real-world example of regenerative transformation can be found in Egypt, where a decades-long initiative demonstrates the capacity of ecological agriculture to restore not only soils but entire landscapes and communities. "Creating out of nothing!" became the guiding motto of an ambitious effort to green the Egyptian desert, turning arid sands into living soil through a holistic approach rooted in biodynamic farming. In 1977, Dr. Ibrahim Abouleish founded the SEKEM Initiative on 173 acres of untouched desert land, about 37 miles northeast of Cairo. Inspired by a vision of renewal—of drawing water from a well, planting trees, and witnessing life returning to barren earth—Dr. Abouleish imagined a space where ecological, spiritual, and human flourishing would be inseparable. He described it this way:

> In the midst of sand and desert I see myself standing before a well drawing water. Carefully I plant trees, herbs and flowers and wet their roots with the precious drops. The cool well water attracts human beings and animals to refresh and quicken themselves. Trees give shade, the land turns green, fragrant flowers bloom, insects, birds and butterflies show their devotion to God, the creator, as if they were citing the first Sura of the Qu'ran. The human, perceiving the hidden praise of God, care for and see all that is created as a reflection of paradise on earth.[70]

Through biodynamic agriculture, an ecological method integrates soil, plant, and animal care with spiritual and energetic principles, SEKEM turned desert into fertile land.[71] What began as a farm evolved into a thriving ecosystem of businesses, educational institutions, and cultural initiatives. Today, SEKEM is home not only to diverse flora and fauna but also to a dynamic human community where people of varied backgrounds live, work, and learn together. Over the past four decades, SEKEM has grown into a constellation of enterprises producing organic food, textiles, herbal medicine, and sustainable clothing, while also housing NGOs focused on education, social innovation, and community development. Through the Egyptian BioDynamic Association, it has supported over 700 farmers in transitioning from conventional to organic farming, scaling ecological practices across the country.[72]

Central to SEKEM's long-term vision is the mainstreaming of regenerative agriculture. By 2057, the initiative aims to make organic and biodynamic practices the norm throughout Egypt.[73] One strategy has been to produce rigorous research that demonstrates the ecological, health, and economic benefits of organic agriculture. When SEKEM encountered challenges with pesticide drift from nearby cotton fields, where 36,000 tons of chemicals were being sprayed by air, they responded not with resistance but with evidence. Through field studies, they proved that ecological pest control could be just as effective, convincing the Egyptian government to ban aerial pesticide spraying and adopt integrated pest management methods. The result: Pesticide use in Egyptian cotton production has dropped by more than 90 percent.[74]

In 2009, SEKEM expanded its educational vision by founding Heliopolis University for Sustainable Development,[75] a pioneering institution that blends scientific research with systems thinking and social responsibility. One of its key contributions was a 2016 study applying true cost accounting analysis to compare organic and conventional agriculture. The research found that although organic farming had slightly higher upfront costs, primarily due to labor and compost inputs, it significantly reduced environmental and health-related damage. In the long run, organic systems proved more cost-effective and beneficial for society as a whole.[76] SEKEM exemplifies how regeneration is not just a method of farming but a holistic paradigm—one that aligns ecological restoration with economic viability, spiritual renewal, and community cohesion. It reveals what becomes possible when agriculture is guided by the principles of care, reciprocity, and life-centered design.

MEASURING SUCCESS

For decades, the dominant measures of agricultural "success"—focused narrowly on yield per acre and labor efficiency—have locked us into an industrial model of food production. These metrics, used to shape research priorities, allocate funding, and inform policy decisions, reinforce a paradigm that privileges output over resilience, efficiency over equity, and scale over care. Academic and governmental comparisons between farming systems often rely on simplistic cost–benefit analyses that fail to account for the broader ecological, social, cultural, and economic implications of agriculture, such as soil erosion, water pollution, biodiversity loss, and community health.[77]

These narrow framings have also distorted public understanding. The prevailing narratives of industrial agriculture not only obscure its harms but create the illusion that alternative approaches are unviable or regressive. As Dr. Andrea Beste of the Institute for Soil Conservation and Sustainable

Agriculture notes, ecological proposals have long been dismissed, derided, or described as impractical.[78]

But this narrative is losing ground. Around the world, the regenerative movement is gaining momentum, showing that agroecological systems can not only work—they can thrive. In fact, the majority of food grown for direct human consumption, not for livestock feed or biofuels, comes from what has been called the Peasant Food Web. This diverse, decentralized network includes smallholder farmers, pastoralists, fisherfolk, foragers, and urban growers, many of whom are women.[79] These communities have sustained food security for centuries by working with, rather than against, local ecologies. Data supports what these communities have long known: that diversified, agroecological systems can maintain, and sometimes increase, yields[80] while improving food and income security, reducing dependency on synthetic inputs, and enhancing ecological resilience.[81] A study by the Institute for Sustainable Development and International Relations (IDDRI) projected that a transition to regenerative farming across Europe, paired with reduced food waste and a shift toward more plant-based diets, could meet the nutritional needs of all Europeans, maintain export capacity, dramatically cut emissions, and restore biodiversity.[82]

A meta-analysis examining nearly 42,000 comparisons between conventional and diversified regenerative systems found overwhelmingly positive outcomes. Farms practicing crop diversification through rotation, intercropping, or agroforestry reported gains in pollination, pest control, soil fertility, nutrient cycling, water retention, and carbon sequestration, all without sacrificing productivity. These are not just ecological wins; they are integrated, systems-level benefits that industrial agriculture cannot replicate.[83]

In North America, the Rodale Institute's Farming Systems Trial, the longest continuous comparison of organic and conventional agriculture, has provided over four decades of data.[84] It compares a conventional chemical-based system with two organic models: one based on legume rotations and another on integrated livestock. Across these systems, corn and soybeans—the mainstay of US grain farming—are used as benchmarks. The results speak for themselves.[85] In years of drought, organic corn yields were 31 percent higher than conventional. Total operational costs were lower across the organic systems, and profitability—especially when organic premiums were factored in—significantly outpaced the conventional model. Even without premiums, the organic manure-based system proved the most profitable overall. In contrast, conventional systems showed a trajectory of soil degradation, requiring more inputs for diminishing returns over time.[86]

While these trials prove that regenerative systems can compete on the industrial system's own terms: yield, they also challenge the very notion that yield should be our primary measure of value. Diversified farms produce a mosaic of outputs: animal fodder, compost materials, nutrient-dense crops, and food for household or community consumption. These products often remain off the books, shared, bartered, or used within closed nutrient loops—yet they generate profound value: nourishment, autonomy, resilience, and reciprocity. This is why per-acre or per-worker productivity metrics miss the point. Agroecological systems offer a host of underrecognized benefits: healthier soils, more nutrient-diverse diets, greater adaptive capacity to climate disruption, and stronger local economies.[87] These are the real measures of thriving food systems, and they cannot be captured by commodity calculus alone. In the midst of global crises, there are beacons of hope: movements, projects, and communities around the world that are not merely critiquing industrial agriculture, they are embodying a new paradigm rooted in relationship and regeneration.[88] Farmers and food growers across the globe are already showing what's possible: that we *can* restore degraded ecosystems, nourish communities, and build truly transformative food systems. This transformation is not only necessary; it is underway. And it must be deepened, protected, and accelerated.

BEACONS OF HOPE

One of the most compelling examples of regeneration in action is unfolding in Territorio AlVelAl, a vast mixed landscape of steppe and forest spanning over 1 million hectares (2.5 million acres) in southeast Spain. This region, home to the world's largest organic rainfed almond production, also cultivates olive oil, honey, and wine.[89] Yet, it has faced severe ecological degradation, marked by increasing desertification driven by climate stress, over-tillage, agrochemical use, and the erosion of traditional grazing systems like sheep herding. These ecological shifts have been compounded by social fragmentation, with high unemployment and youth outmigration threatening the continuity of local agricultural life and culture. In response, local communities came together to form the AlVelAl Agroecology Association in 2015, supported by the Commonland Foundation, regional governments, local businesses, and research institutions.[90] With over 490 members, including 280 farmers, AlVelAl has built a thriving network grounded in regeneration. It has launched three regenerative business ventures: an almond company, an olive oil cooperative, and a food platform that connects regional regenerative producers with broader markets.[91]

AlVelAl supports farmers through training and hands-on assistance in transitioning from extractive monoculture to diversified, ecologically integrated systems. These include planting aromatic herbs and fruit and nut trees, reintroducing heritage crops like saffron, integrating sheep and beekeeping, and restoring degraded natural areas through native reforestation. Terracing, hedgerows, swales, composting, and cover cropping are among the soil-building practices promoted. These efforts not only enhance water retention, sequester carbon, and prevent erosion but also revitalize biodiversity and support rural livelihoods.[92] In 2021, a collaborative study involving local farmers and researchers demonstrated that regenerative practices on AlVelAl's almond farms significantly improved soil health and enhanced the resilience of Mediterranean dryland agroecosystems, all while maintaining the nutritional quality of the harvest. The findings suggest that regeneration isn't merely a theoretical ideal; it is a viable path to long-term sustainability and food system resilience in vulnerable climates.[93]

Nearly 6,000 miles away in California's Central Valley, where 80 percent of the world's almonds are grown, a parallel initiative is emerging: the Almond Project.[94] This collaboration unites farmers, researchers, technical advisors, food brands, and processors to identify farming practices that improve soil health and restore biodiversity. Multi-species cover cropping is at the heart of their approach, used to build organic matter, improve water infiltration, and nurture microbial life, all while enhancing ecosystem stability. California's almond industry, the state's largest crop by acreage and revenue, faces a convergence of ecological threats: deepening droughts, rising temperatures, and declining winter chill hours that disrupt tree dormancy cycles.[95] Industrial almond farming, practiced on over 99 percent of almond acres, relies on exposed soils, routine herbicide applications, synthetic fertilizers, and chemical pest controls. These inputs damage soil structure, eliminate ground cover, and weaken the orchard's ecological resilience both above and below the surface.[96]

Water, too, is a flashpoint. It can take one to three gallons to produce a single almond, raising urgent concerns as water scarcity worsens and irrigation restrictions tighten.[97] Regenerative practices offer a pathway forward. Healthy soil acts as a sponge and buffer, storing more water and moderating extreme temperatures. This resilience not only supports crop health but also reduces irrigation demand—critical in a warming, drying climate.[98]

Joe Gardiner, vice president of business development at Treehouse Almonds, helps lead regenerative experimentation across 50,000 acres of almonds in California. On his own family farm, four 75-acre test plots were designated to compare conventional, organic, and two regenerative approaches focused on soil health. Techniques include rotational sheep grazing, diverse cover crops (such as dill to attract pollinators and scallions to deter pests), spot irrigation, and compost integration. This system reduces reliance on synthetic fertilizers and pesticides while enhancing nutrient cycling and ecosystem services.[99]

Outcomes from these efforts are promising. A recent study comparing regenerative, organic, and conventional almond orchards found regenerative systems outperformed conventional ones across nearly every key metric: higher soil carbon, better water infiltration, richer micronutrient profiles, greater biodiversity, and improved overall soil health. Notably, regenerative orchards also produced nearly twice the profit, bolstered by market premiums for regenerative organic almonds.[100] These findings affirm what many regenerative advocates already know: that transitioning agriculture toward ecological principles can address global crises—climate instability, biodiversity collapse, water scarcity, and rural economic decline—while creating food systems rooted in care, justice, and renewal.[101]

THE POWER OF POLICY TO SUPPORT FARM TRANSITIONS

The far-reaching benefits of regenerative farming are beginning to be realized at scale, where bold policies, strategic investments, and community-led vision align in service of ecological health and food sovereignty. One such breakthrough has taken place in the Indian state of Sikkim, nestled in the Himalayan foothills. In 2003, Sikkim became the first state in the world to

commit fully to organic agriculture, laying out a comprehensive plan to preserve soil fertility, protect ecosystems, promote public health, and eliminate
dependence on chemical-based inputs. To bring this vision to life, the state
enacted a robust policy framework: halting the import of synthetic fertilizers
and pesticides, banning their use and sale, and supporting farmers with a
publicly funded transition strategy.[102]

In 2010, the Sikkim Organic Mission was launched to supply farmers
with organic seeds, manure, and hands-on training in ecological farming
methods. The result? Today, all cultivable land in Sikkim operates under organic practices.[103] More than 66,000 farming families have been positively
impacted, contributing to not just ecological restoration but also rural development and sustainable livelihoods. Organic agriculture has improved public
health outcomes and become a magnet for tourism, with visitors drawn to the
state's vibrant landscapes and nutrient-rich foods. A similar story of systemic
transformation is unfolding in Andhra Pradesh, another Indian state chart-
ing a bold course away from industrial agriculture. Vijay Kumar Thallam, a
retired civil servant in the agricultural sector, captured the crisis plainly: "Soils
have eroded, the productivity has plateaued, the costs of cultivation are increasing year after year. . . . The riskiness of agriculture has increased." In response, Thallam helped establish Rythu Sadhikara Samstha (RySS), a
government-linked nonprofit that supports a shift to "natural farming"—a
holistic, regenerative approach that works *with* rather than *against* nature.
Natural Farming in Andhra Pradesh incorporates many regenerative practices: maintaining soil cover with crops or residues year-round, intercropping
up to 15–20 crop species (including trees), minimizing tillage, using indigenous seed varieties, replacing synthetic inputs with botanical pest controls,
and reintegrating livestock. These are not just technical shifts; they are expressions of a deep relational ethic: farming in harmony with local ecologies.[104]

In 2018, the Andhra Pradesh government, in partnership with the UN
Environment Programme, launched an ambitious initiative to transition all
6 million farmers to natural farming by 2031, covering more than 19.7 million acres. UN Environment's then-chief Erik Solheim called it "an unprecedented transformation towards sustainable agriculture on a massive scale,
and the kind of bold change we need to see to protect the climate, biodiversity, and food security."[105] By 2023, the Andhra Pradesh Community-Managed
Natural Farming Program had reached 3,730 villages and enrolled 850,000

farmers across twenty-six districts. A longitudinal study from 2020 to 2022 showed impressive gains: Net incomes rose by 49 percent, use of chemical inputs declined by up to 73 percent, average yields were 11 percent higher than conventional farms, crop diversity nearly doubled, and dietary diversity jumped by 88 percent on regenerative farms. Inspired by this success, other Indian states and international bodies have sought RySS's guidance to replicate the model elsewhere.[106]

These stories make clear: There is no single blueprint for regenerative agriculture. Its forms must be rooted in local ecosystems, cultural contexts, and community needs. Yet across geographies, the core impulse remains the same—honoring life systems and restoring the interdependence between food, land, and people. From Indigenous knowledge systems to innovative farmer-led movements, regenerative agriculture draws on both ancient wisdom and contemporary adaptation. A growing body of proposals outlines how states and international institutions can support this transition. These include redirecting subsidies away from harmful industrial practices; investing in participatory research on agroecology and regenerative systems; strengthening chemical regulations; incentivizing practices that nourish ecosystems and communities; and safeguarding farmers' access to diverse, locally adapted seeds through open exchange systems. International frameworks on genetic resources and intellectual property must also be aligned with principles of justice and sovereignty for small-scale producers.[107]

Agroecology has already been endorsed by multilateral organizations like the UN's Food and Agriculture Organization as a viable solution to intersecting ecological and social crises.[108] Scientific studies continue to highlight its power to restore soil health, sequester carbon, protect biodiversity, and reduce water use. The question is no longer whether regenerative agriculture works. The question is how to make it work across different contexts—ecological, social, and economic—and how to support those doing this work with the resources, policies, and solidarity they need.[109] Ultimately, regenerative agriculture is not just a new technique; it's part of a broader transformation. To truly sustain the regenerative transitions already underway, we must reweave the underlying assumptions of our political, economic, and cultural systems.[110] This calls for a shift in worldview, a reorientation from extraction to reciprocity, from domination to relationship. It is to that deeper systemic reckoning that we now turn.

5

An Ecological Economy

The phrase "The food system is broken" has become a common refrain to describe the intertwined social and ecological crises explored in previous chapters. Yet, as food justice scholar Eric Holt-Giménez aptly observes, this framing may be misleading. The industrial food system is not malfunctioning—it is operating exactly as designed within the context of late-stage capitalism. "It's supposed to concentrate land and resources and power in the hands of the few," Holt-Giménez explains, "and it's supposed to offload all of the social and environmental externalities onto the rest of society. And that's what it's doing."[1] This is not a system in crisis; it's a system in extraction. The foundational architecture of capitalism—private ownership of production, wage labor, profit maximization, commodification, and market exchange—shapes every aspect of the industrial food system.[2] From monoculture plantations and factory farms to the widespread application of chemical inputs and the relentless conversion of living ecosystems into commodities, today's dominant food model reflects capitalism's core logics: control, efficiency, and accumulation.

At the heart of this model lies a fundamental disconnection. Food, rather than being recognized as a sacred source of nourishment, community, memory, and relational interdependence, is reduced to a tradable object.[3] When food becomes just another commodity, its cultural, spiritual, ecological, and nutritional meanings are erased—replaced with metrics like price point, shelf life, packaging, and caloric value. This commodification is not neutral; it actively obscures the structural roots of hunger, land theft, labor exploitation, and ecological collapse. Efforts to fix food system "failures" without challenging this economic architecture are likely to reproduce harm. The system is operating on the logics it was built upon. To transform food systems, we must confront the deeper problem: an economic worldview that treats land, labor, and life itself as resources to be optimized and extracted. Capitalism appears inevitable not because it is natural or beneficial but because we have been conditioned to believe that no alternative exists. As Mark Fisher wrote in *Capitalist Realism*,[4] this ideology functions like atmospheric pressure, shaping what feels possible, acceptable, or thinkable. Margaret Thatcher's infamous claim that "there is no alternative" wasn't just a political assertion; it was a cultural enclosure. Like industrial agriculture, capitalism frames itself as the only rational way forward, masking its contradictions with stories of growth, innovation, and efficiency. Even our language reflects this indoctrination. The term "supply chain," for instance, conjures the logistical machinery of global trade, but it also echoes the legacy of colonial violence—of enslaved people chained into extractive commodity systems.[5] "Stakeholder," now used broadly in business and policy contexts, originates from the colonial practice of driving stakes into land to claim ownership. These terms reveal a deeper epistemology: one of domination, enclosure, and dislocation.[6]

The entrenchment of this worldview can be traced to the post–World War II era, when economic growth became a national mantra and neoliberalism began its rise. Neoliberalism, a doctrine emphasizing deregulation, privatization, individualism, and market supremacy, was not an organic evolution but a calculated project.[7] In 1947, a group of intellectuals led by Friedrich Hayek and Ludwig von Mises gathered in Mont Pelerin, Switzerland, to counter the growing influence of social democracy and collective economic planning. Their network, later known as the Mont Pelerin Society, laid the ideological groundwork for what would become the global neoliberal consensus.[8] Supported by powerful business interests, this movement strategically invested in academic institutions, think tanks (like the Cato Institute and Heritage Foundation), media outlets, and school curricula. Their goal was to normalize a vision of society where market forces—not public values or ecological limits—determine what counts as success, health, or progress.[9]

This legacy lives on in our food system's obsession with GDP, yield, and "efficiency," while masking the cost borne by workers, communities, and ecosystems. Today, this extreme version of neoliberalism sustains monopolistic agribusinesses, widespread hunger amid abundance, and environmental degradation framed as economic growth. It continues to shape how we measure value, allocate resources, and imagine the future. From an ecofeminist lens, this economy is fundamentally disconnected from the cycles and limits of the Earth. It ignores seasonality, discards waste into the bodies and lands of marginalized communities, and lionizes "economic man"—a disembedded, self-interested consumer with no accountability to the life cycles he disrupts.[10]

As explored earlier, today's economy and industrial food system are propped up by a techno-optimist worldview that assumes innovation will inevitably outpace any energy or resource limits. This belief is central to the myth of endless growth. Yet, as economist Kenneth Boulding wryly noted, "Anyone who believes that exponential growth can go on forever in a finite world is either a madman or an economist."[11] Ecological economists like Herman Daly have long challenged this logic. They remind us that the human economy is not autonomous but a subsystem embedded within the biosphere, powered by solar energy and governed by ecological limits. As the economy expands, it does so by converting living systems into capital—generating carbon emissions, destroying habitats, depleting finite resources, and producing waste that chokes the air, poisons waterways, and erodes the Earth's regenerative capacities. This pattern also plays out in corporate climate pledges. A 2023 *New York Times* investigation into the climate disclosures of major food corporations revealed a stark gap between rhetoric and reality. For instance, Starbucks publicly committed to reaching net zero emissions by 2050. But by 2022, its emissions had increased by 12 percent compared to 2019,[12] coinciding with a 23 percent revenue jump. A company spokesperson described the rise in emissions as "inevitable."[13] Starbucks is not an outlier: More than half of the twenty large food and restaurant companies reviewed had made no real progress toward their emissions targets or had regressed.[14]

Daly understood that unbounded material consumption inevitably undermines ecological stability. He argued that economic thinking must start with the premise that the economy is nested within, and constrained by, the biosphere.[15] This insight was shared by the Club of Rome, a collective of economists, scientists, and systems thinkers, including Donella Meadows, that published the landmark *Limits to Growth* in 1972. Their research warned of ecological and economic collapse unless humanity shifted away from extractive development and toward a steady-state global equilibrium.[16] That message remains relevant today. In 2019, over 11,000 scientists signed a statement from the Alliance of World Scientists declaring a global climate emergency. They called for a rapid halt to overexploitation and material extraction in order to preserve the biosphere's long-term viability. Rather than prioritize GDP or affluence, the scientists urged governments to center ecological health, meet basic needs, and reduce inequality.[17]

At the root of our current crisis lies an economic logic built to serve and preserve wealth. The system is designed to extract—channeling resources from ecosystems and communities to maximize financial capital for the few. This orientation is institutionalized through financialization:[18] the growing influence of speculative capital in food systems via commodity markets, land grabs, and corporate consolidation.[19] Social theorist Marjorie Kelly describes this as *wealth supremacy*, a cultural and economic bias that treats the accumulation of wealth for the wealthy as sacrosanct, even at the expense of the common good.[20] This bias underpins how we protect property rights with near-religious zeal while treating labor protections, environmental safeguards, and community rights as negotiable. The dominant measure of economic success, whether for nations or corporations, is the return to shareholders.[21]

Corporate governance has become synonymous with maximizing shareholder value, sidelining long-term ecological stewardship, worker wellbeing, and community resilience. Financialization encourages short-termism: investing in startups to extract their innovations, merging with competitors to consolidate power, and prioritizing profit over diversity: biological, cultural, or economic.[22] These strategies hollow out the food system's ability to adapt, recover, or regenerate. Many leaders and institutions treat this system as natural and immutable.[23] But as management professor Sandra Waddock notes, neoliberalism now operates as an invisible atmosphere, so pervasive we no longer recognize it as an ideology. We are so firmly in the grip of this narrative that "it is easier to imagine an end to the world than an end to capitalism."[24] This fatalism keeps us locked into a system that prizes capital over life.

Even our resistance often betrays this captivity. When we frame our efforts as "fighting corporate power," we concede that corporate capitalism is the only game in town—that our best hope is reforming it from within. As Marjorie Kelly reminds us, if that's our premise, "then we've lost before we begin."[25] The task, then, is not just to resist capitalism's harms but to remember that it, too, is a human invention. As the author Ursula K. Le Guin once said, "We live in capitalism. Its power seems inescapable. So did the divine right of kings. Any human power can be resisted and changed by human beings."[26] If we are serious about building a viable future, we must not only redesign our food systems. We must reimagine an economy grounded in reciprocity, ecological limits, and the vitality of all life—not just the demands of capital.

TRANSFORMATION OR CO-OPTATION

Over the course of my work as a food systems educator and activist, I've witnessed a proliferation of well-intentioned proposals aimed at fixing the food system. But without fundamentally reimagining the economic structures and value systems that underpin it, these efforts risk being diluted, co-opted, or rendered ineffective in the face of mounting ecological and social crises. The corporatization of the organic movement in the United States stands as a cautionary tale of how capitalism neutralizes transformative potential.

Originally rooted in the counterculture movement of the 1960s and 1970s, the organic movement emerged as a direct challenge to industrial agriculture's reliance on toxic chemicals and its ecological harms.[27] It advocated for food systems that nourish both people and the planet. Over time, however, the movement was steadily absorbed into the mainstream market. Rather than serving as an alternative to the dominant model, organics increasingly became a lucrative sector within it, one now controlled largely by multinational food corporations.[28] Today, more than one-third of North America's top hundred food processors, including General Mills, Tyson, ConAgra, Cargill, Coca-Cola, and Pepsi, own popular organic brands. These ownerships are often intentionally obscured.[29] Consumers may associate brands like Cascadian Farm, Kite Hill, or Muir Glen with small-scale, values-driven origins, but few realize they sit under the umbrella of conglomerates like General Mills, which chooses not to display its corporate branding on organic product labels.[30]

In a system driven by the pursuit of capital above all else, even markets born from ethical resistance, like organics or fair trade, can morph into new arenas for profit extraction.[31] As organic products have become more widespread, they've also become more expensive, reinforcing class-based disparities in access to healthy, sustainably grown food. Meanwhile, critics point to a "watering down" of organic standards. Under USDA regulation, organic certification often amounts to little more than swapping synthetic inputs for natural ones, without meaningfully challenging the monocultural, high-input, industrial production model. Practices like large-scale monocropping, animal agriculture resembling factory farming, and soil-less hydroponics now fall under the USDA Organic label, eroding the movement's original integrity.[32] What once posed a structural critique of industrial farming has largely been absorbed into it.

This trajectory raises serious concerns about the future of regenerative agriculture. Increasingly, agribusiness giants are adopting the language of regeneration—but on their own terms. Just as organics were reduced to a checklist of inputs, corporate approaches to regenerative agriculture often distill complex, relational, and place-based farming practices into a single measurable outcome: carbon sequestration. While soil carbon is important, this narrow metric overlooks broader ecological, social, and cultural dimensions. Corporate-led regenerative efforts frequently promote a handful of technical interventions, such as no-till farming and cover cropping, implemented in ways that retain core features of industrial agriculture. Cover crops, for instance, are often terminated with herbicides like glyphosate, undermining their potential ecological benefits.[33] Although companies may invoke principles like biodiversity or circularity, they continue operating large-scale monocultures dependent on synthetic chemicals and genetically engineered seeds. These systems often supply crops for animal feed or biofuels, not for direct human nourishment, thereby reinforcing the very model that regenerative agriculture seeks to transform.

Adding to this concern, a 2023 analysis by the investor network FAIRR evaluated corporate commitments to regenerative agriculture and found a stark gap between rhetoric and action.[34] Of the 50 companies that referenced regenerative agriculture in their public statements, only 36 percent had set any measurable targets. The vast majority offered vague expressions of intent without concrete benchmarks, pilot initiatives, or credible mechanisms for accountability. In fact, just four companies had established outcome-based goals tied to their regenerative claims.[35]

These findings underscore a broader pattern: Voluntary corporate sustainability pledges, whether aimed at curbing emissions, halting biodiversity loss, or staying within planetary boundaries, have consistently failed to deliver substantive change. The idea that free-market forces alone can solve ecological crises is a dangerous illusion. Worse still, the co-opted narrative of "regeneration" now being promoted by industrial agriculture offers the comforting, but false, assurance that the solutions to the climate and ecological breakdown lie within the current extractive system. By focusing narrowly on technical changes at the farm level, these narratives sidestep deeper structural questions: What kinds of food should we be growing? How should land, water, and seed systems be governed? And who decides? That these questions are avoided is no coincidence. A genuine shift toward diversified, agroecological food systems threatens the business model of the agribusiness giants who dominate the current system.[36]

As detailed in Chapter 4, regenerative food systems minimize dependence on external inputs—many of which, under industrial models, are produced and sold by corporate entities. These systems rely instead on locally adapted seeds, farmer-led innovation, and decentralized production and distribution models. Such a transformation would reduce demand for genetically engineered commodity crops, synthetic fertilizers, and pesticides, all of which are cornerstones of corporate profit. Because of this, truly regenerative agroecological approaches are fundamentally at odds with conventional industrial agriculture. Despite corporate attempts to absorb and rebrand them, these approaches cannot be integrated into business-as-usual without losing their transformative power. Yet mainstream science, policy, and sustainability agendas—shaped by the dominant growth-based economic paradigm—continue to legitimize industrial agribusinesses, offering them a seat at the table while sidelining systemic critiques. As a result, many so-called solutions fall short. They fail to address the entrenched institutions, policy frameworks, and economic logics that perpetuate inequality, ecological harm, and unsustainable practices in the food system. Without confronting these foundations, efforts at transformation risk becoming yet another exercise in sustaining the unsustainable.[37]

True transformation toward a regenerative food future demands more than changes in farming techniques; it calls for a fundamental reimagining of how food systems are governed. At the heart of this shift is the need for community-based, democratic control over food, land, seeds, and water. In such systems, governance is rooted in relational values rather than trade agreements or investment returns, and the rights of people and ecosystems are prioritized over corporate profit and global market logic.[38] This is precisely the vision advanced by agroecology—not just as a body of ecological farming practices, but as a worldview and social movement aimed at systemic change. Agroecology is grounded in food sovereignty, the right of communities to define their own food systems in ways that are ecologically sound, culturally meaningful, and socially just. Rather than measuring success by productivity or economic gain, agroecological frameworks emphasize human well-being and ecological integrity as the central indicators of a healthy food system. They are a direct challenge to corporate-controlled industrialization and a call for re-centering the autonomy of food-producing communities.[39]

This movement gained global momentum in the early 1990s with the formation of La Via Campesina (LVC), the International Peasants' Movement. Born in response to deepening agrarian crises and the rollback of agricultural support across the Global South, LVC now represents over 200 million peasants, small farmers, landless workers, Indigenous peoples, and allied movements. Their rallying cry—food sovereignty—emerged as a collective counter-narrative to the dominant food regime.[40] It was given global expression at the 2007 Nyéléni Forum in Sélingué, Mali, where more than 500 delegates from over 80 countries gathered to articulate a shared vision. The Nyéléni Declaration outlines six core principles that continue to shape the global food sovereignty movement:[41]

1. The right of all people to access sufficient, nutritious, and culturally appropriate food
2. The defense of the rights and dignity of small-scale food producers
3. The revitalization of localized and resilient food systems
4. Community stewardship and democratic governance over natural resources
5. Recognition and protection of traditional knowledge and agroecological wisdom
6. Commitment to agroecological production as a pathway to sustainability and resilience

In the years since, the concept of food sovereignty has moved from grassroots vision to legal recognition. Today, at least fifteen countries have adopted food sovereignty into law, and seven—Bolivia, Venezuela, Ecuador, Nicaragua, Mali, Senegal, and Nepal—have enshrined it in their constitutions. This shift is not symbolic. It reflects a growing political commitment to reorganizing food systems from the ground up. One striking example of this commitment came in December 2023, when Mexican President Andrés Manuel López Obrador banned genetically modified corn for human consumption and began phasing out the herbicide glyphosate. Invoking the constitutional right to food sovereignty, health, and environmental protection, his administration framed the decision not as anti-science but as a defense of Indigenous corn varieties, ecological integrity, and public health.[42] In contrast, the US government, invoking provisions under the US–Mexico–Canada Agreement (USMCA), challenged the ban, accusing Mexico of violating trade agreements and threatening the commercial interests of US biotechnology firms. But as Tania Monserrat Téllez, an organizer with the coalition Sin Maiz, No Hay Pais ("Without Corn, There Is No Nation"), made clear, "We are challenging an entire model of production that threatens not just Mexico, but the world."[43] This confrontation reveals a deeper tension between competing paradigms: one rooted in place-based care and sovereignty, and the other in extractive, borderless profit.

Food sovereignty has emerged as a powerful counter-response to the dominant forces of globalization, corporate consolidation, and market fundamentalism. It directly challenges the "feed the world" narrative long used to justify industrial food regimes and extractive trade practices. Through persistent organizing and advocacy, grassroots movements have pushed global institutions, such as the United Nations Food and Agriculture Organization (FAO), to acknowledge the central role of food sovereignty in shaping equitable and ecologically sound food systems.[44] This shift is reflected in the UN Declaration on the Rights of Peasants and Other People Working in Rural Areas. Article 15.4 asserts that "peasants and other people working in rural areas have the right to determine their own food and agriculture systems, recognized by many states and regions as the right to food sovereignty. This includes the right to participate in decision-making processes on food and agriculture policy and the right to healthy and adequate food produced through ecologically sound and sustainable methods that respect their cultures."[45]

In affirming these rights, the declaration lays a foundation for transformative governance rooted in self-determination, ecological stewardship, and cultural integrity. The recent trade dispute between Mexico and the United States over genetically modified corn further illustrates the global reckoning with corporate control in agriculture. By asserting its right to protect public health, biodiversity, and Indigenous food traditions, Mexico has not only defended its sovereignty but positioned itself as a global leader in the rising agroecology movement. Its stance underscores a growing resistance, particularly across the Global South, to the power of agritech conglomerates and the homogenizing logic of free-market agriculture.[46]

What makes food sovereignty especially potent is its ability to integrate democratic governance of food systems with the recognition of food as a commons: something to be cared for, shared, and governed collectively.[47] It moves beyond calls for local food or support for smallholders and points toward a systemic reorientation: from linear, extractive "supply chains" to regenerative food webs grounded in mutual nourishment. In this vision, value is not extracted and centralized but generated and recirculated throughout the web of life—among farmers, eaters, ecosystems, and communities. Agroecology, then, becomes more than a set of practices; it is a covenant, a renewed "social contract" based on equity, dignity, and interdependence among people, and a "natural contract" that honors our entwinement with the rest of the living world. It invites us to co-create food systems that do not merely sustain life but regenerate the conditions for life to thrive.[48]

NEW ECONOMICS AND THE NEXT SYSTEM

The landscape of emerging economic thought is rich with diverse visions, theories, and practices that seek to enable regenerative agriculture and food systems. These frameworks, often referred to under the umbrella of "new economics," encompass concepts such as ecological, solidarity, circular, regenerative, restorative, degrowth, post-growth, steady-state, Indigenous, and wellbeing economics. They also draw from diverse knowledge traditions and Indigenous wisdom, including buen vivir.[49] While differing in strategies and visions, these movements converge in recognizing the unsustainability of our current capital-focused economy and offer new answers to fundamental inquiries: What is an economy and what is it for? Tracing back to its ancient Greek roots, "economics," shares the same origin as "ecology." Aristotle defined it as the pragmatic science of living virtuously as a member of the polis, or community, through wise household management. In our interconnected world, "household" extends to encompass the entire Earth, and "wise management" necessitates consideration of the interrelatedness of living systems.[50] Simply put, an economy is how we care for our common home—the Earth.

New economic movements generally share foundational principles underpinning ecological, life-affirming economies.[51] The Global Assessment for a New Economics (GANE) identifies ten cross-cutting principles categorized into four areas: holistic, ecological, social, and political economy.[52] A regenerative new economy begins with recognizing social-ecological embeddedness, acknowledging that economies are nested within societies and ecosystems. The fundamental purpose of economics, then, is to support human and planetary wellbeing. This approach embraces complexity and necessitates interdisciplinary action in addressing economic problems. Reliance on singular metrics such as GDP, return on capital, and calories per acre, is contributing to the polycrisis, manifesting as carbon tunnel vision, dismissal of agroecology for its perceived inefficiency in monocrop yields, and corporate definitions of regenerative agriculture that implement isolated practices divorced from context and place.

Building an ecological economy requires reimagining what truly underpins systemic social, economic, and ecological health and wellbeing. Emerging indicators move beyond using the conventional measure of GDP, adopting more holistic approaches to national economies. The Better Life Index[53] compares wellbeing across countries based on factors like housing, community, education, civic engagement, and work–life balance. The Genuine Progress Indicator[54] accounts for environmental and social factors in evaluating national wellbeing. The Human Development Index assesses achievement based on life expectancy, education, and standard of living.[55] Bhutan's Gross National Happiness Index measures progress through psychological wellbeing, health, education, time use, cultural diversity and resilience, good governance, community vitality, ecological diversity and resilience, and living standards.[56] Bhutan's model has led to a significant reduction in poverty rates over a decade while achieving climate neutrality.[57]

The concept of moving beyond GDP is gaining traction at the highest level of international policy discussions and inside governments from New Zealand to Wales. The Well-Being of Future Generations Act of 2015,[58] embedded in the Welsh Constitution, is transforming policymaking from short-term to long-term perspectives. It mandates all public bodies to design and publish goals encompassing social, economic, environmental, and cultural wellbeing,[59] including resilience, health, equality, cohesion, and global responsibility. The act requires government agencies and departments to consider the long-term impact of their decisions and collaborate with communities, ensuring actions reflect the diversity of the populations they serve.[60] To safeguard the wellbeing of future generations, the role of a Future Generations Commissioner has been established.[61] A regenerative food system is central to this vision, with the Wales Agriculture Act of 2023 supporting the expansion of agroecological farming practices and strengthening local supply networks for institutions like schools and hospitals.[62]

As highlighted by the Club of Rome, there are limits to growth, and as a sub-system of the planet, the economy is subject to biophysical and biochemical constraints. New economy theories recognize that human-derived capital fundamentally depends on nature, acknowledging the limited substitutability of natural capital. Regenerative design, applying living systems principles such as circularity, becomes central to developing a new economy. In the current capital-focused economic system, financial capital is emphasized to the exclusion of all others. But the survival of humans, businesses, or societies primarily depends on living capital and the material capital that it yields: food, water, energy, and shelter. The Eight Forms of Capital model recognizes additional forms, including social, material, living, intellectual, experiential, cultural, and spiritual capital.[63] Tools like true cost accounting, discussed in Chapter 2, make the interconnections among different forms of capital visible by systematically measuring and valuing the positive and negative environmental, social, health, and economic impacts to guide policy, business, and investment decisions.[64]

As Indigenous, feminist, and postcapitalist activists and communities have long illuminated, transforming the food system requires more than technocratic reform; it demands an undoing of the growth paradigm itself.

This shift involves reimagining food practices and lifeways guided not by extraction, speed, or accumulation but by values that center sufficiency over efficiency, regeneration over exploitation, distribution over hoarding, the commons over enclosure, and care over control.[65] These are not modest pivots; they are radical departures from the industrialized logic that underpins the global food system—a system that leaves 733 million people in acute hunger,[66] many of them children, while simultaneously enabling the waste of up to 160 billion pounds of food annually in the United States alone.[67] A food economy rooted in sufficiency doesn't just feed bodies, it attends to the nourishment of cultures, ecologies, and relationships. It asks different questions: not "How do we scale up to meet global demand?" but "What kinds and quantities of food honor local ecosystems, cultural wisdoms, and the limits of the Earth's generosity?"

Farmer-poet Wendell Berry reminds us that agrarian values begin with a sense of finitude. On such farms, decisions emerge from a lived awareness that there is "this much and no more"—only so much water, so much hay, so much strength in the back and arms.[68] This grounded realism is not a scarcity mindset but a form of reverence for natural rhythms and relational interdependence. This stance contrasts sharply with the industrial delusion that abundance is born from the violation of limits: through machines, mobility, and technofixes. When bees collapse, we build robot pollinators. When soils degrade, we invent new chemicals. This orientation treats Earth as a warehouse to be plundered, not a living entity to be in relationship with.[69]

In response, a new ecological economy and the practice of nature governance are emerging—not as solutions to be implemented, but as frameworks that restore relationality.[70] Nature governance is not just policy innovation; it's a different philosophical foundation.[71] It invites human governance to be reshaped by the wisdom and agency of the more-than-human world. It offers a path toward eco-responsible cultures of stewardship, where planetary health and collective well-being are not afterthoughts but animating principles.[72] Faith In Nature, a UK-based company known for its environmentally conscious skin and hair care products, made headlines not just for what it sells but for how it governs. In 2022, it became the first company in the world to appoint Nature as a non-executive director on its board.[73] This was not a marketing stunt; it was a profound reframing of corporate accountability. The move emerged from a recognition that decisions with ecological consequences are routinely made without any formal representation of the natural world.[74] To implement this shift, the company amended its Articles of Association to include "hav[ing] a positive impact on Nature as a whole" and minimizing the harms of its operations.[75] They then appointed Nature as a legal presence on the board, not in abstract, but through the practical designation of a human guardian empowered to speak and vote on Nature's behalf.[76] This guardian is currently represented by Lawyers for Nature, and their role is not merely consultative.[77] If the board acts contrary to Nature's stated position, they are required to publicly justify their decision, with detailed reasoning recorded in the board minutes.[78] This creates an accountability mechanism not just to shareholders, abstract ESG metrics, or consumers, but to the planetary commons.

This is not merely symbolic. Since 2022, Faith In Nature's action has catalyzed interest across sectors. Companies, foundations, and even some public institutions are now exploring how nature proxies—trained stewards who understand both ecological science and systems of law and governance— might transform the ethics and logics of decision-making. The Earth Law Center aims to train 100 such proxies, cultivating a new form of leadership that aligns fiduciary responsibility with planetary thresholds.[79] And the questions this provokes are neither utopian nor rhetorical. What would Nature say about the extraction of palm oil or the depletion of aquifers for bottled water? How might Nature vote on decisions about pesticides, monoculture, or regenerative agriculture? This isn't about consensus or convenience. It's about learning to stay with complexity when the stakeholders include rivers, forests, mycelial networks, and future generations.

Designing an economy aligned with principles of relational accountability requires more than policy reform; it demands an ontological shift. It calls for moving beyond the reductionist worldview that positions humans as apex agents, entitled to extract whatever Earth offers without consequence. In place of this supremacist logic, we are invited to reinhabit a worldview rooted in living systems—one that recognizes humans not as masters of the Earth but as participants in intricate webs of interdependence.[80] In this way, what some call new economics, including regenerative agriculture and post-growth paradigms, is less an innovation than a remembering. It echoes the worldviews held by many Indigenous societies, which never severed economics from ethics, ecology, and kinship. As

Dr. Ronald Trosper explains, "For most indigenous societies, the economy isn't a separate, material thing as it is in general economic theories. It is an aspect of a social system, and it is embedded in their social relations. They see each other and the land as existing in social relations."[81] In these traditions, land is not property; it is relative. Wealth is not hoarded; it is circulated. The health of the people and the vitality of the land are not separate indicators—they are mutually enfolded. This worldview is not merely philosophical; it is being codified into law. In the Andean nations of Ecuador and Bolivia, Indigenous cosmologies that center harmony with the Earth have reshaped national constitutions.

Ecuador's 2008 Constitution explicitly commits to building a society rooted in diversity and in "living in harmony with nature," declaring this path as essential to achieving *el buen vivir*—the good way of living. Notably, it was the first constitution in the world to enshrine the Rights of Nature. Article 71 recognizes that "Nature, or Pacha Mama, where life is reproduced and occurs, has the right to integral respect for its existence and for the maintenance and regeneration of its life cycles, structure, functions and evolutionary processes."[82] This wasn't mere aspiration. In 2021, Ecuador's Constitutional Court invoked these rights to halt mining in the Los Cedros Protected Forest, affirming that the survival of endangered species and the integrity of forest ecosystems outweighed extractive economic interests.[83]

Similarly, Bolivia's 2009 Constitution embraced buen vivir as a principle guiding state action, and in 2011, it passed the Ley de Derechos de la Madre Tierra—Law of the Rights of Mother Nature—the first national law to recognize Nature as a rights-bearing entity.[84] These legal frameworks mark a significant divergence from Western models of development that prioritize GDP over planetary health. Rather than seeing these interventions as exotic exceptions or symbolic gestures, they offer a glimpse into what post-extractive economies might require: legal, cultural, and spiritual infrastructures that center the Earth not as a resource but as a living relation. As explored in more depth in Chapter 8, the expanding global movement for the rights of nature is not just a legal innovation; it's a profound invitation to reconfigure the very terms of value, progress, and belonging. It could become a crucial lever for interrupting the spread of industrial agriculture and restoring cycles of reciprocity and regeneration.

IMAGINAL CELLS OF THE NEW ECONOMY

Just as ecosystems hold wisdom for regenerating land and reweaving our relationship with food, they also offer guidance for transforming our economic systems. One of nature's most poetic teachings comes from the metamorphosis of the caterpillar.[85] In preparation for its radical transformation, the caterpillar encases itself in a chrysalis, dissolving into a nutrient-rich soup. Within this dissolution lie imaginal cells, dormant during the caterpillar's earlier life, yet always present, holding the blueprint for a radically different being. At first, the caterpillar's fading immune system treats these imaginal cells as threats. But eventually, these cells find each other, cluster together, and begin to cooperate. They differentiate, specialize, and cohere into wings, legs, eyes, and other parts of a new body. From this integration emerges a creature that is not a modified caterpillar but something entirely new: a butterfly.[86]

This metaphor mirrors what is stirring in the realm of economic transformation. The solidarity economy—an expansive movement of democratic, post-capitalist initiatives—acts like imaginal cells, already embedded in the body of capitalism but often dismissed or suppressed by dominant logics. As Emily Kawano of the US Solidarity Economy Network observes, we do not have to wait for a future revolution; these regenerative practices are already here, taking root across communities and bioregions.[87] They include worker and producer cooperatives, community land trusts, public banking, credit unions, time banks, participatory budgeting, mutual aid networks, collective childcare, and food justice initiatives like community gardens and community supported agriculture (CSAs).[88] Making these initiatives visible interrupts the presumption that capitalist systems are the only viable economic option. These diverse experiments offer tangible proof that another economy is not only possible; it's already unfolding.[89]

The solidarity economy, first named in Latin America and Europe in the 1990s, rests on principles of pluralism, solidarity, equity (across race, class, gender, and species), ecological stewardship, and participatory democracy. It resists universalist blueprints and instead supports a patchwork of localized solutions rooted in relationship and reciprocity.[90] In this framework, economic activity is not driven by competition and growth but by the shared responsibility to meet collective needs without causing harm. It reorients the purpose of infrastructure—housing, education, food, transportation, health-care, governance—away from profit and toward community wellbeing and ecological regeneration.[91] This is not a utopian fantasy but an emergent reality. These projects exist not just in theory but in neighborhoods, villages, and urban enclaves across the globe.[92]

Food, in particular, has become a catalytic site for this transformation. Regenerative food movements are experimenting with how food is grown, gathered, prepared, and shared—not as a commodity but as a commons. These efforts are unsettling the dominance of industrial food systems and cultivating alternatives that nourish bodies, lands, and relationships. Food becomes a lever for social, ecological, and economic justice.[93] The pressures of converging crises—climate collapse, global pandemics, inflation, food in-security, ecological destruction, and widening inequality—have destabilized the illusion of capitalism's inevitability.[94] What once seemed permanent now feels precarious. In this space of rupture, imaginal initiatives are stirring. The dormant is awakening. But what emerges from the chrysalis is not a reformed caterpillar. As author and educator Carol Sanford reminds us, regeneration is not about incremental improvement; it is about transformation.[95] A re-generative economy will not arise from regulating capitalism or softening its edges. It will emerge from composting its failures and seeding new para-digms of ownership, value, and exchange—ones that redistribute power and wealth and are accountable to the flourishing of all life.[96]

Across the world, and likely close to home, communities are already enacting these futures. They are engaging regenerative finance, rethinking ownership, revitalizing local governance, and reclaiming food sovereignty. These are not merely alternatives; they are seeds of resilience, glimpses of a future economy grounded not in extraction but in care, reciprocity, and interdependence.

REVIVAL OF REGIONAL GRAIN ECONOMIES

While the Midwest is now often referred to as the "breadbasket" of the United States, this designation obscures a deeper and more decentralized history. Before industrial agriculture centralized grain production and commodified it, particularly following the mid-twentieth-century shift westward, grain was a local affair. Most communities had their own grain supply chains, intimately tied to local ecologies, knowledge systems, and food cultures. The industrialization of staple grains like wheat came with a heavy toll: diminished flavor, loss of seed and varietal diversity, environmental degradation, and the erosion of human and animal health. It also dismantled regional economies and relational food infrastructure, as local mills shut down and farmers ceased growing grains for direct human nourishment.

In Maine, this trajectory began to shift in the 1990s. Jim Amaral, baker and founder of Borealis Bread in Wells, noticed the absence of Maine-grown wheat and began asking why. His inquiry led him to Matt Williams, then a small grains specialist with the Aroostook County Extension Service. At Amaral's urging, Williams planted a crop of hard red winter wheat, harvested in 1998. At the time, Maine had no milling infrastructure, so the grain had to be trucked to New Brunswick, Canada. After 9/11 made border crossings more difficult, Williams decided to mill the grain onsite.[97] With his daughter Sara and son-in-law Marcus, Williams transformed Aurora Mills & Farm into a model of mid-scale, regenerative organic grain production[98]—now growing wheat, oats, rye, spelt, legumes, and buckwheat on 420 acres.[99]Aurora Mills is part of a revitalized "grainshed" in the US Northeast, an emerging regional ecosystem of grain cultivation, processing, and distribution. This local grain economy now extends across New England, supporting a web of seed growers, food businesses, farmers, bakers, maltsters, and eaters, while restoring value to rural communities and ecological systems alike.[100]

Yet, the revival of regional grain economies doesn't depend on farmer enthusiasm alone. Rebuilding requires relational infrastructure—physical and social—that was dismantled during decades of industrial consolidation. As a founder of the Sustainable Sourcing Initiative, which connects plant-based food companies to regenerative farmers, I've encountered these obstacles firsthand. Early in the project, I believed that increasing demand for regeneratively grown crops could redirect farmland away from extractive corn and soy monocultures. But the barriers turned out to be far more structural. Midwestern farmers often want to diversify—to grow oats, lentils, millet, and other nutrient-rich crops—but they are blocked by a lack of accessible processing infrastructure. There are few facilities to clean, sort, dry, or package these crops. Transportation is a major bottleneck, with shortages of trucks and drivers. Without reliable local markets or post-harvest handling systems, even the most willing farmers cannot transition. This gap is what food systems advocates refer to as "the missing middle." It is the infrastructural void between small-scale direct-to-consumer models and large-scale industrial supply chains. Filling this gap is essential for scaling regional, regenerative food systems and creating viable markets for mid-sized farms.[101]

Today's systems of storage, transport, and processing were built for the industrial model, optimized for efficiency, volume, and uniformity. These systems are now deeply consolidated, as detailed in earlier chapters. When it comes to building resilient grainsheds, the Artisan Grain Collaborative has identified a key constraint: the lack of accessible infrastructure for diverse, small-batch, organic, or regionally adapted grains. Essential needs include certified processing facilities willing to work with small quantities; temperature-controlled drying and storage; market development for products made with local grains; and support with packaging, food safety, and labeling. Beyond the physical infrastructure, farmers also need access to arable land, appropriate equipment, skilled labor, and—perhaps most critically—capital in the form of grants, loans, and cooperative investment.[102] Rebuilding this infrastructure is not merely about logistics; it's about reweaving economic life around place-based resilience, ecological care, and community nourishment.

Building on the early efforts of Amaral and Williams, the revitalization of Maine's local grain economy was further catalyzed by the vision and tenacity of Amber Lambke and baker Michael Scholz. Recognizing the need to restore regional processing capacity, they co-founded Maine Grains, a gristmill housed in a repurposed county jail in downtown Skowhegan. This project wasn't just about flour; it was about rebuilding an entire ecosystem of relationships: between farmers and bakers, soil and seed, economy and place. Maine Grains collaborates with organic growers to expand markets for grains and rotation crops, supplying stone-milled flours and oats to bakeries, breweries, food cooperatives, and restaurants across the Northeast corridor from Maine to New York City. Farmers who supply Maine Grains pledge not to use chemical fertilizers or pesticides on their grain crops, and the mill itself is certified organic.[103] But its significance is not merely technical—it serves as a community anchor, hosting a farmers' market, local café, radio station, and other place-based enterprises that reflect a shared ethic of regeneration.[104]

Lambke, who also founded the Maine Grain Alliance in 2007, envisioned this work as both economic and cultural repair. Through efforts like the Kneading Conference, an annual gathering of grain farmers, millers, bakers, researchers, and oven builders, she helped reweave the social fabric that makes a grain economy not just functional but alive. After five years of careful planning, research, and business development, Maine Grains began milling Maine-grown grain in 2012.[105] Yet, transforming a jail into a mill, like many regenerative infrastructure projects, did not attract conventional investors. Most traditional lenders remain tethered to industrial metrics of risk and return, unable or unwilling to recognize the broader social, ecological, and relational value such projects generate.[106]

The Covid pandemic starkly revealed the fragility of centralized food systems and the urgent need for diversified, place-based infrastructure. In response, the USDA launched programs like the Resilient Food Systems Infrastructure Program (RFSI), distributing up to $420 million in American Rescue Plan funds to bolster regional processing and distribution.[107] While public investment is essential, the scale of transformation needed far exceeds current commitments. Estimates suggest that transitioning to agroecological and regenerative models globally would require between $250–430 billion annually, yet only $44 billion is currently invested across philanthropic, public, and private sectors. This funding gap signals the need for a tenfold increase in regenerative investment, and an equally significant shift in how value is defined and capital is deployed.[108]

Regenerative food enterprises often require a different kind of capital: patient, relational, and rooted in place. While conventional financing is structured around quick returns and high margins, regenerative systems generate slow, layered returns over time.[109] A study of Kansas wheat farmers transitioning to regenerative practices found that while profits dropped 30 percent to 60 percent during a three to five year transition period, returns rebounded to 15–25 percent over a decade.[110] These timelines and trade-offs are ill-suited to extractive capital models focused on quarterly profits or venture-scale exits.[111] To finance Maine Grains, Lembke assembled what she calls "a smorgasbord" of funding, blending philanthropic support, individual investments, and bank loans.[112] Among the early contributors was the Maine chapter of Slow Money, a grassroots movement that mobilizes local capital for local food systems. Rather than chasing profit, Slow Money's approach foregrounds community resilience, ecological health, and food sovereignty. Since 2010, more than $100 million has flowed to over 1,000 local food enterprises[113] through decentralized networks.[114]

One innovative offshoot,[115] Beetcoin, provides o percent loans to small organic farms and local food businesses through donor-funded, democratically governed lending circles.[116] Groups like the one in Boulder, Colorado, operate on a "one member, one vote" basis, regardless of donation size.[117] When loans are repaid, funds are recycled into new loans, creating a regenerative loop that mirrors the cycles of the ecosystems these farms depend on.[118] These funding ecosystems don't fit the mold of philanthropy or finance; they blend elements of crowdfunding, micro-lending, mutual aid, and cooperative governance to reimagine capital as a relational force rather than an extractive one.

RELATIONAL CAPITAL

Where capitalism rewards efficiency, uniformity, and scale—often at the expense of diversity, resilience, and relationship—the new economy is seeded in symbiosis. It flourishes not through consolidation but through interdependence. In place of extractive linearity, it embraces regenerative complexity. Farmers, millers, maltsters, bakers, brewers, distillers, and food system activists are increasingly stitching together networks that value quality over speed, local reciprocity over distant profits, and nourishment over commodification. In the Northeast, this collaborative web extends beyond Maine through the Northeast Grainshed Alliance, which connects stakeholders across New England, New York, and New Jersey. Their efforts amplify the ecological, nutritional, and cultural value of regional grains while building supply chains that can withstand global disruptions.[119] They work in tandem with initiatives like the Artisan Grain Collaborative in the Midwest, founded by bakers and environmentalists who saw local grains as a way to heal soil and diversify diets,[120] and Cascadia Grains in the Pacific Northwest, which links farmers, processors, and markets across Washington State's specialty grain economy.[121]

These regional grainsheds do not mimic industrial production. Instead of beginning with anonymous seed contracts dictated by commodity markets,[122] they start with relationships between land, climate, culture, and community.[123] At Aurora Mills, for example, grain is grown in dialogue with specific partners: oats for Allagash Brewing, buckwheat for Portland restaurant Yosaku's soba noodles,[124] and rye for Borealis Breads. The Maine Grain

Alliance is reviving seed strains like einkorn, black emmer, flint corn, and rare Estonian wheats such as sirvinta—reconnecting past and future through embodied plant knowledge.[125] Regional grain systems do more than feed people; they revitalize economies, support farm viability, and embed wealth in place. They allow farmers to adapt to climate realities, rotate crops to build soil health, and avoid the rigidities of industrial monoculture. Importantly, these systems enable communities to respond to crises like pandemics, war, and supply chain collapse by leaning into networks of mutual aid, social media storytelling, and word-of-mouth resilience.[126] Unlike the "food from nowhere" model of global supply chains, regional food economies foster flexibility, adaptation, and relational accountability. Farms growing for their region can experiment with ecologically attuned varietals. Processors like Maine Grains negotiate prices directly with growers, bypassing commodity pricing to create equitable and transparent trade agreements. These are not just transactions; they are relational commitments.[127]

Audre Lorde's reminder that "the master's tools will never dismantle the master's house"[128] remains instructive. We cannot expect to build regenerative economies using the extractive logics of dominant finance. As former JPMorgan executive-turned-regenerative-economics-thinker John Fullerton argues, finance must be reclaimed as a tool, not an end, for cultivating systems that center life.[129] This is the core premise of the regenerative finance, or ReFi, movement. Organizations like RSF Social Finance articulate this shift through four core pillars:[130] integrated capital, equitable deployment, trust relationships, and holistic analysis.[131] Integrated capital blends grants, loans, guarantees, technical assistance, and social network connections in ways tailored to the unique needs of community-rooted enterprises.[132] Equitable deployment challenges structural bias in funding, inviting practitioners, especially from historically marginalized communities, into decision-making roles. Trust and context, rather than metrics alone, become the scaffolding for financial engagement.[133]

This philosophy is operationalized by groups like Mad Capital, one of the few impact-driven lenders committed solely to regenerative organic agriculture. Mad Capital provides integrated financing for transition costs, land access, infrastructure, and operations, while co-developing transition plans and market linkages with each farmer. Their model affirms that financial health, ecological integrity, and community vitality are inseparable.[134]

Similarly, the Equitable Food Oriented Development (EFOD) Fund embodies an integrated capital approach grounded in racial justice, food sovereignty, and community leadership. Rather than applying one-size-fits-all investment terms, EFOD practitioners co-design capital structures and support systems tailored to their lived realities.[135] Their portfolio includes projects like the Detroit People's Food Co-op;[136] the Black Food Sovereignty Coalition co-packing hub in Portland, Oregon;[137] Boston Farms Community Land Trust's urban farm development on vacant public land;[138] and Oakland Bloom's[139] cooperative kitchen for immigrant and refugee chefs.[140] Each of these projects does more than challenge extractive systems; they prototype post-capitalist forms of ownership and governance. Their emphasis on community control, collective stewardship, and relational capital reflects core tenets of the new economy: that wealth is not just material but relational; that value is not just measured in profit but in healing, reciprocity, and the capacity to regenerate life.

RETHINKING OWNERSHIP

Alongside the need to reimagine finance, a regenerative economy demands a fundamental transformation of ownership. The extractive logic of concentrated control, whether by corporations, shareholders, or private equity firms, must give way to forms of ownership that are distributive, relational, and rooted in stewardship rather than speculation. Cooperatives, employee-owned firms, community land trusts, and purpose-driven enterprises are not fringe experiments; they are foundational building blocks of the new economy. They shift economic power from the few to the many and redefine enterprise as a tool for sustaining life rather than maximizing profit. In regenerative enterprises, workers are not a "countervailing force" negotiating with management—they *are* the enterprise. Take Organic Valley, the largest farmer-owned organic cooperative in the United States. With over 1,600 member-farmers, Organic Valley is not just a supplier of organic products; it's a model of stakeholder ownership embedded in ecological and community wellbeing. Rather than serving distant investors, wealth generated through the co-op is reinvested into the communities and ecosystems that sustain it.[141] Another vivid example is Cooperation Jackson, which in May 2024 celebrated a decade of organizing for solidarity economies in

Jackson, Mississippi. Their ecosystem of worker cooperatives includes an urban farm,[142] a co-op incubator, a cooperative financial institution, and a community land trust–based eco-village.[143] These initiatives are not animated by GDP growth but by the regeneration of relationships between people, land, and livelihoods.[144]

Unlike publicly traded companies trapped in a cycle of inflating shareholder returns, regenerative enterprises are structurally designed to resist the pressures of extractive finance. One innovative path that has emerged is stewardship ownership, which decouples profit extraction from governance power. The story of Organically Grown Company (OGC) illustrates this evolution. Founded in 1978 by a collective of gardeners and small-scale farmers, OGC has grown into one of the largest independent organic produce distributors in the United States.[145] Over four decades, it evolved from a nonprofit to a cooperative to an S-Corp with an employee stock ownership plan (ESOP).[146] But this structure came with risks: Under ERISA (the federal law governing retirement plans), the company could be forced to accept buyout offers that maximized retirement plan assets, even if such offers undermined its values or mission.[147] To escape this trap, OGC restructured.[148] In 2018, with a $10 million loan from RSF Social Finance and $1 million in working capital,[149] it created the Sustainable Food & Agriculture Perpetual Purpose Trust.[150] This new legal framework enables OGC to remain permanently independent, mission-aligned, and immune to shareholder takeover. The trust holds the company "in perpetuity" for its stated purpose: to deliver environmental, social, and economic benefits while centering the needs of its broad stakeholder community: farmers, employees, customers, investors, and local communities. As Natalie Reitman-White of OGC explains, "You can't fix a broken food system with a broken finance system. You have to create new models for running a business." That is exactly what they did. Under this model, profits become tools, not goals. They are reinvested to serve the mission, not extracted as dividends. Governance is grounded in long-term stewardship rather than short-term gain. Control rests with those generating relational value over time, not just capital inputs. The Perpetual Purpose Trust aligns mission continuity with economic resilience and community accountability.[151]

OGC's transformation has inspired others.[152] In January 2023, organic food wholesaler Hummingbird Wholesale transitioned ownership to an employee-owned purpose trust, securing its values-driven mission—offering locally sourced, sustainable, and nutritious food—for future generations.[153] These shifts in ownership are part of a growing movement challenging the supremacy of shareholder capitalism. Steward ownership, community trusts, and mission-locked governance mechanisms reflect the logic of regeneration: that value must circulate, that enterprises must be held in service to life, and that economic institutions must be accountable to the living systems they inhabit.

FARMLAND FOR THE FUTURE

Even as we reimagine business structures and regenerate finance, no meaningful transformation of agriculture or food systems is possible without confronting the foundational theft of land and labor upon which capitalist economies were built. Regeneration cannot be abstracted from repair. The soil carries the weight of dispossession—of Black and Indigenous communities whose access to land has been systematically denied through enslavement, violence, policy, and profit. In the United States, Black farmers have lost millions of acres of farmland through legal trickery, structural discrimination, and forced displacement. A 2022 study conservatively estimates the financial loss at $326 billion.[154] Today, 97 percent of farmland is owned by white people, much of it passed down within families or snapped up by investors before it even hits the open market. Farmers of color face additional barriers: student debt, language access issues, racist lending practices, and discriminatory land transactions.[155] The problem is not just personal, it is systemic, and young, low-income, and BIPOC farmers are most impacted.[156]

Despite these barriers, a new generation of land stewards is emerging—fiercely committed to growing nourishing food, restoring soil health, and repairing relationships with land. According to a 2022 survey by the National Young Farmers Coalition, 86 percent of young farmers identify their practices as regenerative.[157] Yet, without secure access to land, their vision remains precarious. As older farmers retire, over 40 percent of the agricultural land in the United States, up to 370 million acres of farmland, may change hands in the next two decades.[158] This "great transfer" presents an unprecedented opportunity to redress land inequities and embed regenerative principles in land tenure and policy. If mishandled, it will deepen ecological crisis, food insecurity, and sprawl.[159]Already, we are losing farmland at a staggering rate—2,000 acres per day, according to American Farmland Trust (AFT).[160] Rising sea levels, soil salinization, and speculative energy infrastructure further threaten the land base.[161] As John Piotti, president of AFT, notes, "We need farmland not just to feed a growing population, . . . but to heal an environmentally degraded planet."[162] The loss of farmland is not just a threat to food; it's a rupture in our ability to live relationally with Earth.

Reclaiming land from the speculative economy is central to the new ecological economy. This means moving from commodification to caretaking, from ownership to stewardship. It means supporting BIPOC land return, community land trusts, and rematriation, such as that practiced by the Sogorea Te' Land Trust, which describes its work as restoring people "to their rightful place in sacred relationship with their ancestral land."[163] Policy levers already exist, or could be created, to enable such shifts:[164] to prioritize sales to Black and Indigenous farmers, tax incentives for landowners who sell to BIPOC and beginning farmers, restrictions on non-operator investor land ownership, and access to integrated capital for land acquisition. These are not just technical solutions; they are ethical obligations.[165]

Regenerative land relationships also require interpersonal repair. As Donna Bransford and Jocelyn Wong write, healing the land includes healing the relationships among people who have been torn apart by settler colonialism and capitalist alienation.[166] Jubilee Justice, based in Louisiana, embodies this ethic.[167] Founded in 2018 by Konda Mason, the organization supports Black farmers through regenerative rice cultivation, reparative capital, and racial healing.[168] Their cooperatively owned, solar-powered rice mill—on land once part of a cotton plantation—represents both an ecological and spiritual return.[169] By vertically integrating production, milling, and distribution, Jubilee Justice ensures that Black farmers retain value and agency.[170] The project, supported by the Kataly Foundation's Restorative Economies Fund, weaves together land, race, money, and spirit into a model of restorative economics[171] grounded in repair, relationship, and redistribution.[172]

Nationally, organizations like Agrarian Trust are also reimagining land tenure. Through the Agrarian Commons initiative, they remove farmland from speculative markets, placing it in community-held legal structures that ensure long-term access and stewardship.[173] One such commons, Little Jubba Central Maine, supports Somali Bantu farmers displaced by war.[174]

Through Liberation Farms, families cultivate plots for themselves and form traditional Iskashito cooperatives for collective growing and shared profit.[175] Their white flint corn, brought from Somalia, now grows in Maine soil and is milled through Maine Grains: an act of diasporic continuity and place-based sovereignty.[176] With a 99-year rolling lease and self-determined governance, the Little Jubba Agrarian Commons ensures that agricultural knowledge and cultural traditions will endure across generations.[177] It's not merely land access; it's land belonging.

All of these initiatives—from Jubilee Justice to Agrarian Commons—remind us that a regenerative food system cannot be built on extractive foundations. The new ecological economy is not just about growing different crops or changing financial tools. It's about re-grounding our lives in reciprocal relationships with land, community, and each other. It requires radical imagination, deep accountability, and a collective commitment to transforming the root structures of harm. If we are to meet the challenges of an uncertain future, we must design food systems that are not only ecologically sound but socially reparative and spiritually grounded. This means breaking from the logic of endless growth and returning to the deeper purpose of economy: to care for the Earth and one another in ways that honor life, restore justice, and regenerate possibility.

6

A New Lens on Technology, Science, and Innovation

To meaningfully confront the challenges of today's food systems, we must begin by interrogating the narratives we've inherited—not only about agriculture and justice, but about *technology* itself. As botanist and storyteller Robin Wall Kimmerer reminds us, technology is not confined to digital devices or Silicon Valley start-ups. The word originates from the Greek *tekhne*: an art, craft, or skill. In this broader sense, an ear of corn, an irrigation system, and a grinding stone are as much technologies as artificial intelligence or gene editing.[1] The development of high-fructose corn syrup and the Indigenous processes of domesticating and cultivating corn are both technological expressions, each embedded with different cosmologies, intentions, and consequences.[2]

Today, "technology" is a polarizing word in food system discourse. Some regard it as an essential lever for a sustainable future, while others view it as inseparable from the logics of industrialization and ecological harm. These opposing positions often harden into caricatures: techno-utopians promising innovation at scale versus agroecologists accused of nostalgic idealism. But such binaries obscure more than they reveal. The deeper question isn't whether technology is good or bad, but what kinds of technologies are we developing, in service of whom, and grounded in what values?

Over the past decade, I've moved between worlds—spaces of hyper-optimistic ag-tech conferences and grounded, back-to-the-land regenerative farming initiatives. This journey has brought me to a central inquiry: Can technological innovation be reconciled with food justice and agroecological principles? Or more provocatively: Can technology be re-rooted in relational values rather than extractive imperatives? This tension surfaced clearly

during the 2019 Tufts University Food Systems Symposium, where I partici-
pated in a conversation that shaped my understanding of this divide. The
central question: Can food tech and food justice co-exist, or even collaborate?[3] Opening the session was Neftalí Durán, a chef, advocate, educator,
and organizer committed to building equitable food systems. A member of
the Ñuu Savi (People of the Rain) from Oaxaca, Durán spoke with clarity and
urgency about how today's industrial food system is inseparable from legacies of colonialism: land theft, enslavement, and erasure. He named the
commodification of coffee and corn not just as economic shifts but as spiritual and cultural displacements. For many Indigenous communities, these
are not "crops" but relatives, woven into their origin stories. Central to
Durán's message was the question of power: Who controls the means of
nourishment, and who decides what counts as knowledge or progress?
Agroecology, he argued, is not just a set of practices; it's a reclaiming of the
right to feed ourselves without dependence on corporate systems designed
around extraction and control. In contrast—and in conversation—was
Dr. Pam Ronald, a plant geneticist and distinguished professor at UC
Davis. Ronald, co-author of *Tomorrow's Table* with her husband, organic
farmer Raoul Adamchak, advocates for an "all-of-the-above" strategy that integrates genetic engineering with organic agriculture.[4] For Ronald, biotechnology is not the enemy; it's a tool that, when governed responsibly, can help
address urgent agricultural challenges. She offered examples like the virus-
resistant papaya in Hawaii and pest-resistant eggplants in Bangladesh,
technologies she sees as both practical and justice-oriented, particularly in
alleviating toxic labor burdens borne by marginalized communities.

During the moderated dialogue, the conversation turned to the ethics
of ownership and control. Ronald suggested that technologies should be developed with farmer input and made accessible through public seed systems. Durán countered with a deeper critique: The very premise of
ownership over seeds or nature is rooted in colonial assumptions.
Indigenous approaches, he noted, are not about improving nature but responding to it: listening, adapting, and stewarding rather than modifying.

What emerged for me was not a resolution but a reorientation. This
debate is not merely about productivity, efficiency, or even sustainability. It is
about worldviews. For some, farming is a burdensome necessity to be

optimized. For others, it is a sacred relationship that requires humility. The push for food technology is often framed as moral—an imperative to relieve suffering. Yet, this same framing can pathologize agroecology as outdated or unjust, casting refusal of certain technologies as a failure to care. Ronald's claim that "food tech is food justice" is not without merit, but it must be held in tension with Durán's reminder that justice includes not just outcomes but origins, relationships, and ways of knowing. The challenge before us is not to pick a side but to ask: What kind of food futures are we cultivating, and at what cost? Whose voices shape the direction of innovation? Whose cosmologies are centered? As we move into this chapter, we explore whether it is possible to reclaim and repurpose technology, not to dominate nature but to align with living systems. Can we craft tools that regenerate rather than deplete, and design technologies that are embedded in place-based wisdom rather than abstracted from land and life? This chapter does not offer a singular answer. Instead, it extends the invitation to navigate these tensions with humility, curiosity, and the courage to hold contradiction.

PUTTING ETHICS BACK ON THE TABLE

Within the industrial food and agriculture system, the prevailing ethos around technological development has often been: If we *can* do it, then we *should*. But capability alone does not justify application. The symposium made clear that technology is never value-neutral; it is always shaped by, and shapes, the worldviews, priorities, and power structures in which it is embedded.[5] While Ronald represents a technologically driven approach to food system challenges, Neftalí Durán foregrounds a relational paradigm—one that centers care, autonomy, and ecological integrity. These are not merely opposing opinions; they reflect fundamentally different understandings of what it means to meet human needs within the limits of a living Earth. Technology theorists like Lewis Mumford and Langdon Winner warned decades ago that the consequences of technological innovation ripple far beyond their intended use. Every tool not only solves a problem but subtly reshapes the social and ecological systems it enters.[6] Consider the plow: Designed to increase efficiency and enable surplus, it also transformed landscapes, catalyzed urbanization, and altered religious cosmologies—from animistic to hierarchical.

As animals became beasts of burden, violently conscripted into agricultural systems, their perceived sacredness diminished. Thus, the plow not only restructured food production; it reshaped our relationship to nonhuman life and laid groundwork for the extractive logic that underpins the Industrial Revolution.[7]

This illustrates a critical point: Technologies never emerge in isolation. They are part of what some call an "ecology of innovation," a web of machines, infrastructures, social norms, economic imperatives, and cultural narratives. Whether it's a simple tool or an AI-powered robot, every technology operates within a network of human and nonhuman actors and is shaped by behavioral patterns, institutional structures, and market logics.[8] A robot moving through a field isn't just a machine; it is participating in a dance with farmer practices, scientific assumptions, crop markets, and policy incentives. In turn, it reshapes that dance: shifting roles, reducing contact, redefining labor.[9]

This becomes especially clear with so-called "smart farming" technologies. Designed to optimize efficiency and yield, these tools often reduce the sensory, relational dimensions of farming.[10] Take, for example, automatic milking systems (AMS). Where farmers once physically milked cows, learning to read subtle behavioral cues and bodily rhythms, AMS now translates animal wellbeing into data points: udder hygiene, temperature, production metrics.[11] Farmers consult dashboards rather than make contact. Over time, the very meaning of "animal welfare" shifts, co-evolving with the technology.[12] As with overreliance on GPS dulling our sense of direction, digital mediation in farming may widen the gap between humans and the ecosystems they depend on.[13]

Technological design is always a reflection of its creator's vision of what the world *is* and what it *ought* to be.[14] These visions determine how problems are framed, and thus what solutions are considered legitimate. Agroecologists, for instance, see ecological and human health crises as outcomes of colonial, capitalist structures that lock in corporate control and monocultural thinking.[15] Scientists focused on techno-efficiency, by contrast, often frame the crisis as one of optimization—too much water, too many pesticides, not enough data. This framing leads to technical "fixes": gene editing crops to reduce water use or increase yield, refining inputs rather than transforming systems.[16]

Consider the case of wheat cultivation in the UK. In 2021, Rothamsted Research received permission to field-test wheat that had been gene-edited to contain less asparagine, a naturally occurring compound that transforms into the carcinogen acrylamide during high-heat cooking.[17] While this may seem a sound solution, agroecologists point out that low-asparagine wheat varieties already exist[18] and that high asparagine levels are often a symptom of soil stress, nutrient imbalance, and industrial overproduction.[19] Gene editing enables the current system to persist. Agroecologists would instead advocate for soil restoration, diversified cropping, and the reduction of nutrient stress, shifting the very conditions under which asparagine accumulates.

This pattern recurs. In drought-prone California, biotech advocates suggest engineering water-resistant crops.[20] Agroecologists counter: Why grow water-intensive crops there at all? Why not revisit the hydrological reengineering of colonized landscapes, or consult Indigenous communities who once cultivated food adapted to that ecology?[21]

Technologies, then, are not just tools—they are active participants in reshaping social, political, and ecological life. They are vehicles for value transmission, shaping how we think, what we prioritize, and how we relate.[22] The challenge is not simply *which* technologies we use, but *how* we develop and govern them, and *whose* values they encode. To navigate this complexity, we need frameworks that surface values *upstream*, before the technology is widely deployed. The Consilience Project offers one such approach through axiological design: designing technologies that make explicit the values they embody and assess their broader impacts not only physically but behaviorally and psychoculturally.[23] Another approach, responsible research and innovation (RRI), emphasizes participatory ethics, urging that value-laden decisions be deliberated early in the innovation process. RRI recognizes that the design of a tool shapes society, and that shaping tools must therefore be a collective, accountable act.[24]

The RRI framework offers a practical way to apply these reflections at the point of design. It urges us to ask not only *what* a technology can do, but also what it *ought* to do and for *whom*.[25] Rather than treating ethical and social impacts as afterthoughts, RRI embeds them at the front end of innovation. It invites scientists, funders, policymakers, and communities into collective deliberation about the values, risks, and alternatives that shape technological development.[26] This isn't just about ticking boxes or drafting regulatory frameworks; it's about cultivating a culture of inquiry that holds space for complexity and humility. Questions at the heart of RRI include:

- *Who defines the risks and benefits, and how will they be distributed?*
- *Why is this technology being developed, and by whom?*
- *Who will benefit, and who might be harmed or excluded?*
- *What other pathways have been considered, and why were they rejected?*
- *What might we never fully know or foresee?*
- *What values are being embedded into the technology itself?*
- *Who takes responsibility if harm arises?*[27]

These inquiries anchor four principles of RRI: anticipation (of both intended and unintended risks), responsivity (to evolving contexts and feedback), reflexivity (on the assumptions, values, and methods shaping research), and inclusion (of diverse publics and knowledge systems in decision-making).[28] This framework is especially important in food systems, where complexity, uncertainty, and high stakes are the norm. Global agriculture is a nonlinear system shaped by ecology, labor, markets, weather, culture, and historical trauma. Anticipation in this context is less about prediction and more about iterative learning—staying responsive to feedback and course-correcting when things go awry. Crucially, anticipation must also extend to alternatives. What if we choose not to implement a new technology? What if the best path isn't innovation at all, but restoration, preservation, or redistribution? These counterfactuals should be part of the deliberation, not only to avoid harm but to expand our sense of what is possible.

The history of agricultural technology offers sobering lessons.[29] Enclosure, the privatization and consolidation of land, seeds, and knowledge, has long-undermined food sovereignty. Technologies like genetically engineered (GE) seeds often intensify this trend, with intellectual property laws concentrating control in the hands of a few. Identifying and mitigating rights-based harms must be part of any ethical tech development process. Another pattern is over-innovation, or what some call tech saviorism, the assumption that high-tech solutions are inherently superior. The Green Revolution exemplified this logic: Outside experts imposed technical fixes onto local food systems without understanding their ecological or cultural context, often causing new forms of dependency and ecological degradation. When innovation is driven by cultural bias or prestige rather than grounded need, it risks displacing more effective, locally rooted approaches. Then there's the Jevons paradox: the tendency for efficiency gains to increase, rather than

decrease, resource use. In agriculture, this means that even when tools reduce water, fertilizer, or pesticide use per acre, total usage may still rise as
production scales up. The real issue isn't just the technology; it's the growth
paradigm that frames more as always better.[30]

Rather than reactively managing these outcomes after the fact, RRI calls
for proactive, participatory foresight. Mechanisms like the precautionary
principle, temporary moratoria, and codes of conduct can provide guardrails,
but only if backed by ongoing public engagement and a willingness to question dominant assumptions. Ultimately, the RRI approach is not about
rejecting technology but about situating it *in service to life*. It offers a pathway
toward a regenerative ethic of innovation, one that attends to the deeper
structures of power and meaning and that holds scientists, developers, and
communities accountable to the futures we are co-creating. As technologies
embed themselves into the fabric of food systems, their influence extends
far beyond efficiency or productivity gains. They not only produce unintended
consequences, such as those exemplified by the Jevons paradox, but can also
generate new, unforeseen value systems that reshape how society understands labor, land, care, and control.[31] This is why reflexivity in innovation
isn't a luxury; it is a necessity. When those advancing new technologies are
unwilling to engage with dissenting perspectives or learn from harmful
outcomes as they emerge, they are not practicing responsible innovation—
they are reproducing extractive systems under the guise of progress.[32]
Responsiveness, then, requires more than agile engineering or product it-
eration. It involves the ongoing capacity to revise goals, strategies, and design assumptions as technologies interact with living socio-ecological
systems. In a world where agricultural conditions, climate patterns, and
cultural contexts are in constant flux, this kind of adaptivity must be built
into the innovation process itself.

One of the most effective ways to build such responsiveness is to center
knowledge of local conditions, relationships, and lived experience.[33]
Technologies, no matter how promising in abstract, do not land on neutral
terrain; they are filtered through political priorities, funding mechanisms,
and the assumptions of their developers. A robot designed to harvest fruit,
for instance, will likely reflect the investment goals of its financiers, privileging speed and cost-efficiency over the embodied wisdom or livelihoods of
farmworkers. This creates what scholars call a "technological frame": a set

of guiding logics that shape design and deployment, often without scrutiny.[34] When these frames are dominated by venture capital metrics or industrial logics, they foreclose alternative futures before they can emerge. This is why inclusion, not as a token gesture but as a foundational principle, is central to the RRI framework. Without it, we risk entrenching a singular, techno-optimist narrative that equates innovation with inevitability, excluding those most impacted by these shifts: farmers, land stewards, Indigenous communities, and food workers.[35]

Around the world, participatory approaches are offering models of how technology deliberation can be made democratic, pluralistic, and grounded. In Mali, the l'ECID—Espace Citoyen d'Interpellation Démocratique (Citizen's Space for Democratic Deliberation) initiative convened a fifteen-month process involving farmers from every district in the Sikasso region, global witnesses, independent observers, and media representatives to deliberate the future of genetically engineered cotton. This citizen jury blended Western deliberative structures with traditional African methods like the palaver, a communal dialogue practice rooted in relational accountability. The result was not only a decision about GE crops but a space where diverse cosmologies, values, and visions of agriculture could be held in shared conversation.[36]

In Canada, Living Labs[37] are fostering co-creation between farmers, scientists, and policymakers to develop place-based responses to climate change and biodiversity loss.[38] In 2022, nine new Living Labs were launched, including the first Indigenous-led Living Lab by the Mistawasis Nêhiyawak and Muskeg Lake Cree Nation. Similarly, the Diversity by Design project connects small-scale farmers with technologists and public institutions to examine how digital tools impact biodiverse farming, land access, and equity—particularly for women, BIPOC farmers, and Indigenous nations.[39]

In the United States, Semillero de Ideas (Nursery of Ideas) offers another powerful model, this time led by farmworkers themselves.[40] Rather than being displaced by robotic harvesters,[41] workers like Josefina Luciano are reclaiming space in the innovation conversation. Semillero creates workshops where workers can not only learn new tools but shape them, imagining how mechanization might reduce bodily strain while preserving the dignity and wisdom of human labor.[42] As Luciano puts it, "These large corporations may be able to replace us where we work, but they'll never replace our human touch, sacrifice, and essence."[43] These examples remind us that technologies are never just about tools; they are about relationships. Who gets to define the problem? Who determines the solution? Who benefits, who is left out, and who gets to decide?

Responsible innovation means asking not just what a technology will do, but how it will transform our ways of living, relating, and stewarding the Earth. It demands that we widen the aperture of our inquiry not only to anticipate outcomes but to examine the deeper ethics of how we coexist with each other, with more-than-human life, and with the systems upon which we all depend.

LEARNING FROM TRADITIONAL ECOLOGICAL KNOWLEDGE

These questions—about the purpose, design, and consequences of technology—cannot be disentangled from deeper inquiries into democracy, power, and food sovereignty. At the heart of debates around agricultural innovation lies a fundamental issue: Who has the right to define, access, and produce culturally meaningful food? Food sovereignty is not just about access; it's about self-determination. And self-determination, in turn, depends on whose knowledge counts and whose science is legitimized in shaping our shared futures. The field of science and technology studies (STS) has long emphasized that science, like technology, is not neutral. It is a social practice, one that emerges from particular worldviews and institutional arrangements. It does not simply describe the world; it constructs it, through frameworks that reflect dominant assumptions about value, method, and evidence.

As the Tufts symposium highlighted, debates over who controls food systems are also debates over epistemology: the ways of knowing that guide how we engage with land, life, and one another. Western science, for all its strengths, is not the only legitimate pathway to understanding. A 2017 "Letter from Indigenous Scientists in Support of the March for Science in 2017" affirmed that the concept and practice of science exists across thousands of Indigenous languages and lifeways.[44] While often excluded or devalued by institutional research systems, traditional ecological knowledge (TEK) offers a robust, relational, and time-tested paradigm rooted in observation, reciprocity, and reverence for the Earth.

Marine ecologist Fikret Berkes describes TEK as "a cumulative body of knowledge, practice, and belief, evolving by adaptive processes and handed down through generations by cultural transmission, about the relationships of living beings (including humans) with one another and with their environment."[45] Unlike Western scientific models, which tend to separate fact from meaning, TEK holds that knowledge cannot be detached from ethical responsibility. It aligns human action with ecological principles and natural law, emphasizing kinship rather than control.[46] Where Western science may analyze a soil's mineral composition, TEK invites us to ask whether the land is being honored. It does not reject observation, experimentation, or innovation; it situates them within cultural frameworks of care. This distinction matters. Western resource management, shaped by assumptions of scarcity and supremacy, tends to view land and species as units of production. TEK, by contrast, begins with the recognition that humans are part of, not above, the web of life.[47]

These different approaches to science also produce different definitions of evidence. In Western science, evidence is that which can be isolated, replicated, and generalized. In TEK and agroecological traditions, knowledge is situated and relational, emerging from lived experience, oral history, community science, and long-term engagement with specific places.[48] Here, the farmer is not a passive recipient of research but a co-creator of knowledge. This is not less scientific; it is simply a different science, one that embraces complexity, context, and humility.

The story of the Ojibwe people and *manoomin*, wild rice, offers a powerful case study in these divergent worldviews. The Ojibwe, along with their Anishinaabe relatives the Potawatomi and Odawa, followed a prophecy to journey west to a land where "the food grows on water." That prophecy led them to Gichigami (Lake Superior), where they encountered manoomin, a sacred grain with deep nutritional, cultural, and spiritual significance. Every late summer, during *manoominike giizis* (the wild rice moon), families gather rice from canoes, guided by ancestral protocols of care and restraint.[49] But this lifeway is increasingly threatened. Manoomin populations have declined due to sulfates from mining, damming, invasive species, climate disruption, and agricultural encroachment.[50]

When the Minnesota Pollution Control Agency proposed loosening water quality protections, ostensibly in response to industry pressure, the Ojibwe resisted, framing manoomin not merely as a resource to be managed but as a relative to be defended.[51] Their opposition was not just environmental; it was epistemological. It challenged the notion that science divorced from land-based relationships should govern decisions about ecosystems and livelihoods. This conflict illuminates the deep incongruities between worldviews: between commodification and kinship, ownership and stewardship, domination and interdependence. It also reminds us that questions of technological development, food systems, and environmental policy are inseparable from deeper reckonings about whose knowledge is respected, whose sovereignty is recognized, and what forms of life are deemed worth protecting. In navigating the future of food and technology, we need more than interdisciplinary expertise; we need *inter-epistemic* humility. We need to recognize that no single science can solve the crises we face.

Whereas the Ojibwe relationship to manoomin is grounded in collective stewardship and sacred responsibility, Western institutions have approached it through a framework of privatization and commodification, often governed by the logic of intellectual property and extractive science.[52] Efforts to domesticate and industrialize wild rice in the United States can be traced back to *The Wild Rice Gatherers of the Upper Lakes*, a 1900 Smithsonian publication in which anthropologist Albert Jenks derided traditional Ojibwe harvesting methods. He claimed wild rice cultivation had reached its apex with the Ojibwe, and that their techniques—communal, spiritual, and nonmechanized—were inherently limited and backward: "Wild rice . . . held them back from further progress, . . . for with them it was incapable of extensive cultivation."[53] This colonial judgment seeded decades of research at the University of Minnesota (UMN), where scientists launched a wild rice breeding program aimed at increasing yields and developing characteristics like sturdy stems to facilitate mechanical harvesting.[54] These strains, grown in diked paddies and reliant on chemical synthetic fungicides and herbicides, were framed as modern upgrades to an outdated system. As California rice producers began to outpace Minnesota's cultivated rice production by the mid-1980s, the pressure to "modernize" increased.[55]

But to the Ojibwe, these interventions represented a rupture, an attempt to sever manoomin from its web of ecological, cultural, and treaty-based relations. Tensions deepened in 1999 when the California-based company NorCal patented a process for breeding wild rice that enhanced yields and mechanization.[56] Around the same time, UMN researchers completed mapping the wild rice genome, a project launched in 1987, without consultation, consent, or benefit-sharing with the Ojibwe.[57] The extraction of genetic information without tribal involvement was seen not just as a scientific overreach but as a continuation of colonial appropriation.

In a 1998 letter to UMN, then-Minnesota Chippewa Tribal President Norman Deschampe called for an end to the "exploitation of our wild rice for pecuniary gain." He emphasized that wild rice is not merely a plant or a product but a living relative, protected under treaty and federal trust obligations:

> We were not promised just any wild rice; that promise could be kept by
> delivering sacks of grain to our members each year. We were promised the
> rice that grew in the waters of our people, and all the value that rice
> holds . . . a sacred and significant place in our culture.[58]

Yet, from the perspective of David Biesboer, longtime director of UMN's wild rice research program, the conflict was distilled into a matter of cultural difference: "It's culture versus industrial uses, Native American uses and farmer uses. . . . Europeans might say it belongs on a plate while Native Americans say there are spirits involved. It's really rooted in that cultural difference between Native Americans and Europeans. And I'm a scientist; to me, it's an aquatic weed, and you get a little foodstuff out of it."[59] Such statements lay bare the epistemological rift between reductionist science and relational worldviews. But they also reveal a deeper issue: who gets to define what counts as knowledge, and what is dismissed as "myth" or "belief."

In response to this legacy of erasure, a growing coalition of Indigenous scholars, students, and allies at UMN began organizing symposia starting in 2009 under the banner "Nibi [Water] and Manoomin: Bridging Worldviews." These gatherings called for ethical research practices and honored the sovereignty of tribal knowledge systems.[60] Out of this momentum, a more reparative project emerged in 2018: *Kawe Gidaa-Naanaagadawendaamin Manoomin* (First We Must Consider Manoomin). Distinct from UMN's genetics work, this interdisciplinary initiative brought together tribal resource managers and university researchers from a wide range of fields, including hydrology, law, ecology, and Indigenous studies, united by a shared commitment to understanding manoomin through the guidance of tribal knowledge holders and relational accountability.[61]

This effort reflects a broader shift in scientific communities: an increasing recognition that Western science alone is insufficient to address the complex socioecological challenges of the Anthropocene.[62] At the same time, many Indigenous leaders acknowledge that respectful collaboration can be generative, particularly as climate change accelerates threats to long-held land-based practices. Manoomin, for example, is highly sensitive to changing hydrological conditions. Its floating leaves and barely anchored roots make it vulnerable to the intensified storms and fluctuating water levels brought on by global warming.[63] One framework that helps navigate this complexity is Two-Eyed Seeing (*etuaptmumk* in Mi'kmaw), a concept introduced by Mi'kmaq Elder Dr. Albert Marshall. It invites us to learn "to see from one eye with the strengths of Indigenous knowledges and ways of knowing, and from the other eye with the strengths of mainstream knowledges and ways of knowing, and to use both these eyes together, for the benefit of all."[64] Importantly, Two-Eyed Seeing is not a merger or assimilation of Indigenous knowledge into Western science. It is a call for epistemic equity, where both knowledge systems are held with respect, their distinctiveness acknowledged, and their collaboration guided by protocols of consent, reciprocity, and shared purpose.[65] It offers a model for navigating not just technical challenges but relational ones: how we learn to coexist across difference, listen beyond dominance, and reimagine science as a practice of care, humility, and collective survival.

SEEING WITH TWO EYES

In contrast to the top-down promises of agri-tech and the Green Revolution, a growing recognition is taking root: Addressing hunger and ecological breakdown will require more than technological fixes; it will require rethinking our diets, reducing waste, and revitalizing food systems through diversification.[66] This includes the reintegration of underutilized crops and their wild relatives, many of which have been pushed to the margins by industrial agriculture's narrow focus on a handful of commodity species. The consequences of this narrowing are profound. As monocultures spread, many edible plant species have been forgotten or rendered endangered, diminishing agrobiodiversity and the genetic options needed to adapt to climate disruption. Losing this diversity not only limits our nutritional possibilities but weakens our resilience.[67] Efforts to reverse this trend must be led by those who have long stewarded these species— Indigenous and local communities whose knowledge systems have preserved them for generations.[68]

Take the banana, for example. While the globally traded Cavendish variety is under threat from a devastating fungal disease, the same disease has not taken root in Africa and southeast Asia, where hundreds of varieties are grown, many with natural resistance. In Uganda alone, nearly a hundred varieties support millions of livelihoods.[69] Yet, attempts to commercialize or "rescue" such crops without community partnership risk replicating extractive patterns, as seen when a Dutch company patented processes for storing and processing *teff*—an ancient Ethiopian grain—without consultation or benefit-sharing.[70]

To engage in truly regenerative research, we must disrupt not only colonial patterns of resource extraction but also the epistemic hierarchies that privilege academic science over Indigenous and local knowledge. The notion of "discovering" new plant species, when local communities have long known their uses and significance, is a particularly egregious example of this epistemic violence. Approaches like participatory action research and community-based participatory research offer alternatives: frameworks grounded in relational accountability, shared decision-making, and community-led inquiry.[71]

The framework of Two-Eyed Seeing can guide this shift. It invites us to bring together the strengths of Indigenous knowledges and Western science—not to merge or dilute them, but to work with both, side by side, for the benefit of all. For instance, TEK can guide the identification of nutrient-rich or climate-resilient food sources, while AI and genomics might help predict gene functions and enhance understanding of crop traits. But the direction and purpose of this work must be determined in collaboration with the communities it affects.[72]

Promising efforts are already underway. Since 2021, Africa's open-access publishing platform has made scientific research freely available to support local knowledge production. The African BioGenome Project, involving over 100 African scientists and twenty-two organizations, is sequencing the genomes of 105,000 endemic species, including plants, fungi, and animals essential to regional food security. Importantly, the project centers traditional custodians of knowledge in its design and governance, affirming that knowledge must be co-created, not extracted.[73] For example, a 2021 genome study of 245 Ethiopian native chickens revealed the genetic basis of resilience to cold and drought, traits long observed by local farmers but only recently recognized by formal science. By sequencing crops like fluted pumpkin or marama beans, which are not traded internationally but are vital for local food systems, the project supports biodiversity conservation grounded in regional priorities and relational ecologies.[74]

Yet for these efforts to have integrity, academic collaboration with Indigenous and local knowledge must be supported by institutional shifts in governance, funding, and research ethics.[75] A hopeful step came in 2022, when the White House Council on Environmental Quality (CEQ) and the White House Office of Science and Technology Policy (OSTP) issued the first US government-wide guidance on recognizing and including Indigenous knowledge in federal decision making.[76] International platforms like the Intergovernmental Science-Policy Platform on Biodiversity and Ecosystem Services (IPBES) and the Convention on Biological Diversity (CBD) have similarly affirmed the importance of multiple knowledge systems in shaping biodiversity policy.[77]

The application of Two-Eyed Seeing has already transformed relationships on the ground. One such example comes from the Saugeen Ojibway Nation (SON) in Ontario, where the community and the provincial Ministry of Natural Resources and Forestry (MNRF) have co-governed commercial fisheries in Lake Huron. Historically, these entities operated in silos, with Western scientists guiding policy and Indigenous concerns marginalized. Mistrust grew as external experts mediated communication and Indigenous insights were undervalued.[78] Two-Eyed Seeing shifted this dynamic. It replaced extractive consultation with reciprocal knowledge exchange.[79] When SON members raised concerns about the decline of lake whitefish, pointing to an increase in lake trout as a potential factor,[80] many scientists initially dismissed the connection as speculative.[81] But the community's observations, grounded in generations of land-based knowledge, became the basis of a joint research proposal.[82] Rather than debating whose evidence was "valid," the partnership embraced both perspectives: One eye attending to the data, the other to lived experience and ecological memory. SON members, including fish harvesters and elders, had long noticed behavioral and phenotypic changes in stocked lake trout, as well as increased predation on whitefish. Their two hypotheses: Lake trout were either competing with whitefish for habitat or preying on them directly.[83]

The research was co-designed from these community hypotheses. Through shared framing, data collection, and interpretation, the team worked in what has been called an "ethical space": a relational field that honors difference without subsuming it, and where decision-making is rooted in equity and respect.[84] The Anishinaabe principle of *Mino-bimaadiziwin* (the good life) guided the collaboration, helping ensure that process—not just outcomes—aligned with relational values.[85] This ethical grounding also influenced language. When researchers proposed that the project aimed to "protect" whitefish for future generations, Elder Sidney Nadjiwon noted that "protection" is not an Anishinaabemowin concept. The team instead reframed their goal: not to manage whitefish as a resource, but to learn *from* whitefish as teachers and participants in a living system, not objects of scientific inquiry.[86] Such shifts from extraction to relation, from dominance to dialogue, signal what is possible when knowledge generation becomes an act of shared stewardship. Two-Eyed Seeing is not merely a methodology; it is an ethic of co-existence, reminding us that truly regenerative futures can only emerge from the plural, grounded, and reciprocal wisdom of the many.

WHAT WOULD NATURE DO?

Learning from living systems was at the heart of the New Alchemy Institute's vision to design "ecologically derived human support systems," practical, integrated approaches to food, energy, shelter, waste treatment, and ecosystem repair.[87] Emerging in the 1970s amid oil shocks and the rise of environmental consciousness, the New Alchemists sought not to "innovate" in the traditional industrial sense but to observe and align with the inherent intelligence of nature.[88] Their systems mimicked ecological relationships: wind pumps powered water systems that nourished gardens, which fed rabbits, whose waste enriched the earthworms, which supported aquaculture, which in turn fed humans. As co-founder John Todd explained, "Each time we make a connection, as in nature itself, the whole becomes more stable, more strong and more healthy." This approach was both experimental and deeply scientific. The New Alchemists collected data on everything from soil chemistry to fish feed quality, adjusting their designs through ongoing observation and adaptation. Over time, their organically grown crops rivaled, and in some cases surpassed, the yields of conventional industrial farms.[89]

Inspired by the principles of E. F. Schumacher's "Small Is Beautiful," the New Alchemists championed "appropriate technology," tools and systems that sustain human life with minimal ecological impact. From low-cost windmills for Indian farmers to solar herb dryers in Costa Rica, their innovations emphasized accessibility, local materials, and decentralized autonomy. Their critique of "supertechnology" wasn't about rejecting all modern tools but about selective, frugal use, grounded in the belief that industrial systems built on fossil fuels, mass production, and pollution are not only unsustainable but fundamentally dehumanizing.[90] For Todd and his collaborators, nature itself became the blueprint for redesigning how we live. "We don't have to invent anything," he once said. "We just have to pay attention to what's been learned."[91]

This ethos lives on in the practice of *biomimicry*, a field that studies nature's strategies and applies them to human design. Like TEK, biomimicry isn't new; Indigenous cultures have long turned to nature as teacher. From polar bears guiding Alaskan hunters to bees inspiring architecture, humans have always sought to live in dialogue with more-than-human wisdom.[92] Biomimicry deepens this tradition into a formalized discipline rooted in three pillars: *emulate, ethos,* and *reconnect.*[93] Emulation involves studying nature's patterns—how ecosystems regulate, how organisms adapt—and translating them into regenerative design. The ethos is grounded in humility and reciprocity, affirming our responsibility to support life rather than extract from it. Reconnection calls us back into relationship with the Earth, not as dominators but as kin.[94]

In practice, this looks like asking a landscape: What grows well here? What relationships already exist between species? These were the questions that led Wes Jackson to rethink the foundations of agriculture on the Kansas prairie. He noticed that modern farming, with its monocultures and shallow-rooted annual crops, was stripping the soil and destabilizing ecosystems. In response, Jackson co-founded The Land Institute in 1976 to explore how prairies, with their perennial polycultures, might offer a template for sustainable food systems. The Land Institute's research showed that perennials, with deep roots and year-to-year resilience, could maintain or even increase grain yields while building soil health, sequestering carbon, and improving water infiltration. Unlike annual crops that require constant plowing and herbicides, perennials protect and regenerate. Considering that over 70 percent of our global calories and croplands are tied to grains, the implications are enormous.[95]

One of the most promising examples is *kernza*, a perennial grain developed from intermediate wheatgrass (*Thinopyrum intermedium*), a Eurasian forage plant identified in 1983 by Rodale Institute breeders. Building on early USDA research, The Land Institute began selecting and breeding the plant in 2003 under the guidance of lead scientist Lee DeHaan. Through traditional breeding, not genetic modification, DeHaan and his team analyzed thousands of plants, selecting for traits like seed size and yield, using genetic tools only to speed selection, not alter DNA.[96] Now grown in 16 US states on more than 2,400 acres,[97] Kernza is making its way into breads, beers, and cereals. And it's only the beginning. The Land Institute is also developing perennial versions of wheat, rice, sorghum, and oilseeds—reimagining grain production as a restorative rather than extractive process.[98] In both the New Alchemy Institute and The Land Institute, we see the same pattern: working *with* nature, not against it. And like biomimicry and TEK, these approaches begin not with control or conquest but with careful listening. As biomimicry advocate Janine Benyus reminds us, "When we ask nature, first we quiet our human cleverness. Then we ask, and then we listen. The answer is the echo that bounces off the land herself."[99]

Echoing Wes Jackson's land-based epiphany, former USDA scientist Dr. Rick Haney arrived at a similar realization that reshaped his scientific approach. A widely respected figure in sustainable agriculture, Haney developed what later became known as the Haney Test, a soil health assessment that uses water-based extraction to evaluate available nutrients and microbial activity in the soil. Unlike conventional lab analyses detached from the field, the Haney Test allows farmers to engage directly with the living conditions of their land.[100] In 2022, I witnessed this test in action on a farm in Arkansas: a compact, handheld device gauging microbial biomass, nutrient availability, and soil habitat conditions to identify problem areas, guide cover crop choices, and reduce reliance on synthetic fertilizers. For Haney, the transformative insight was simple but profound—soil is not inert; it is alive. The microbes in the soil breathe like we do, he explained. They take in oxygen and emit CO_2. By measuring this CO_2, the test provides a snapshot of microbial vitality, a proxy for the soil's overall health. But more than a technical tool, the Haney Test is an invitation to shift our relationship to soil, asking not, "What can I extract?" but "What are you experiencing?" "Are you in balance?" "What do you need?"[101] This orientation—treating soil as a dynamic, responsive system—requires relational attentiveness. It recognizes that soil health cannot be prescribed universally; it must be understood contextually. Each field is its own living community, with its own needs and responses. In this way, the Haney Test exemplifies a broader shift in agroecological science: toward practices that listen to the land, work in partnership with life, and treat technology not as a master's tool but as a humble translator between humans and ecosystems.

RESTORING RELATIONSHIPS

"Convivial technologies"[102] are not simply tools; they are relational infrastructures. Adapted to specific ecological and cultural contexts, these technologies support cooperation, shared stewardship, and interdependence among people and the land. They foster social connectivity between rural and urban communities and provide platforms for mutual aid, skill-sharing, and decentralized innovation. In this way, convivial technologies help us re-learn how to be in community, an essential capacity for cultivating a resilient, regenerative food future. Across movements for food sovereignty and ecological justice, networks of farmers, scientists, engineers, and activists are co-creating technologies that reject proprietary enclosures and engage with communities historically excluded from shaping food system tools. Initiatives like Farm Hack, farmOS, and the broader open-source hardware and software communities, including the right to repair movement and the seed sovereignty and cooperative movements, are challenging the structures of corporate competition, data hoarding, and technocratic control. Take farmOS, for example: an open-source platform designed by and for regenerative farmers, supported by a decentralized collective of public-sector researchers, unconventional growers, and community tech workers.[103] The platform operates according to a "development methodology" rooted in openness, accessibility, and co-creation.[104] All of the code is freely available, contributions are welcomed from anyone, and farmers are not just users; they are co-developers. This reverses the usual hierarchy of expertise and centers the needs and wisdom of those closest to the land.[105]

OpenTEAM (Open Technology Ecosystem for Agricultural Management) and its innovations like the Ag Data Wallet extend this approach.[106] The Ag Data Wallet allows farmers to move their data securely across platforms without losing ownership or control. This enables peer benchmarking, collaborative modeling of greenhouse gas emissions and soil health, and community dialogue about best practices, all without corporate surveillance or extraction.[107] Meanwhile, initiatives like the Farm Hack community are building open-source agricultural equipment designed for small and mid-scale farms. Through its Tool Library and collaborative workshops, Farm Hack offers blueprints for everything from solar-powered greenhouse tables to DIY cultivation tractors and humidity sensors for chicken coops.[108] These designs are freely shared, collectively improved, and rooted in the belief that open knowledge, like open seed, is a commons for collective flourishing. In Europe, L'Atelier Paysan takes this further by training small-scale farmers, bakers, brewers, and others to build their own agroecological tools and structures.[109] Their advocacy challenges the dependency fostered by industrial agricultural systems, which condition farmers to rely on external expertise and high-cost technologies.[110] L'Atelier's approach returns autonomy and skill to those working the land, reconnecting technological practice with lived experience and place-based wisdom. In a context where venture-funded agri-tech increasingly seeks to capitalize on data and privatize tools, ensuring that innovation remains in the commons is urgent. The Bionutrient Food Association's Bionutrient Meter, a handheld device that measures nutrient density in food and carbon content in soil, is one example of reclaiming data sovereignty. By making this information openly available through a public, noncorporate platform, the association shifts the narrative from quantity to quality—from extractive yield to nutrient and soil vitality—while keeping the knowledge generated accessible to all.[111]

Though growing, many of these open-source movements remain scattered. The Gathering for Open Agricultural Technology (GOAT) exists to weave these efforts together, offering convenings, forums, and web-based communities to reduce redundancy, identify gaps, and increase the relevance and effectiveness of open tech for farmers, producers, and eaters.[112] As we return to the central inquiry—can food tech serve food justice?—the answer depends not only on the tools themselves but on the worldviews, values, and power structures that shape them. When technology is created within frameworks like RRI, Two-Eyed Seeing, biomimicry, and conviviality, it can become more than a tool; it can be a practice of relationship. These approaches ask us to design with humility and with care, embedding equity, ecological reciprocity, and local wisdom into our innovation processes. In doing so, we reorient technological power toward the flourishing of all life. And in this, the seeds of a truly regenerative future are already sprouting.

7

A Regenerative Mindscape

Scaling and accelerating agroecology isn't only about expanding access to healthy, diverse, and culturally meaningful food across urban and rural land-scapes.[1] It also requires scaling deep: transforming the inner landscapes of our values, worldviews, and relationships.[2] A truly regenerative food future calls not just for technical shifts but for profound cultural and psychological change: a remembering of how to live *with* the Earth rather than from above it. More and more, leaders across traditions, disciplines, and movements are naming the limits of purely external, mechanistic responses to our interwoven crises.[3] As the 2020 UN Human Development Report affirms, "Nothing short of a wholesale shift in mindsets, translated into reality by policy, is needed to navigate the brave new world of the Anthropocene, to ensure that all people flourish while easing planetary pressures."[4] We are not just in a food crisis or a climate crisis—we are in a *relational* crisis, where the dominant ways of knowing and being that structure society have become misaligned with the conditions for life.

Farmers engaged in agroecology, along with the practitioners who walk alongside them, understand that transitioning to regenerative systems is not only a matter of technique or policy. It also involves inner transformation: shifts in how we see the land, ourselves, and our place in the web of life. These changes unfold across individual, communal, and societal scales and are both shaped by and shaping global systems.[5] Soil conservationist Ray Archuleta, after decades with the USDA, came to define regenerative agriculture as a journey to "change the heart and mind" by learning to mimic nature.[6] Similarly, Mark Biaggi, a ranch manager, speaks of the need to shed linear, mechanistic thinking and embrace nature's nonlinear, ever-changing complexity—a move that demands humility, not hubris.[7]

Many of the destructive patterns in our food systems are upheld by deep-rooted narratives, particularly from the Global North, that normalize extraction, domination, and disconnection. But both ancient wisdom and contemporary science remind us that we are capable of evolving: of remembering how to listen, relate, and regenerate.[8] And we don't have to imagine this transformation from scratch. It's already underway in the gardens of resistance, the cooperative farms, the Indigenous land rematriation movements, and the many living experiments that are building life-affirming food systems right now.

Within the field of sustainability transitions, a growing body of thought recognizes that outer systems change must be accompanied by inner development.[9] Systems theory points to worldviews, values, and paradigms as deep leverage points, places where change can ripple across levels because they touch the root.[10] These inner dimensions form a kind of "mindscape" that coexists with, and co-creates, the material landscapes we inhabit.[11] Mindsets shape actions, just as actions reinforce or disrupt mindsets.[12] Practices of care—for land, for one another—can strengthen our sense of interdependence. But habitual disconnection can keep us from seeing new possibilities, even when our values begin to shift.[13]

Indigenous ways of knowing have long carried this recognition of inner–outer reciprocity.[14] Today's frameworks for transformative systems change are increasingly indebted to such wisdom. But to engage Indigenous teachings without reenacting extractive logics or colonial harm requires deep accountability and humility. Efforts to realign food systems must center not only Indigenous knowledge but Indigenous sovereignty and leadership.[15]

This interdependence between inner and outer change is vividly illustrated in the model of three nested spheres of transformation: practical, political, and personal. At the core is the *practical*—what we can see and measure: seed choices, farming techniques, consumer behavior. This is encircled by the *political*: the systems and structures that constrain or support practice, including policies, markets, and institutional norms. But surrounding both is the *personal*—our beliefs, values, and cultural paradigms.[16] Lasting change does not begin at the center; it ripples from the outermost ring inward. Transforming the personal, our relationships to land, to time, to each other, is what enables shifts in the political, which then allows new forms of practice to emerge and take root.[17] As we step further into this chapter, we explore what it means to nourish the personal sphere, not in isolation from systems, but as a vital terrain of transformation. This is a question at the heart of scaling deep. And it is essential to the future of food, and the future of life.

MOTIVES FOR CHANGE

To truly center regeneration as the guiding principle for the future of food, we must listen deeply to those already walking this path: farmers and land stewards who have long practiced agroecology, often outside the spotlight and beyond institutional support. At the same time, we must also attend to the complex realities of those working within conventional agriculture, many of whom are already grappling with the accelerating impacts of ecological disruption. Farmers stand at the frontlines of climate volatility and ecological collapse. They feel, in immediate and embodied ways, the destabilizing effects of extreme weather, droughts, floods, and soil degradation. Their proximity to the land gives them a vital perspective, and a potentially transformative role, in shaping food systems that nourish rather than exhaust. But shifting from extractive to regenerative approaches isn't just a technical pivot; it's a profound cultural, economic, and emotional reckoning. What will it take for farmers to move away from dependence on fossil fuels, synthetic fertilizers, and industrial models of production toward practices that build soil health, enhance biodiversity, and restore resilience? Many dominant strategies for catalyzing food systems change still rely on familiar levers: Provide financial incentives ("pay them"), enforce regulations ("stop them"), or deliver information ("tell them"). But extrinsic interventions often fail to reach the scale or depth needed for transformation.[18] They may spark compliance, but not commitment.

Behavioral science, particularly self-determination theory developed by Richard Ryan and Edward Deci offers a different lens: Lasting change arises not from coercion or reward but from alignment with a person's intrinsic values and sense of self.[19] Top-down mandates, subsidies, and taxes can influence short-term behavior, but they rarely shift the deeper motivations that govern how people relate to land, work, and community. Farmers who comply with a rule or chase a payment may still seek to maintain their core objectives—and often those objectives are shaped by their worldview, their cultural inheritance, and the economic pressures they face. In contrast, when change is driven from within, rooted in identity, values, and purpose, it tends to endure, even amid shifting external conditions.[20] Supporting the transition to agroecology, then, means creating conditions that nurture intrinsic motivation.[21]

GETTING OFF THE TREADMILL

Flowing from the previous exploration of intrinsic motivation, it becomes clear that enduring change in agriculture is less about external incentives and more about internal awakenings: shifts in identity, purpose, and worldview. The long-dominant productivist logic, which treats yield as the primary metric of farming success, is beginning to loosen its grip. For decades, agroecologists have emphasized that soil is not merely a substrate but a vibrant living system, and that farming must be understood in relational, not extractive, terms. Now, even within conventional sectors of the food and agriculture industry, there's a growing openness to more holistic frameworks, ones that center ecological interdependence and social wellbeing alongside productivity. This emerging receptivity is not just philosophical; it's practical. Farmers are increasingly confronted by the ecological consequences of industrial farming: depleted and flood-prone soils, extreme and erratic weather patterns, and soaring input costs, particularly fossil-fuel-based fertilizers. These cascading pressures are pushing many toward a precipice, creating conditions ripe for change not through coercion but through epiphanic disruption.

Dr. Hannah Gosnell, a professor of geography at Oregon State University, has been studying exactly these pivot points. Her research delves into the personal and relational dimensions that shape farmers' decisions to embrace regenerative agriculture, not just as a technical fix but as a life-altering reorientation. Unlike earlier studies that focused mainly on policy levers or economic incentives, Gosnell's work reveals how transformation often emerges through moments of crisis that rupture established patterns and invite deeper reflection.[22] Such "induced epiphanies," whether triggered by financial stress, environmental catastrophe, or health-related burnout, have historically accompanied major agroecological transitions around the world.[23] In

Latin America, India, and Cuba, agroecological movements often arose in response to geopolitical shocks, food insecurity, or ecological collapse. In each case, the failures of industrial agriculture did not simply invite critique; they provoked radical imagination and relational experimentation.[24]

For most of the regenerative farmers whose personal stories have been documented by researchers, the trigger that opened their minds to the possibility of a different way of farming was some form of personal or business crisis, be it environmental, economic, health, or psychological. The crises that farmers experienced, often a series of events, led to feelings of desperation and vulnerability, forcing them to critically reappraise themselves and their farming practices. Economic problems were frequently linked to environmental stressors such as drought and land degradation. These persistent challenges and the growing realization that their current approaches were insufficient created opportunities and motivation for change.[25] This was the case for Adam Chappell, who grows cotton, soy, rice, and corn on his 8,000-acre farm in Cotton Plant, Arkansas.

On a recent visit to Arkansas with members of the Soil & Climate Alliance, I witnessed this kind of transformation firsthand. Adam Chappell, who stewards an 8,000-acre operation in Cotton Plant, had once followed the conventional playbook: heavy tillage, chemical inputs, and yield-maximizing strategies. But as debt mounted and pigweed took over his fields, the sustainability of that model began to unravel. "We were spending more than we could ever hope to make," he recalled. Staring down financial ruin, Chappell took to YouTube, and discovered a pumpkin grower managing weeds with cover crops. That small spark led to a radical shift. Within a year of planting cereal rye, Chappell saw results that decades of chemical dependence hadn't delivered. What changed wasn't just his farming method; it was his sense of what was possible and what was worth striving for.[26] This is what intrinsic motivation looks like in practice: not compliance with policy mandates but a reorientation of purpose anchored in lived experience. When farmers like Chappell begin to see the land not as an input–output system but as a living partner, it catalyzes a deeper relational accountability—one that neither subsidies nor scolding can produce.

Building on Adam Chappell's story of transformation under pressure, Australian sheep farmer and author Charles Massy offers another poignant example of what it takes to fundamentally shift course. During the droughts of the 1980s and 1990s, Massy found himself trapped in a downward spiral, taking on debt to buy feed for his animals as his land deteriorated into dust. In 1982, his farm resembled a barren wasteland, plagued by grasshoppers and stripped of vitality. Emotionally and financially drained, he came close to giving up. "My mental health was atrocious," he recalls. "I was depressed for a year or more because I was in this trap." That breaking point, however, became a turning point. As Massy puts it, it was the rupture that "cracks the mind open"—the moment of disillusionment that opened the door to regenerative thinking. Rather than retreat, Massy leaned into change. He sold half the farm, enrolled in a PhD program in human ecology in his 50s, and began to study the deep narratives driving land degradation. As part of his research, he interviewed 80 leading regenerative farmers, listening to story after story of land and life renewal. By 2000, Massy had fully transitioned his own operation to regenerative methods, including the planting of over 60,000 trees.[27] His transformation wasn't just technical; it was psychological, even spiritual. It marked a departure from domination toward reciprocity, from extraction toward care.

Massy's journey mirrors what many farmers describe: the slow unraveling of the productivist paradigm and a yearning for sovereignty—over their time, labor, land, and livelihoods. Research from the US Great Plains reinforces this trend. Farmers increasingly see regenerative agriculture not merely as an environmental choice but as a path to autonomy and economic resilience. While agribusinesses promote themselves as indispensable partners, many farmers view their soil—healthy, resilient, and alive—as a way to exit extractive systems that demand ever-higher yields and ever-deeper debts.[28] One young Nebraska farmer, returning home after years away, described how conventional agriculture had begun to feel both irrational and dangerous. Propped up by subsidies and price supports, the system demanded heavy inputs, enormous loans, and continual expansion.[29] His family began planting cover crops out of financial necessity, hoping to reduce their escalating input costs.[30] That initial step opened a door to broader transformation, rooted in both economic pragmatism and ecological care. As the cost of synthetic fertilizers and other fossil fuel-based inputs continues to rise, more farmers may find themselves questioning not just their practices but the entire architecture of industrial agriculture.[31]

For many, the appeal of regenerative farming lies in the promise of independence, not just from chemicals or market fluctuations but from the entire mindset of control and compliance.[32] Researchers have documented numerous similar accounts of farmers embracing regenerative practices during periods of acute pressure—times when, in the words of a middle-aged Kansas cattle farmer, "your back's against the wall."[33] These farmers are not romanticizing hardship. Rather, they are translating crisis into clarity, and reclaiming agency in the face of systems that often reduce them to passive implementers of someone else's model. In doing so, they are redefining success, not by yield alone but by quality of life, ecological integrity, and intergenerational resilience.

RECOVERY AND HEALING

Building on the financial motivations discussed earlier, many farmers also cite health and ecological concerns, particularly regarding dependency on chemical inputs like glyphosate, as powerful drivers for transitioning to regenerative agriculture. One Kansas farmer voiced a common sentiment: "You can walk outside in the morning and smell chemicals. . . . When your eyes are watering and you see [and] smell that stuff for ten minutes, I can't help but think that it's doing something to you. That's what we're trying to get away from."[34] A California almond grower echoed this concern, sharing, "Our home is in the middle of the orchard. We drink the same water as our trees and we don't want to subject our family to health concerns that 'conventional' farming operations can present. We also strongly believe in being good stewards of our land and grow our almonds using sustainable processes."[35]

In Ohio, farmer Les Seiler shifted to regenerative practices out of concern for the watershed and a desire to build soil health through cover cropping and minimal fertilizer use. His turning point came during the 2014 Toledo water crisis (as described in Chapter 3), when toxic algae blooms linked to agricultural runoff made the city's municipal water undrinkable and unsafe to touch. With 80 percent of the Maumee River Basin comprised of farmland, including Seiler's own 1,700 acres, he acknowledged, "No producer around here can say they didn't play a role in that. And that really resonated with me—we were part of that problem, and I just look at that as a bad deal. We can't do that to people."[36]

While the catalysts for change vary widely, a recurring thread in farmers' stories is a deepening awareness of how industrial practices not only harm ecosystems but create a kind of chemical dependency—one that depletes both land and livelihood. As one agricultural consultant described, the first use of synthetic fertilizers and pesticides can feel like a miracle cure. But over time, as soils degrade and crops grow weaker, more inputs are needed to achieve diminishing returns. This escalates into a harmful cycle that reduces yield stability and long-term viability.[37] One central Kansas farmer compared his regenerative transition to helping both people and soil "get off drugs," while another described it as a recovery process from years of industrial exposure, an agricultural detox that begins with acknowledging harm and choosing a different path.[38]

NEW WAYS OF SEEING

Expanding on the shift away from chemical dependence, many farmers describe how positive, life-affirming experiences, particularly around the soil microbiome, further catalyze their transition toward regenerative agriculture. Learning about the living dynamics beneath their feet and seeing firsthand how chemical-free practices can enhance the vitality of their land often becomes a powerful motivator to not only begin but to stay on this path. Encounters with alternative agricultural visions—whether through peer-to-peer exchanges, public talks, or hands-on demonstrations, often trigger epiphanies that transform not just farming practices but worldviews. Central to this transformation is a shift in belief: the realization that not only is it possible to farm without chemicals but that doing so can restore relationships with land, life, and community. Ray Archuleta, widely known as "The Soil Guy," refers to this as a kind of "de-programming" that allows farmers to "see" the land differently.[39] His soil demonstration tests—comparing lifeless, chemically treated soil with rich, biodiverse forest soil—illuminate this difference with startling clarity. The contrasts in texture, color, smell, and life spark new mental models, challenging farmers to rethink what health and productivity really mean.[40] In Gosnell's study, this awakening is deepened through the use of microscopes, giving participants a direct glimpse into the hidden microbial worlds within soil. Such encounters mark the beginning of a deeper learning journey, one in which farmers begin to see their work as part of a much larger web of interconnections between soil, water, animals, people, and place.[41]

This holistic perspective runs counter to dominant industrial narratives that frame farming as a battle against nature, in which success is measured narrowly by yield and the use of chemical "solutions." Regenerative agriculture instead invites a new relationship, one grounded in stewardship rather than extraction. As Steve Gliessman, a pioneering agroecologist, explains: "The idea of farming with nature rather than against her is a foundational component of how we as agroecologists think."[42] It calls for seeing the farm as part of an ecological whole and recognizing the farmer not as a controller of nature but as a participant within it.[43] As one Kansas farmer put it, "I very much view myself as a part of nature, and I want to be . . . I want to live my life in contrast to the tradition I was raised in where you have dominion over nature, and you fight nature."[44] This emerging ethic extends beyond the farm's boundaries—considering planetary health, future generations, and even the agency of nonhuman beings such as the soil.[45] In New Zealand, regenerative farmers described becoming attuned to the presence and needs of pollinators, birds, and even inanimate entities like water and soil, cultivating a sense of responsibility that transcends human-centered stewardship.[46]

The emotional and psychological shifts that accompany this transformation are profound. One farmer captured this change succinctly: "Before I changed my farm and started caring about my soil and thinking about the soil as an ecosystem, it was a very depressing life. Every day I'd wake up deciding what I was going to kill next, whether that was a pest or a weed. It was a life revolved around death. But when I completely changed the outlook of my life—when I started trying to grow things and bring life to my farm—it gave me a positive attitude . . . It's less draining to work around life instead of death every day."[47] This sentiment echoes what a long-time regenerative practitioner in Nebraska described as his pivotal realization: "I heard Gabe Brown say one time, 'I used to wake up every day and say, what am I going to kill today?' I thought, that's exactly what I do." Now, instead of seeking to eliminate problems, he looks to enhance vitality by collaborating with pollinators and cover crops to get "more life into the system."[48] In this way, regenerative agriculture becomes not just a method but a mode of being, an invitation to reorient life and work around renewal rather than control.

Continuing the shift in perception that characterizes regenerative agriculture, many farmers begin to see so-called weeds not as nuisances to be eradicated but as allies in disguise—potential sources of forage, nutrient cycling, and soil improvement. Rather than relying on synthetic fertilizers, which often exacerbate weed issues by disrupting soil biology, they come to understand that weeds can be indicators of soil health and even contributors to it. Herbicides, once considered essential, begin to appear not only unnecessary but actively harmful—costly tools that diminish the resilience of the ecosystem they're meant to "manage." This transformation in thinking extends to how pests and water are understood as well. Pests are no longer cast as enemies to be wiped out but as part of a complex ecological web that, when functioning well, self-regulates. Insecticides become a last resort rather than a first response.[49] Similarly, the narrative around drought begins to shift. Rather than seeing drought solely in terms of precipitation deficits, regenerative farmers recognize that the capacity of soil to absorb, hold, and cycle water is just as critical. A dry spell becomes not simply a weather event but a reflection of how well, or poorly, the soil ecosystem has been stewarded.

This broader, systemic view is accompanied by a deepened sense of time and seasonal rhythm. Regenerative farming is not a quick fix; it's a long-term commitment to working with, rather than against, the land's natural cycles and limits. Early years in the transition are often marked by uncertainty, as biodiversity takes time to reestablish and soil health builds slowly. The full benefits of practices like cover cropping with rye, millet, or winter peas typically become evident only after several seasons. Research shows that while farmers may implement one or two regenerative practices within the first couple of years, it often takes another five or six years before a broader set of principles becomes integrated into their farming systems.[50] As Di Haggerty, an Australian sheep farmer, puts it, "you've got to work within the natural processes of that land, and it needs time. Sometimes it needs time just to rest, so it means it might not have a crop in it. It might not have animals on it. It might just be plants . . . sitting there if the seasonal conditions aren't conducive to growth at that particular time."[51] This understanding of rest, limits, and patience flies in the face of industrial models that demand constant productivity and short-term gains.

The regenerative mindset also resists the universalizing tendencies of industrial agriculture. What works in Arkansas may not work in Iowa. Even within the same farm, different fields may require different approaches. The soil is not an inert substrate but a living, evolving system that must be listened to and learned from. As Dr. Rick Haney reminds us, regenerative agriculture is less about following a rigid prescription and more about cultivating a relational practice: "What are your conditions? Are you in balance? What can I do to help?" These are not just technical questions but relational ones, questions that reframe the farmer's role from master of the land to humble caretaker within it. In this way, regeneration is not merely a strategy for producing food; it's a deep reorientation of how we relate to land, time, and life itself. It challenges the monocultural mindset of uniformity, efficiency, and control, offering instead a pathway rooted in attentiveness, reciprocity, and adaptability.

THE SOCIAL CONTRACT

Building on these stories of transformation and intrinsic motivation, my visit to Chappell's farm left me with a lingering question: What will it take for a regenerative worldview to supplant the dominant extractive model? Over time, I've come to see that the cultural ideals surrounding what constitutes "good farming" are a powerful, yet often overlooked, key to this transition. These ideals don't just influence farming practices; they underpin the entire social architecture that sustains industrial agriculture. Even in the face of direct exposure to the harms of agrochemicals—toxins in pesticides and fertilizers, noxious gases from manure storage and silage, deteriorating soil and water quality—many farmers remain reluctant to name these realities. In certain rural communities, there exists an unspoken social contract: a shared silence that protects the status quo. Take, for example, the finding from Great Lakes Protection that nearly 80 percent of farmers underestimate the level of soil erosion on their own farms.[52] This isn't just a data gap; it reflects a deeper dissonance between lived experience and prevailing norms. Being recognized as a "good farmer" is central to social standing in many farming communities. It's more than a compliment; it's a badge of cultural belonging and a sign of respect. This label carries real weight; it designates someone whose knowledge and skills are trusted, someone others turn to for help or advice.[53]

Importantly, this identity is deeply entangled with conventional indicators of success: a "clean" field free of weeds, a landscape shaped by plows and inputs, and yields that speak to productivity above all else.[54] This

aesthetic of order and control—plowed fields, manicured rows, visible applications of agrochemicals—is not simply about preference. It reflects generations of embedded assumptions reinforced by family legacy, land-grant
universities, agribusiness advisors, and federal policies. In this paradigm,
productivism isn't just an economic strategy; it's a moral framework.
Maximizing yield becomes synonymous with being a responsible, successful farmer.[55] What's striking is that even when new practices like cover
cropping or reducing tillage offer clear financial or ecological benefits,
many farmers hesitate to adopt them, not because they don't work but
because they challenge these deeply held cultural norms.[56] As one researcher
noted, social status often outweighs economic logic.[57] Trying something different can be read as a critique of your neighbors, your parents, or even
yourself.[58] And in tight-knit communities, that can mean social isolation—
from church pews to local diners to family gatherings.

One Kansas farmer shared a painful story about his father-in-law, who
rejected him as a tenant after he began transitioning to regenerative
methods:

> My father-in-law . . . worked the soil eight times: he plowed it, disked it, and
> field cultivated it over and over . . . to look perfect. I'd have 15 percent better
> yields every year with my minimum tillage . . . and my input costs were
> significantly less. But all he could hear in his head was the coffee shop crowd
> bragging about how clean his fields were. I helped my father-in-law the last
> 20 years of his career. One day I found out that he rented the farm to another
> guy. Why? Because I was a "crazy cover crop, minimum tillage guy."[59]

The stakes are even higher for farmers who don't own the land they cultivate. In the United States, nearly 40 percent of farmland is rented, much
of it under short-term, informal agreements.[60] These arrangements often
come with implicit expectations: maintain the visual aesthetic, don't challenge tradition, avoid anything "weird." Without longer-term leases or cost-
sharing agreements, tenant farmers are unlikely to take the financial or
relational risk of experimenting with regenerative methods, especially if those
changes might be misread as neglect.[61]

Ultimately, stepping off the industrial treadmill is not just about farming differently; it's about being willing to reimagine one's role, values, and

identity in the world. The transition to regenerative agriculture involves more than changing tools or techniques. It requires a profound shift in world-view—a new moral compass grounded in care, interdependence, and stewardship. And that kind of transformation can create as much tension with peers and kin as it does with the land.[62]

SOCIAL LEARNING

Given the deep relational ties that farmers maintain with peers, crop advisors, seed and chemical vendors, and others embedded in the agricultural system, it's no surprise that regenerative practices spread, or stall, through patterns of social learning. To understand how regenerative agriculture might move from the margins to the mainstream, we can turn to social learning theories, particularly those rooted in the sociology of innovation. One foundational framework, the diffusion of innovation theory developed by E. M. Rogers in 1962, is a social science model that describes how new ideas, technologies, or practices are adopted over time by segments of a population,[63] beginning with innovators, followed by early adopters, the early majority, late majority, and finally, laggards.[64] Crucially, both the diffusion of innovation theory and social network theory[65] underscore that people rarely adopt new behaviors in isolation. Social learning happens through trusted relationships and within the reference frames of familiar networks. Especially in tight-knit farming communities, the spread of innovation depends less on access to information and more on how that information is received: who delivers it, whether the messenger is trusted, and whether the practice aligns with the group's shared values and identity. This is why norms around what constitutes a "good farmer" carry such weight: They serve as moral and cultural touchstones that shape what is seen as legitimate or deviant, successful or suspect.[66] Identifying as a "regenerative farmer" or "agroecologist" isn't just a label; it's an identity marker that shapes how one perceives their role in a community and how they are perceived in return.[67] Group identities operate as reference points for behavior and influence how individuals frame what is meaningful, doable, and worth pursuing.[68]

This dynamic becomes especially powerful when new knowledge is introduced peer-to-peer rather than from top-down institutions.[69] A neighbor's successful cover crop or a fellow farmer's insight about reducing synthetic

inputs often carries more persuasive power than a university extension bro-
chure. Social learning, understood as a process of learning through dialogue, observation, experimentation, and reflection within a community,
has proven to be one of the most effective mechanisms for transforming
farming practices. Rather than simply transferring information, it invites
relationship-building and co-creation of meaning. When farmers gather in
shared spaces—whether pasture walks, field days, conferences, or informal
kitchen table conversations—they engage in a kind of collective inquiry that
helps them explore alternatives, troubleshoot challenges, and imagine
new possibilities.[70]

Evidence from California and beyond shows that these farmer-led learning environments outperform traditional top-down models of agricultural extension.[71] Movements like *campesino-a-campesino*, farmer field schools, and
participatory on-farm research have long demonstrated that farmers are not
just passive recipients of knowledge but co-producers of it. Through these
horizontal exchanges, they gain confidence to test regenerative methods, reflect on their outcomes, and iterate based on local conditions and collective
wisdom.[72] These networks not only support technical learning but also help
shift cultural narratives, destigmatizing regenerative practices and challenging entrenched notions of what makes a farmer "good" or "successful."[73]

Even when a farmer's first exposure to regenerative principles comes
through digital sources, such as Dr. Elaine Ingham's Soil Food Web School,[74]
Regen.ag Academy,[75] the Rodale Institute's organic transition training,[76] or
YouTube, the deeper transformation often happens in community. Farmers
seek out events and networks where they can meet others on a similar journey. These gatherings provide not just information but affirmation—spaces
where they don't feel alone in questioning the dominant model. Some farms
and regions have become known as "agroecology lighthouses," places that
model the viability of regenerative systems and anchor local networks of peer
learning. These exemplars not only share techniques but foster shared identities and values, building collective momentum. They serve as living counter-
narratives to industrial agriculture, showing that another way is not only
possible but already unfolding in the hands of ordinary people.[77]

One of the most impactful lighthouses helping to guide US farmers
through the fog of industrial agriculture is the grassroots network Practical
Farmers of Iowa (PFI). Notably, its name was deliberately chosen to invite a

broad spectrum of participants, including those not yet ready to fully embrace a radical departure from conventional systems. In a study of agroecological transitions across the Mississippi River Basin, nearly every farmer interviewed cited PFI as a critical, often primary, influence on their decision to shift practices.[78] Since its founding in 1985, PFI has supported farmers in reducing chemical inputs while remaining economically viable, encompassing a wide spectrum of practices from precision nutrient management to organic production and rotational grazing. A cornerstone of PFI's work is its collaborative, farmer-led research model. By establishing on-farm trials and test plots, members learn through experimentation, helping to fill the significant void left by underfunded public agroecology research.[79]

While regenerative practices are deeply shaped by place, the broader potential of the movement rests on building bridges across place-based communities. Transformation deepens when localized experiments cohere into trans-local solidarities.[80] In regions like the Great Plains, where departing from dominant agrarian norms can invite social penalties, farmers often seek out like-minded peers beyond their immediate context.[81] These extended networks offer both technical learning and emotional refuge. Mitch Hora, a multigenerational farmer from Washington County, Iowa, articulates the importance of finding such alternative communities of practice:[82]

> Many farmers live in communities and are surrounded by other farmers
> who have done things a certain way and anything outside of that system is
> seen as ludicrous and unconventional. . . . . Farmers who are learning new
> ways of farming do it because they see another farmer doing it and say to
> themselves, if that farmer can do it, so can I.[83]

This peer-to-peer form of solidarity and witnessing is being cultivated through field days, pasture walks, and multiday gatherings hosted by initiatives such as Beyond the Yield in Nebraska[84] and Fuller Field School in Kansas.[85] For many, these encounters are catalytic. One attendee, Daniel Deepe, called the Fuller Field School "life-changing," describing how a single presentation on glyphosate shifted his entire relationship to chemical farming: "The sprayer has been parked ever since.[86]" The cultural and epistemic diversity at these events is itself a form of transformation. As Didi

Pershouse noted, it's not uncommon to witness "conservative Christian farmers in cowboy hats stand next to permaculture folks in rainbow bandanas, and investors who made it big in the dot-com boom in California next to Mennonite women, all brought together by a shared passion for soil microbes, human health, photosynthesis, and grazing animals."[87]

Across the United States these networks, formal and informal, are nurturing new social learning ecosystems. No Till on the Plains, which began in the 1990s as a grassroots gathering of farmers interested in natural systems farming,[88] has grown into an influential node, drawing hundreds to its conferences and farm tours.[89] As one former board member recalled, the most valuable outcome of those early events wasn't just information exchange but the sense of not being alone—of meeting others "who were open minded and looking down the same path they were looking down."[90] Similarly, Green America's Soil & Climate Initiative (SCI) and the Women, Food, and Agriculture Network (WFAN) are weaving dense webs of support. SCI has convened over a hundred farmers across twenty-three states, connecting them with soil scientists, food system stakeholders, and advocacy groups.[91] WFAN, which began in 1994 as a regional effort, now supports women farmers and landowners across the country through mentorships, community-building, and conservation programming.[92]

These initiatives create relational infrastructure, what some might call "communities of courage," that help farmers reimagine what's possible and reshape what it means to be a "good farmer." For many, attending their first regenerative agriculture event becomes a portal. One farmer described walking into a session on soil health and leaving with his "mind blown."[93] These kinds of encounters aren't just about acquiring new techniques; they're about recovering a different sense of self, land, and possibility. They help coalesce not just new practices but new identities and cultural imaginaries. And in doing so, they erode the isolation that often silences dissent within dominant systems. These relational and peer-led transformations mirror global movements as well. Although the US-based regenerative movement often emphasizes soil health, its emphasis on farmer-led knowledge creation resonates with international agroecology efforts that center food sovereignty, land justice, and ecological self-determination. Whether through Iowa field trials or Cuban farmer-to-farmer trainings, what emerges is a shared ethic: Transformation begins in relationship.

Despite minimal institutional investment in agroecology, whether in the form of research funding, extension services, or coherent policy support, millions of smallholder farmers across the globe have adopted and expanded agroecological practices through peer-based knowledge sharing.[94] One of the most remarkable and instructive examples comes from Cuba, where a widespread transformation of the food system emerged not through top-down mandates but through farmer-to-farmer solidarity and necessity. Following the collapse of the Soviet bloc in 1991, Cuba faced an abrupt and profound crisis. Deprived of access to imported agrochemicals, fertilizers, and industrial equipment previously supplied by the USSR, the island nation was forced into a period of intense austerity. This rupture, what Cubans call the "Special Period," demanded new approaches to survival. In response, Cuban farmers turned toward organic and ecological methods, drawing both on traditional wisdom and innovations from Cuban scientists. The Farmer-to-Farmer Agroecology Movement (Movimiento Agroecológico de Campesino a Campesino, or MACAC), housed within the Cuban National Association of Small Farmers, became the beating heart of this transition. Rather than relying on technical experts or outside institutions, MACAC embraced a horizontal pedagogy rooted in the campesino-a-campesino model, a Central American method of peer learning and participatory experimentation. Through mutual aid, demonstration farms, and shared problem-solving, farmers became each others' teachers. This approach not only restored agroecological practices but also cultivated new forms of dignity and collective agency. From just 200 farmers in 1999, the movement grew to include 200,000 by 2018, representing more than half of all smallholder farmers in the country.[95]

A similarly powerful transformation is unfolding in India through the Andhra Pradesh Community-Managed Natural Farming (APCNF) initiative. Here, women, who comprise over three-quarters of agricultural workers, are at the forefront. Organized into self-help groups, women farmers are spreading regenerative practices, supporting one another through transitions, and mobilizing resources to fund seeds, composting materials, and training.[96] At the heart of APCNF is the recognition that transformation is as much about relationship and care as it is about technique. Veteran farmers, known as "champions," mentor those just beginning their journey.[97] Massive training camps, some with thousands in attendance, demonstrate that farming without chemical inputs is not only possible but economically and ecologically sound.[98] This initiative, which began with 40,000 farmers in 2016, expanded to over 700,000 by 2021. With active support from the state's agricultural department, the goal is to reach millions more.[99] Importantly, this scale is not built on central control but on distributed wisdom—on networks of trust, mentorship, and iterative learning. In 2020, members of the Alliance for Food Sovereignty in Africa (AFSA) traveled to Andhra Pradesh and witnessed this model firsthand. Inspired by what they saw, AFSA initiated a transcontinental peer-learning exchange, connecting farmers across twelve African countries and their counterparts in Andhra Pradesh. Through this collaboration, the knowledge-rich, input-light ethos of natural farming is being shared across continents, not as a prescriptive solution but as an invitation to adapt and co-create locally resonant forms of agroecological regeneration.[100] These stories remind us that deep change rarely begins in the halls of power. It emerges from the margins, from those already cultivating different relationships to land, to each other, and to the future.

REGENERATIVE CHAMPIONS

Over the past decade, the National Wildlife Federation's Sustainable Agriculture team has been exploring how to shift prevailing perceptions of "good farming" by applying the diffusion of innovation theory. Recognizing that most farmers are not early adopters but rather "wait-and-see'ers," as Nebraska farmer Richard Oswald describes, the team focused on understanding the motivations of middle adopters.[101] To this end, they collaborated with Robyn Wilson, a professor in risk analysis and decision science at the Ohio State University. Wilson emphasized that simply providing information is insufficient to change behaviors.[102] Through engagement with Great Lakes soil conservationists and farmers, they identified four key outreach strategies: tailoring initiatives to individual farmer needs, offering cost-sharing programs to mitigate risks, promoting farmer-to-farmer mentoring, and emphasizing long-term soil health over immediate economic gains.[103] These insights led to the creation of the Grow More program in 2019, equipping farm innovators and outreach professionals with effective messaging tools to engage middle adopters.[104]

Central to the Grow More initiative is the Conservation Champions program, which offers grants up to $15,000 to teams of innovative farmers and local outreach partners. These teams develop creative outreach strategies aimed at engaging new farmer audiences and promoting soil health practices in the Mississippi River Basin.[105] Recognizing that traditional outreach methods often only reach already interested farmers, the Champions program seeks to craft new themes and messaging that resonate with those yet to adopt regenerative practices.[106]

The Rare Center for Behavior & the Environment (BE.Center), the world's first institute dedicated exclusively to behavioral science for environmental challenges, has also conducted qualitative research into farmers' decision making. They identified two primary drivers: social proof and social pressure. In Colombia, the BE.Center applied these insights to the Lands for Life regenerative agriculture program.[107] Previous efforts in the region faltered due to a lack of consideration for social variables influencing farmers' decisions. Lands for Life adopted a nuanced approach, first engaging "innovador" farmers eager to adopt new practices.[108] Their successes provided social proof, encouraging early and late majority farmers to follow suit. Peer-to-peer workshops and media engagements further bolstered this transition, reshaping the community's perception of a "good farmer" as one who embraces regenerative methods.[109]

In Mexico's Chiapas state, the BE.Center supported the Centro de Investigación y Servicios Profesionales A.C. (CISERP) in promoting traditional milpa intercropping of beans, maize, and squash, known as Las Tres Hermanas or the Three Sisters. The decline of these ancestral methods, due to the promotion of commercial seeds and synthetic inputs, adversely affected biodiversity and public health.[110] CISERP's campaign in Tojtic encouraged composting, reduced chemical fertilizer use, and seed sharing.[111] Collaborating with the community, they developed materials in the native Tsotsil language, including a slogan, comic book, puppet show, and mural celebrating ancestral practices. Children played a pivotal role, bringing messages of pride in local agriculture to their families.[112] Field Schools facilitated hands-on training, while seed and nutritional fairs promoted local seed conservation and milpa-based dishes.[113] Within a year, 90 percent of the 324 milpa farmers in Tojtic adopted the agroecological practices promoted by the campaign, with 65 percent committing to participating in a five-year seed conservation program.[114]

While initial adoption often hinges on simple practices yielding quick results, the overarching goal is a transition to integrated agroecological systems that restore biodiversity, enhance soil fertility, and reduce reliance on external inputs.[115] This journey requires a holistic understanding of living systems. As more farmers and communities embrace such systems, supported by policies, research, and equitable markets, agroecology can regenerate entire territories, as exemplified in Andrah Pradesh. For agroecology to effect meaningful change in local and global agricultural systems, a collective effort from farmers, policymakers, scientists, investors, educators, and others is essential. The path forward involves each of us becoming regenerators, a focus explored in the next and final chapter.

8

Regeneration Rising

We stand at a pivotal threshold, a collective rite of passage, confronting the urgent call to reimagine and redesign our relationship with Earth. The shift this book has explored—from a mechanistic, extractive worldview toward one that recognizes life as interconnected, dynamic, and sacred—can feel daunting when viewed against the backdrop of ecological collapse and economic instability. And yet, as systems thinker Donella Meadows reminds us, "there's nothing physical or expensive or even slow in the process of paradigm change. In a single individual it can happen in a millisecond. All it takes is a click in the mind, a falling of scales from eyes, a new way of seeing."[1] A paradigm shift begins not with new tools but with new eyes, an inner reorientation, a falling away of inherited certainties. This is the invitation at the heart of regeneration. My intention in writing this book has not only been to reveal the hidden costs, the so-called externalities, of the global food system but to offer a pathway toward wholeness. A way to reclaim our inherent capacity to live in right relationship with Earth. While farmers play a crucial role in this transition, the work of regeneration belongs to all of us. It is not a job title but a relational stance, a way of being in the world.

Regeneration is not a universal formula or checklist. It takes root in specific places, shaped by distinct cultural, ecological, and economic histories.[2] While the scope of global crises might tempt us toward sweeping "solutions," experience shows that abstract, globalized thinking often misses the nuance, memory, and complexity that live in place. In contrast, bioregional regeneration begins where our feet touch the ground. It acknowledges that climate, food security, soil health, and biodiversity loss are not merely "global problems" to be solved but relational imbalances to be tended, community by community, watershed by watershed. This perspective, bioregionalism, has deep roots.[3] In the 1970s, Bay Area environmentalist Peter Berg began to articulate a vision of aligning human identity and lifeways with ecologically meaningful boundaries such as watersheds rather than artificial political borders. When asked where he was from, Berg replied not with a state or nation but with a river: the Sacramento watershed.[4] Bioregionalism recognizes what many Indigenous cultures have always embodied: that identity is shaped in relationship to land, water, and the living systems that sustain us.[5] It is a way of knowing echoed in Thich Nhat Hanh's concept of "interbeing,"[6] Joanna Macy's "ecological self," and Arne Naess's deep ecology, which invites us to see humans not as separate from nature but as intricate knots in the web of life.[7]

Bioregions are not just ecological zones; they are relational territories. Within them, we can cultivate shared meaning, restore ancestral knowledge, and design context-specific responses to complex challenges. Take, for example, the Cascadia bioregion—a vast and diverse ecological corridor stretching from southeastern Alaska through British Columbia and Washington to northern California.[8] Bound together by rivers, salmon migrations, and shared vulnerability to fire, drought, and seismic activity, Cascadia offers a living example of how bioregional identity can inspire cross-border cooperation and holistic governance.[9] In this movement,[10] forests, cities, and coastlines are not separate "issues" but interdependent systems requiring integrative approaches across jurisdictions, communities, and generations.[11] As we turn toward regeneration, not as a technique but as a relational ethic, the bioregion offers a scale that is both intimate and expansive. It invites us to become students of place, to listen for the wisdom embedded in local ecologies, and to weave networks of mutual care across the artificial boundaries that fragment our world.[12] If this final chapter is about "scaling" regeneration, it does not mean scaling up in the industrial sense. It means scaling deep—into hearts, relationships, soil, stories, and place. It means becoming regenerators not just of land but of culture, community, and connection.

CONNECTING TO PLACE

When I accepted a teaching position at Harvard, my family and I relocated from Indiana, where my sons were born, to move to Massachusetts. Though I longed to develop a meaningful relationship with this new place, the rhythms of academic life, a punishing commute, and the demands of parenting young children left little space for such exploration. It wasn't until the relative stillness of the Covid pandemic, when I spent more time at home, that something shifted. One afternoon, I came across a bioregional quiz from 1981. It posed questions that pierced through my routines and certainties:

1. What soil series are you standing on?
2. What was the total rainfall in your area last year?
3. What were the primary subsistence techniques of the culture that lived in your area before you?
4. Name 5 edible plants in your region and their season(s) of availability.
5. Where does your garbage go?
6. How long is the growing season where you live?
7. What primary ecological event/process influenced the landform where you live?
8. Trace the water you drink from precipitation to tap.[13]

I took the quiz, and failed nearly every question. But something had been stirred. What began as a sense of deficiency became a portal into curiosity. Despite years of working alongside farmers, food innovators, educators, and policymakers across the globe, I had overlooked the histories and ecologies of the very land I was now part of. I didn't yet know the names of the original stewards of this place or the movements cultivating relational repair and food sovereignty in my own city. Like a farmer beginning the transition to agroecology, I began to see with fresh eyes. The woods near my home became more than scenery; they became a teacher. My neighborhood began to hum with histories and ecologies I had once overlooked. This was not about mastery but about apprenticeship: learning to live in a place through attention, reciprocity, and reverence. This kind of practice is often called bioregional regeneration. The poet Gary Snyder named it "reinhabitation"[14], living not simply as residents but as accountable stewards of a shared home. This process is experimental, layered, and necessarily incomplete. It invites us to become attentive participants in a living system, not detached observers. It also invites a recognition that the land does not belong to us; rather, we are of the land. As Robin Wall Kimmerer reminds us, to live as if our children's future mattered is to treat land as sacred kin, essential to both our material survival and spiritual integrity.[15] Returning to the quiz as I wrote this book, I realized that some the questions about rainfall, growing seasons, and temperatures no longer have stable answers. The climate crisis is reshaping the contours of every bioregion. Bioregional literacy now means not only understanding what *has been* but listening for what *is becoming*. It asks us to attune to transformation, loss, and possibility in the places we inhabit. Love for place becomes a form of agency. It is this grounded love, not abstract urgency, that moves us to resist overdevelopment, pollution, and erasure. And it often begins in the most intimate of landscapes: our yards, balconies, and window boxes.

The ways we tend these spaces matter. For decades, a narrow vision of order and productivity encouraged both homeowners and farmers to treat land as a surface to be controlled: fertilized, mowed, or plowed into submission.[16] Today, in the contiguous United States, residential land covers 30 percent of the country, four times more than land officially designated as protected.[17] Most of this is carpeted in turfgrass, a thirsty, chemically dependent monoculture that has become the nation's largest irrigated "crop."[18] These lawns offer little to no habitat for the web of life, becoming what scientists have called biodiversity deserts. This simplification is more than aesthetic; it is ecological collapse in slow motion. When native plants are removed and habitat is fragmented, local ecosystems lose their ability to sustain pollination, pest regulation, carbon sequestration, clean water, and oxygen generation. These are not amenities; they are life-support systems.[19] In response, grassroots initiatives are sprouting with the intention of reweaving fragmented ecologies. The Homegrown National Park movement, envisioned by ecologist Doug Tallamy, encourages people to convert portions of their lawns into habitats. The goal is ambitious: 20 million acres of native plantings, roughly half of all private lawn space in the United States.[20] Tallamy reminds us that our official national parks, while vital, are too small and isolated to protect biodiversity at the needed scale.[21] But when our homes, schools, places of worship, and workplaces begin to pulse with native plants and pollinators, we create corridors of resilience—networks of survival that extend beyond the bounds of any single preserve.[22]

Another movement, Food Not Lawns, was founded in 1999 by permaculture designer Heather Jo Flores. It invites people to convert ornamental lawns into food-producing gardens and to share seeds, skills, tools, and land. Since the Covid pandemic, this ethos has gained traction across diverse communities, from apartment balconies to suburban backyards.[23] This impulse to cultivate life at home, on porches, in backyards, or along apartment windowsills, has gained renewed momentum in recent years. Since the onset of the Covid pandemic, home gardens have proliferated across varied landscapes, echoing a deeper ancestral memory and longing. Yet this surge is not a new trend but a remembering of practices long carried by immigrant and Indigenous communities across generations. In the United States, the legacy of home gardens is deeply rooted in the cultural resilience of those who arrived, often displaced, from Mexico, Latin America, Africa, Asia, and Europe. For many immigrant families, gardens have served not merely as food sources but as vessels of memory and cultural continuity—a quiet refusal to assimilate through the keeping of seeds, flavors, and rituals. For many Indigenous communities, gardening has been an enduring practice of cultural determination and relational continuity, part of a long arc of survival and renewal amid ongoing colonial violence.[24]

Urban and peri-urban agriculture has sustained cities throughout history, not as a novelty but as a necessity. In precolonial Mexico, the Aztec practice of *chinampas*, floating gardens woven into shallow lake beds, fed entire populations with an intricate ecological choreography.[25] In nineteenth-century Tokyo, sophisticated systems of urban-fringe agriculture used organic inputs like "night soil" (human excrement) to create nutrient loops that supported dense urban populations without synthetic inputs.[26] Today, this lineage continues. In countries like Czechia, Japan, and Cuba, home and community gardens still provide a substantial share of fresh produce, often outpacing commercial organic sectors in both volume and biodiversity.[27] Especially when grounded in agroecological principles,[28] they do more than feed people. They can weave food security with climate adaptation, reconnect communities fractured by displacement, and foster intergenerational learning. As global supply chains wobble under the strain of overlapping crises, local food systems are not just resilient alternatives; they are lifelines.[29]

There is precedent for this collective turning toward the soil. During both world wars, the US government encouraged "Victory Gardens" to boost food production and morale.[30] By 1944, nearly half of the nation's fresh produce was grown in these gardens, vacant lots and backyards transformed into

vibrant, sustaining spaces.[31] Today, in the face of climate collapse and ecological grief, many are reclaiming this spirit—this time as climate victory gardens, infused with regenerative practices and a commitment to ecological
repair.[32] These gardens are not just utilitarian. They are relational. Professor
Gabriel R. Valle, who has spent the last decade in conversation with Latinx
gardeners in Southern California, shared a teaching passed on to him by
one gardener: The most important thing you can do every day is to greet your
garden. When you do so, "you greet life."[33] This daily ritual reflects a form of
reciprocal kinship, a humble honoring of the aliveness that sustains us. In
urban settings, gardens expand the possibility of wellbeing. They offer
green breathing spaces, host more-than-human species, and become edible commons, spaces of shared nourishment and collective care.[34]
Movements like community sponsored agriculture (CSA), regenerative farming, and neighborhood garden networks invite people to reenter relationship
with food—not just as consumers but as co-stewards of seasonal rhythms,
uncertainties, and labor. They offer an antidote to the industrial food system
by grounding people in the tactile intimacy of soil, seed, and sustenance.

We have the capacity to co-create living landscapes, even in the midst of
urban sprawl, by engaging with the principles of regenerative land care.
This approach, which some refer to as organic land care,[35] centers ecological
reciprocity: enhancing biodiversity, supporting soil vitality, and honoring the
cycles of life rather than disrupting them. By turning away from synthetic
fertilizers and pesticides, we not only safeguard pollinators and subterranean
life but also respect the earth's innate carbon-sequestering rhythms. When
we mulch, plant cover crops, or allow so-called weeds to coexist with intention, we are not imposing control but entering into stewardship, reducing
water dependency, preventing erosion, and minimizing the flow of toxins into
local waterways. Composting kitchen and yard scraps becomes a practice of
transformation: waste becomes nourishment, methane emissions from
landfills are lessened, and the soil becomes more vibrant, alive, and capable
of sustaining nutrient-dense food. By planting perennials, native species,
herbs, and trees, we reduce disruption, receive the gifts of multiple harvests,
and invite resilience into the landscape—resilience that can weather storms,
regulate temperature, and even quiet invasive tendencies.

Amid cascading crises, it is easy to forget our capacity to regenerate. Yet
every seed we plant, every handful of soil we touch, is a small refusal to give

in to despair. When bees buzz again, when butterflies flutter past us, when the night sparkles with fireflies and morning is announced by birdsong, these are not just signs of success but signs of relationship restored. As more of us step into this practice, not as saviors but as humble co-stewards, the effects ripple outward. Regeneration is not an endpoint but a way of being.

BACK TO BASICS

Perhaps even more vital than the new actions we take is our willingness to let go—of illusions of convenience, of inherited habits of separation, of the narratives that frame our entitlements as inevitabilities. Rachel Carson once described the moment we are living in: "We stand now where two roads diverge. . . . The road we have long been traveling is deceptively easy. . . . but at its end lies disaster."[36] She pointed not toward certainty but toward possibility—a path of preservation, not perfection. This is a hard truth to metabolize: Much of what has made modern life "easier" in the Global North has come at the expense of ecosystems, communities, and future generations. The pursuit of ease has cost us our health, our relationships, and our deeper capacities for care. The Slow Food movement, initiated by Carlo Petrini in 1989, emerged as a refusal of this high-speed disconnection. It sought to preserve local food cultures and traditions while resisting the rise of "fast life," a way of living that severs us from food origins, from the labor behind each bite, and from the interwoven ecological webs that sustain us. Over the decades, Slow Food has become a global movement, not only preserving seeds and skills but fostering joy, education, and collective nourishment. Its practices—school gardens, shared meals, and embodied learning—remind us that food is not just fuel but relationship.[37]

Some might view this as nostalgic or outdated. But evidence suggests we are already in the midst of a transformation, whether we acknowledge it or not. The industrial "food-from-nowhere" regime, disconnected from place, community, and consequence, is eroding. What's emerging is a return to "food from somewhere": relational, grounded, bioregional.[38] Taking the bioregional quiz, a simple yet piercing exercise, disrupted my assumptions. Questions like "where does your trash go" and "where does your food come from" exposed how deeply enmeshed we are in systems that obscure origin, process, and impact. The quiz didn't just illuminate my ignorance about place; it highlighted how the health of one bioregion is entangled with countless others. With over half of the world's population[39] and more than 80 percent of the US population living in urban areas,[40] the call is clear: We must reimagine how and where we live. This is not simply a logistical challenge but a cultural reckoning. Our food systems cannot remain concentrated in the hands of a tiny farming minority, currently just 1.4 percent of the population, if we hope to build resilience beyond fossil fuels and extractive supply chains. Agroecological systems require many more hands and eyes: humans attuned to soil health, weather patterns, and the living needs of plants and animals. The very tools designed to eliminate agricultural labor—chemical inputs, machinery, automation—have severed farmers from ecological wisdom and most of the broader population from any relationship with food production. This mechanized distancing has come at the cost of both ecological balance and cultural memory.[41]

Systems thinker Fritjof Capra reminds us that ecological literacy, an ability to live within, rather than outside of, natural systems, is not optional for our survival.[42] Industrial agriculture violates nearly every principle of living systems: It treats waste as external, not cyclical; it relies on synthetic energy rather than solar flows; it sacrifices diversity for control. And crucially, it forgets that life thrives not through domination but through networks of interdependence formed over billions of years.[43] Hope lies not in returning to the past but in reactivating our relational responsibilities. Agroecological systems thrive on observation, care, and humility. Unlike industrial monocultures, they require constant attention—learning to read soil, water, pollinators, weather, and time. This kind of care cannot be outsourced to machines operating on thousands of acres. It must be practiced by people, in place.[44]

The transformation ahead is not only agricultural. It is cultural. It will require us to redefine work, value, and livelihood, not as escape from labor but as deepened entanglement.[45] As M. Jahi Chappell and others have argued, a truly regenerative future may involve a widespread "re-ruralization" of the United States.[46] Ecologist David W. Orr articulated this decades ago: The future will belong to those willing to be embedded again, with integrity.[47] Jason Bradford, organic farmer and author of *The Future Is Rural*, echoes this trajectory. The material basis of modern urban life—fossil fuels, rare minerals, global trade—cannot be sustained indefinitely.[48] As energy descent accelerates and climate disruptions intensify, our capacity to maintain hyper-urban infrastructures will wane. In this light, relocalization is not only desirable; it may be inevitable.[49] Many are not yet aware that energy decline, rather than food shortage, is one of the greatest threats to food security. This is not a problem technology alone can solve. It is an invitation for responsibility, not as heroism but as reinhabitation. In highly urbanized nations, this may mean embracing smaller, slower, more localized ways of living, less dependent on distant labor and invisible extraction.[50]

What seems increasingly certain is that the future of food is not an individual pursuit or a corporate solution. It is a return to relationship. Regeneration does not live in a singular product, person, or place; it lives in the quality of the connections we tend. As Capra reminds us, a sustainable human community is one that fosters the thriving of all communities, both human and more-than-human, through ongoing reciprocity and co-evolution.[51] The question is not just how we will feed ourselves but how we will live in right relationship with the systems that feed us.

TRANSFORMING FOOD SYSTEMS

If the future of food lies in relationship, then transformation must be relational too. Unlike industrial or capitalist models that seek to replicate success through uniform scaling, real systems change is context-specific, rooted in particular ecologies, shaped by diverse stakeholders, and responsive to the unique dynamics of place.[52] There is no universal blueprint, no one-size-fits-all template. Transformation resists being managed, optimized, or scaled in the conventional sense. It is messy, often uncertain, and always entangled. And yet, amid this diversity, there are patterns. Not prescriptions, but principles. Shared values and orientations, like humility, interdependence, care, and justice, that can guide those committed to cultivating systemic change. As Professor Sandra Waddock puts it, what is needed is "a *catalytic* approach to developing systemic change," one that links diverse, distributed, and emerging initiatives into a living web of purpose and resilience, capable of holding the complexity of our socioecological realities.[53] History shows us that the most profound social transformations—the abolition of slavery, the expansion of voting rights, the dismantling of apartheid[54]—did not arise from a single policy, leader, or protest. They were not linear or centralized. They were emergent. They happened when countless local actions and experiments, often invisible to one another, began to intersect, resonate, and coalesce.[55] The change was not orchestrated; it was composted and catalyzed.[56] Transformation toward more just and sustainable futures requires not only building the new but also hospicing the old. This means attending to what must be let go of, and to whom it will impact, with care and dignity. It is not about eradication but about relational reckoning.[57] It involves reshaping how we live with one another and with the ecosystems that sustain us, reconfiguring the deep relationships that have been frayed or forgotten under extractive systems.[58]

For those of us working in the realm of food and ecological justice, this asks us to become weavers rather than builders—catalysts, not commanders. As Waddock notes, our task is to support emergence, not by controlling outcomes but by cultivating the conditions in which new possibilities can root, stretch, and intertwine.[59] In nature, all living systems organize through networks.[60] These networks are not planned; they arise from mutual recognition of interdependence, and they evolve through reciprocity and feedback, not hierarchy.[61] This principle also holds for food system transformation. When small, place-based initiatives—farmers experimenting with agroecology, educators revitalizing seed traditions, youth organizing for food justice—begin to find each other and align, something new becomes possible. A distributed intelligence forms, not a program to be implemented but a pattern of regeneration to be nourished. To catalyze such emergence, we must connect those working in relative isolation. Across the food system, transformation will require shifts in at least four interwoven dimensions:

- *How we grow (agroecological and regenerative production practices);*
- *How we learn (knowledge creation and cultural transmission);*
- *How we relate (economic and social relationships of reciprocity); and*
- *How we govern (policies and institutions aligned with justice and ecological care).*

When changes in these domains reinforce one another, they can begin to destabilize the entrenched power structures of industrial agribusiness. The "lock-ins" that hold the current system in place—economic incentives, cultural myths, regulatory barriers—can start to loosen.[62] In their place, regenerative systems can take root, not through force but through resonance and convergence. This is not a matter of scaling up but of scaling across, inviting different efforts into mutual awareness, shared purpose, and co-evolution. Transformation catalysts do not lead from above; they invite alignment through shared inquiry, co-created agendas, and durable relationships of trust. Their role is to support convergence, not control outcomes.[63]

Like the imaginal cells within a caterpillar, early signals of transformation may seem marginal, incoherent, or threatening. But over time, these cells begin to cluster, communicate, and cooperate. Eventually, they form the body of something entirely new. We can think of three phases in this process: (1) awakening to the reality that another way is necessary, and possible; (2) clustering with others who share a regenerative vision, even if imperfectly; and (3) coalescing into a larger field of emergence, where diverse efforts resonate into a new collective body.[64] When networks of changemakers converge across sectors, geographies, and worldviews—grounded in shared values but open to difference—they can generate the kind of momentum that shifts culture, policy, and practice. This is how we begin to birth food systems that are healthy, equitable, resilient, and biodiverse, not through replication but through relational convergence.[65]

A MOVEMENT OF MOVEMENTS

Across the globe, countless food-related initiatives are sprouting in communities, each seeded with intention, creativity, and care. Yet many of these efforts remain fragmented, dispersed across landscapes and sectors without the connective tissue necessary to respond to the magnitude and interwoven nature of today's socioecological crises. As the agroecology movement reminds us, when food is treated in isolation from the broader systems of economy, power, and culture, even the most promising practices risk reproducing the very patterns of harm they seek to interrupt. Regenerative agriculture, for example, cannot flourish in isolation.[66] A farm can compost its waste, nurture its soil, and diversify its crops, but still suffer from the pollution of neighboring monocultures, the violence of land speculation, or the instability of shifting climate patterns. Regeneration at the level of production will always be constrained by the extractive logic of the systems in which it is embedded. For regenerative agriculture to truly thrive, it must be part of a regenerative culture, a larger shift in how we understand ourselves in relation to each other and to the Earth. This shift calls for more than techni-cal fixes or piecemeal programs. It requires what the Club of Rome has called "the broadest coalition the world has ever seen," a convergence of ef-forts capable of reshaping not only practices but paradigms.[67] And that work is already underway.

In more than forty-eight countries and thousands of towns, villages, cit-ies, universities, and schools, people are coming together to relocalize and reimagine how we live. These grassroots endeavors, often known as transi-tion initiatives, are not just about food or energy; they are about belonging. Over the past two decades, the global Transition Movement has grown as a mycelial network of communities cultivating resilience through locally rooted projects: establishing renewable energy cooperatives, planting urban gardens, creating community spaces, and tending to the relational health of their neighborhoods.[68] In 2024, this movement launched a Practice of Change ini-tiative: an inquiry into how transformation actually unfolds on the ground.[69] By sharing lessons, identifying patterns, and honoring contextual complex-ity, this effort aims to nourish the adaptive wisdom of communities across contexts rather than prescribe singular solutions.

In parallel, the Global Tapestry of Alternatives (GTA) is weaving together a rich constellation of practices and worldviews that move beyond the logic of capitalist modernity, a system built on extraction, growth, and control.[70] These alternatives are not new; they include Indigenous resurgence, agroecology, commons-based governance, solidarity economies, and water and food sovereignty. But they are being revived, reimagined, and re-rooted in place, sometimes in ancient forms, sometimes in emergent ones. Rather than building toward a centralized vision, the GTA cultivates decentralized solidarity.[71] It grows not by *scaling up* but by *scaling out*, deepening horizon-tal relationships and weaving connections across movements, languages, and geographies. It avoids singular leadership or rigid hierarchy, relying instead on networks of trust, reciprocity, and shared commitment to regeneration and justice.[72] This distributed model reflects a principle known as subsidiar-ity, or what some call cosmopolitan-localism. It holds that decisions should be made at the most local level possible, while remaining in respectful dia-logue with broader planetary concerns.[73] This way, communities are not re-duced to nodes in a system but honored as stewards of place, with their own intelligences, rhythms, and traditions. And as these communities begin to speak with one another, not in unison but in polyphony, a different kind of power emerges, one rooted not in domination but in reciprocity.[74]

SCALING OUT

Movements like the Transition Movement and the GTA remind us that influence is not a matter of size but of resonance and connectivity. What begins as a small, localized initiative can become part of a global pattern—not by scaling up in the conventional sense but by acting as a living node within a shared relational field. One such example is the emergence of Ecosystem Restoration Communities (ERC). In 2017, a small gathering of people came together in the arid region of Murcia, Spain, to regenerate degraded land.[75] What began as Camp Altiplano, a 12-acre plot within the La Junquera farm, has since become a generative hub of ecological and cultural restoration. The farm's steward, Alfonso Chico de Guzmán, is converting his 3,700-acre operation from organic monoculture into a regenerative agroforestry system, weaving together almonds, heritage grains, and native biodiversity. Over the past seven years, more than 2,000 volunteers and learners from around the world have passed through Camp Altiplano. Together, they have transformed a once-depleted cereal field into a thriving landscape—planting over 20,000 native trees and shrubs, restoring a wetland now inhabited by toads, frogs, snakes, and dragonflies, and contributing to the vitality of the larger AlVelAl bioregional collective. But the transformation extends beyond the soil. Camp Altiplano has become part of a broader conversation, offering its learnings into a global commons of restoration. From this seed, a global network has grown. Today, ERC functions as a decentralized "living lab," linking over fifty communities in thirty countries across six continents. These communities are not replicating a model; they are experimenting, adapting, and sharing. Together, they aim to engage one million people by 2030 to reverse ecological destruction and restore biodiversity in over a hundred locations worldwide.[76]

A similar spirit animates the Landcare movement, which began in 1986 when a group of Australian farmers chose collaboration over competition to address erosion, water quality, and land degradation across shared watersheds. Landcare has since grown into a global network of communities—rural and urban, Indigenous and settler, civil and state—working across sectors to regenerate their local ecologies.[77] In the Philippines, for example, Landcare efforts began with contour farming to prevent soil erosion and evolved into an integrated community resilience effort. One Landcare group there supports over 5,000 households in growing vegetables, fruit trees, and livestock while protecting the Pilar Malinao Dam's water catchment—a convergence of food security, watershed governance, and communal self-determination.[78]

These examples reveal that small initiatives do not remain isolated unless we keep them so. When open to connection, the local is never merely local. The village garden, the community wetland, the contour farm: they become part of a distributed, living pattern. Through the thoughtful exchange of people, practices, and place-based insights, these efforts create a web of mutual reinforcement. They do not scale by replication but by reverberation, each site deepening its own capacity while amplifying the capacities of others.[79]

The theory and practice of regenerative systems transformation invite us to think in terms of bioregional networks and nested relationships. The future will not be determined by the success of individual projects but by the health of the relationships that connect them.[80] One compelling example of this relational approach is the International Model Forest Network (IMFN). Founded in Canada, IMFN is now the world's largest decentralized network for sustainable landscape governance, grounded in voluntary, multistakeholder collaboration.[81] Rather than managing forests in isolation, the Model Forests approach invites farmers, fishers, Indigenous communities, businesses, governments, and researchers to come together around a shared vision for their entire landscape—spanning forests, rivers, farms, and settlements.[82] In Latin America, this vision has flourished. The Latin America Model Forest Network (LAMFN) now includes thirty-four model forests in thirteen different Latin American countries, covering more than 31 million hectares (76.6 million acres).[83] But LAMFN is not about standardization; it is about relational learning. Member forests engage in continuous internal and cross-regional dialogue, sharing strategies, reflecting on failures, and refining practices in real time. This is not a network of efficiency; it is a network of resonance. These models—ERC, Landcare, IMFN—offer more than technical solutions. They show us how transformation can unfold through interdependence. By creating spaces where local actions are honored,

connected, and amplified, we begin to foster the conditions for planetary healing, not through command-and-control structures but through reciprocal
alliances of care. Though still under-recognized by much of the public, policymakers, and even some food systems actors, there is a growing constellation of grounded, bioregional initiatives demonstrating the far-reaching
potential of regenerative and agroecological practices, particularly when embedded within landscape-scale efforts. These are not isolated agricultural
interventions but complex, relational undertakings that reweave ecological
integrity, community resilience, and economic regeneration.[84] One of the
most promising approaches in this realm is integrated landscape management (ILM), which begins with a simple yet often overlooked truth: Farms
are not separate from ecosystems; they are embedded within them. By honoring this embeddedness, ILM works to restore the ecological commons
surrounding food production. Planting native vegetation, buffering waterways, nurturing pollinator habitats, and rehydrating soils through wind-
breaks and hydrologic repair—all contribute to landscapes that are more
resilient to floods, droughts, and erosion. These practices do not extract value
from the land; they build relationship with it.[85]

A compelling example unfolded in Ecuador in 2020. EcoAgriculture
Partners (EcoAg), a nonprofit facilitating long-term collaborative landscape
partnerships, collaborated with 2,000 farmers across the region to co-design
a bioregional initiative that merged productivity with ecological care. Over
9,000 acres were planted with native species, helping to safeguard agricultural lands, forests, and riparian zones, while protecting the water sources
that sustain forty-one community-managed watershed reserves. By inter-
planting crops like avocado, banana, and papaya with coffee and cacao, and
by nurturing soil fertility through nitrogen-fixing species, these farmers did
more than grow food—they regenerated the conditions for life.[86] What enabled this success was not a technical blueprint but a web of trust and collaboration. Local organizations and farming communities sustained and
scaled their efforts through shared learning and adaptive strategies. This convergence has not only enhanced local food security and ecological stability;
it has also seeded broader economic revitalization and redefined what development could mean in the region.[87]

Landscape-level regeneration is not limited to farmland. It invites us to
think systemically about how we inhabit territories, design settlements,

govern watersheds, and relate to bioregions as living entities. Though each landscape, seascape, or watershed holds its own cultural and ecological specificity, common threads of challenge and possibility weave across them all. This kind of work requires durable, diverse coalitions: communities, Indigenous leaders, researchers, land stewards, planners, and institutions coming together to cultivate a shared sense of place. These coalitions do not impose solutions; they build shared understanding through dialogue, mapping, scenario-building, and iterative planning. Together, they forge visions that hold complexity, develop regenerative investment portfolios, and track impacts not just in metrics but in relationships. Crucially, they remain open to feedback, adjusting their strategies as the landscape responds and teaches.[88] The integrated landscape approach is not a fringe experiment. It has been endorsed by the UN's climate change, biodiversity, and land degradation conventions, the Food System Summit, UN-Habitat, and other multilateral bodies. But its real momentum comes not from top-down endorsements; it comes from the ground.

As this book enters the world, we find ourselves in the midst of the UN-declared Decade on Ecosystem Restoration.[89] Led by the United Nations Environment Programme and the FAO, this global call is amplifying thousands of locally driven restoration projects.[90] But declarations alone are not enough. What is needed now is deeper alignment and relational fidelity. Across the world, territory-based coalitions are taking form, not as technocratic programs but as intergenerational commitments to place. These bioregional collaboratives understand something essential: that the lands and waters they inhabit are not backdrops to human activity but living participants in collective wellbeing. Regenerative landscapes nourish communities with food and fiber, yes, but also with stories, memory, identity, and a renewed sense of custodial belonging. They sequester carbon, yes, but they also hold meaning, kinship, and the future of countless species.[91]

BIOREGIONAL REGENERATION

Traditional landscape restoration efforts have been constrained by short-term thinking—designed as projects rather than processes, confined to specific sectors like forestry, water management, or agriculture. While these efforts may be initiated with good intentions and initial funding, many struggle to endure beyond the lifespan of grants or development cycles. Recognizing this pattern, tropical ecologist and former director of the Netherlands Office for the International Union for Conservation of Nature (IUCN) Willem Ferwerda began to ask different questions: ones that pointed not only to ecological degradation but to the cultural and economic disconnection underlying it. Ferwerda saw that bioregional regeneration offered more than a technical solution. It held the potential to address multiple interwoven crises—biodiversity loss, climate instability, fractured communities, and economic disenfranchisement—by restoring not just landscapes but relationships. Regeneration, in this frame, is not only about soil health or carbon sequestration. It is also about creating livelihoods, reweaving social fabrics, and offering communities a renewed sense of meaning and purpose. But this kind of transformation is complex. As Ferwerda observed, working at a bioregional or territorial scale requires an approach that is cross-sectoral, long-term, and deeply knowledge-intensive: rooted in a deep understanding of ecosystem processes, water and nutrient cycles, land management, and the interrelations among them.[92] This is what Dr. Sara Scherr, agricultural founder of EcoAg, calls "landscape literacy."[93] And importantly, such literacy cannot be rushed. Research suggests that meaningful landscape regeneration requires at least a generation, twenty years or more. To support this kind of long-view work, Ferwerda founded Commonland, an organization designed not to control but to catalyze.[94] Backed by philanthropic partners, investors, and government actors, Commonland works alongside farmers, landowners, entrepreneurs, civil society groups, and policymakers to initiate and sustain large-scale landscape regeneration. At the heart of their approach is a shift in logic: from profit maximization per hectare to the cultivation of multiple kinds of returns across time. Commonland's Four Returns model reframes value into four interwoven domains:

1. Inspiration—rekindling purpose and vision within communities;
2. Social returns—fostering job creation, equity, and education;
3. Natural returns—regenerating biodiversity, soil health, and water systems; and
4. Financial returns—ensuring long-term, just, and resilient livelihoods.[95]

In 2023 alone, Commonland supported regional partners in restoring over 58,000 hectares (143,321 acres), assisting nearly 700 farmers in transitioning to regenerative practices, and catalyzing over 1,800 jobs rooted in ecological repair.[96] But more than these numbers, what Commonland offers is a narrative shift, a reminder that regeneration is not just possible but already happening, and that its deepest value lies in restoring reciprocal relationships between people and place. Their work now spans twenty-three countries, with active landscape partnerships in Spain, South Africa, the Netherlands, Australia, and India, where the Andhra Pradesh Community-Managed Natural Farming initiative is supporting 6 million farmers across 8 million hectares (nearly 20 million acres) to transition toward regenerative land care.[97]

One of Commonland's most enduring collaborations is with AlVelAl, a grassroots bioregional initiative in the Altiplano of southeastern Spain. Sparked by a co-creative workshop in 2014, AlVelAl has since grown into a multigenerational movement bridging five regions: Altiplano de Granada, Los Veléz, Alto Almanzora, Guadix, and Northwest Murcia. Together, they aim to regenerate a million hectares (approximately 2.5 million acres) through a vision that harmonizes ecological, social, and agricultural renewal. AlVelAl's twenty-year vision is not framed as a goal to be achieved but as a shared field to be tended.[98] Their Manifesto for a Regenerative Territory, launched in 2022, calls on seventy-nine towns and villages across the region to declare themselves not as "developed" but as *in process*, actively participating in natural, social, and economic revitalization.[99] This is not performative branding but a collective act of commitment: to reverse desertification, stem rural depopulation, and transform a deeply stressed region into a mosaic of living systems. The Altiplano's drylands, traditional meadows, and organic almond groves (the largest of their kind globally) are not only producing healthy food with a low water footprint; they are also sequestering carbon, supporting biodiversity, and inviting a different kind of future, one that recognizes that regeneration is not about controlling land or returning to a mythical past. It is about reentering relationship with the living systems that hold us, and finding our place within them again. Building upon the vision initiated by AlVelAl and Commonland, a new chapter in landscape restoration emerged in 2023 with the founding of Fundación Aland. This initiative aims to expand the reach and coherence of holistic bioregional restoration across the Iberian

Peninsula. Acting as a connective tissue, the Iberian network now brings together six landscapes, nine local partners, and five strategic collaborators—including Commonland and AlVelAl—all applying the Four Returns framework within their unique ecological and cultural contexts.[100] This network is not working in isolation. Aland and AlVelAl are also active participants in the Bioregional Weaving Labs (BWL) Collective, a vibrant community of practice co-initiated by Ashoka Netherlands and co-led by OpEPA (a Bogotá-based nonprofit focused on reconnecting people with nature) and Commonland. The BWL Collective brings together over twenty-five system-shifting organizations, impact investors, and funders. These groups are collaborating across multiple European bioregions—from Ireland and the Netherlands to Austria, Romania, and Spain—to support farmers, educators, conservationists, and local communities in restoring and protecting ecosystems. Their collective aim is bold yet grounded: to mobilize and support a million changemakers and regeneration of one million hectares (approximately 2.47 million acres) of land and sea across Europe by 2030, measured not only in ecological metrics but through the depth of the Four Returns: inspiration, social cohesion, ecological vitality, and economic resilience.[101]

Globally, this ethos of bioregional restoration continues to deepen and diversify. One such emergent node is 1000 Landscapes for 1 Billion People (1000L), a vast, multipartner initiative launched in 2019 by EcoAg. Co-led by Commonland, Conservation International, the Rainforest Alliance, the United Nations Development Programme, and Tech Matters, 1000L convenes hundreds of organizations working to co-create integrated, place-based solutions to ecological and social breakdown. 1000L is not a centralized program but a distributed field of partnerships. It currently engages with more than 250 active landscape collaborations around the world, linking them to technical expertise, policy leverage, participatory finance mechanisms, and digital tools designed to serve local sovereignty.[102] Rather than imposing a single model, 1000L supports local leaders in crafting their own regenerative paths, drawing from ancestral wisdom, scientific insight, and emergent governance practices.[103]

As emphasized earlier in this book, the majority of the world's biodiversity—over 80 percent—exists within territories stewarded by Indigenous peoples.[104] Their relationships with land, water, and kin are shaped by generations of embodied ecological knowledge, held within

ceremonial, linguistic, and cultural systems. These practices offer a depth
and relational precision that often exceeds the grasp of dominant scientific
paradigms. A recent paper advocating for Indigenous-led restoration acknowledges this directly: "[W]e are globally surrounded by communities of
people that already know how to holistically manage the land, with the restorative and local nuance that still mystifies reductionist science."[105] This
isn't romanticism; it is recognition. The UN echoes this call. One of the
guiding principles of the Decade on Ecosystem Restoration is the integration of multiple knowledge systems—Indigenous, local, traditional, and
scientific—toward more inclusive, effective, and respectful restoration.[106]
Yet this integration must not be extractive. Article 31 of the UN Declaration
on the Rights of Indigenous Peoples affirms the inherent right of Indigenous
peoples to protect and control their cultural and ecological knowledge.[107] In
many places, the colonial wounds are still raw. Trust cannot be presumed.
Where Indigenous peoples are not ready to share their knowledge, it is not
a deficit to be overcome but a boundary to be honored. Regeneration, in
such contexts, begins with listening, slowing down to engage in relational
repair and honoring the sovereignty of those whose knowledge has too often
been exploited or erased.[108]

True bioregional regeneration cannot occur without this deeper healing.
It must move at the speed of relationship, centering Indigenous leadership,
protecting rights, and building new patterns of reciprocity. Across Australia,
the United States, and elsewhere, restoration projects guided by a Two-Eyed
Seeing—one that holds Indigenous and Western ways of knowing side by
side without collapsing one into the other—offer glimpses of what this can
look like. This is not just about mending landscapes. It is about mending
the ways we see, relate, and belong. Only then can we begin to restore the
world—not as it was, but as it might become, if tended with humility, integrity, and collective care.

REGENERATING THE WHEATBELT WITH TWO-EYED SEEING

For over 60,000 years, the Noongar people cultivated a sophisticated, relational model of ecological and cultural stewardship in what is now called
Western Australia. Their connection to *boodja*—land or country—is not metaphorical; it is living and embodied. *Boodja* provides nourishment, shelter,

medicine, stories, and the foundation for a strong society. Through practices like cultural burning and place-based tending, the Noongar sustained reciprocal relationships with the land and the spirit beings that dwell within it.[109] They have lived as *moondang-ak kaaradjiny*, the carers of everything.[110]

This continuity was violently disrupted in the nineteenth century as European colonizers cleared vast swaths of native forest to graze sheep and grow wheat. This region is now known as the Wheatbelt. Historian Tony Hughes-d'Aeth describes the process starkly as "a vast and almost total destruction of a pre-existing lifeworld"—a rupture so profound it is visible from space.[111] The region's fragile soils have been subject to relentless tillage and burning. In some areas, over 90 percent of the native vegetation has been lost,[112] leaving soils stripped and ecosystems fragmented. The land itself seems suspended in a liminal space: "life trying to die or death trying to live."[113]

This trauma is not only ecological. It mirrors the violence of settler-colonial agriculture, which extracts from both land and people, severing relational bonds in pursuit of profit. The effects continue. For the Noongar and other Aboriginal peoples, the legacies of colonization remain active—so too for settler farming families, many of whom have seen their farms lost to consolidation, climate volatility, and the economic imperatives of "get big or get out." In 2000, there were roughly 10,000 grain growers in the Wheatbelt. Today, there are around 4,000.[114]

Yet, this seemingly depleted landscape is becoming a site of resurgence, not through external rescue but through "Aboriginal-led Regenerative Land-Management," a term used by Noongar leader and farm-owner Oral McGuire. Elders are now sharing their blueprint for living well with boodja, inviting collaboration with Western-trained scientists and regenerative thinkers through a Two-Eyed Seeing approach, one that honors both Indigenous and Western ways of knowing without reducing either. This work is part of the Danjoo Koorliny Walking Together project, a long-term systems-change initiative led by Aboriginal elders and hosted by the Centre for Social Impact at the University of Western Australia. Danjoo Koorliny is guided by a 200-year horizon, looking beyond the upcoming bicentennial of colonization in 2029 toward a future rooted in Noongar country, shaped by collective responsibility and grounded in place-based healing.[115] Partners include Commonland, Wide Open Agriculture (WOA), and other community and regenerative enterprises.[116]

Commonland's Four Returns framework aligns in practice with Noongar approaches, though its conceptual origins differ. Similarly, the Three Zones approach (natural, combined, and economic zones) reflects Indigenous principles of flexible stewardship across seasonal and ecological boundaries. In practice, this means that landscapes are managed as fluid, interconnected systems—not as fixed, extractable plots. Natural zones prioritize ecological restoration, economic zones incorporate human activity with attention to ecological corridors, and combined zones invite careful integration of biodiversity and food production.[117] In collaboration with Noongar Land Enterprise Group, Commonland has helped plant 23,000 seedlings of more than sixty native species across land managed by regenerative farmers, WOA, and investors. These acts of care are not simply reforestation; they are ceremonies of reconnection. Once barren fields now pulse with life, signaling the beginning of a different relationship between people and land. WOA itself represents a new kind of enterprise: the first publicly traded company in the world explicitly committed to the Four Returns. Rather than viewing profit as separate from ecological and social repair, WOA's business model is designed to serve bioregional vitality. Its ventures span regenerative meat distribution, industrial hemp, a farmland investment portfolio, and Dirty Clean Food[118]—a direct-to-consumer platform offering carbon-neutral, regeneratively grown foods, including the world's first certified regenerative plant-based milk.[119] Beyond business, WOA and Commonland are investing in future stewards. A pilot scholarship program is nurturing the next generation of regenerative farmers, and collaborative research is underway to assess the long-term impacts of regenerative practices, not only on land and climate but on community, culture, and meaning.[120] Here, in the Wheatbelt, something more than restoration is occurring. It is a quiet revolution of belonging, led by those who never stopped listening to the land, now joined by those learning to listen anew.

BRINGING BACK THE BUFFALO

The power of landscape-scale regeneration not only in ecological recovery but in its ability to surface—and reweave—the relationships severed through colonization. It invites us to remember that land health and cultural vitality are inseparable. One potent illustration of this is the Indigenous-led resurgence of buffalo across the North American Great Plains, a living model of what it means to restore an ecosystem, a people, and a reciprocal way of being. For millennia before European invasion, the buffalo—known scientifically as *Bison bison*, but more commonly called "buffalo" by Native nations—moved in immense herds across the continent, from the dry plains of northern Mexico to the snowy grasslands of what is now southern Canada. These animals were more than food sources; they were relatives, teachers, and keystone co-creators of the prairie biome. Indigenous peoples, numbering over 25 million before colonization, relied on buffalo for nourishment, clothing, shelter, tools, and ceremony. Their ways of life were deeply attuned to the rhythms of the buffalo and the ecosystems they shaped.

Buffalo were the ancient farmers of the land.[121] Their movements aerated soil, spread seeds, moderated fire regimes, and fertilized the plains through nutrient-rich dung. Their winter coats offered nesting material for birds like burrowing owls, and their wallows collected spring water, forming micro-wetlands that supported amphibians and other life. Their mere presence orchestrated biodiversity, feeding not only humans but predators, scavengers, and pollinators across vast distances.[122] As Indigenous nations followed and learned from the buffalo, they developed knowledge systems rooted in flux, interdependence, and relational accountability. "[E]very action, no matter how small, has the potential to affect others over great distances."[123] This teaching is as vital now as ever.

This entire system was violently interrupted. By the late 1800s, settler expansion and state-sponsored extermination campaigns had reduced buffalo numbers from tens of millions to fewer than 500. At the same time, the genocide and forced removal of Native peoples confined them to reservations, stripped them of land, and criminalized their foodways, ceremonies, and governance systems. The fates of buffalo and Indigenous peoples became tightly intertwined, both rendered nearly extinct in the name of expansion and control.[124] Today, both are reemerging. There are now over half a million buffalo across the continent, and Indigenous communities, whose numbers had fallen below 250,000 in the early twentieth century, have grown to nearly 7 million, according to the most recent US Census. But this is not merely a demographic resurgence. It is a cultural, ecological, and spiritual renewal.[125]

A pivotal moment came in 2014 with the signing of *The Buffalo: A Treaty of Cooperation, Renewal and Restoration*. Led by the Blackfoot Confederacy and joined by allied First Nations and Native American tribes, the treaty marked the first cross-border Indigenous agreement focused on the return of buffalo to 6.3 million acres of prairie. This effort spans from Montana to Alberta and includes signatories like the Blackfeet Nation, Kainai/Blood Tribe, and Siksika Nation, among other.[126] The treaty, conceptualized by Blackfoot scholars Leroy Little Bear and Amethyst First Rider, articulates a vision grounded in ancestral responsibility:[127]

> To honour, recognize, and revitalize the time immemorial relationship we have with BUFFALO . . . so together WE can have our brother, the BUFFALO, lead us in nurturing our land, plants and other animals to once again realize THE BUFFALO WAYS for our future generations.[128]

This is not restoration as technical intervention; it is rematriation of kin, culture, and responsibility.[129]

More recently, in July 2024 the Tribal Buffalo Lifeways Collaboration was launched by the InterTribal Buffalo Council, Native Americans in Philanthropy, The Nature Conservancy, and the World Wildlife Fund. This alliance aims to deepen Indigenous-led buffalo restoration and strengthen the cultural, spiritual, ecological, and economic lifeways embedded in these efforts.[130] With support from the US Department of the Interior and the USDA, the collaboration seeks to expand co-stewardship arrangements, return land, and provide critical infrastructure such as fencing, herd training, and access to water.[131] Troy Heinert, a member of the Rosebud Sioux and executive director of the InterTribal Buffalo Council, frames this work plainly: "[I]t's also land restoration, water resource restoration, and cultural revitalization."[132] These are not separate aims—they are expressions of one interwoven world.

The central grasslands of North America, one of the planet's most endangered ecosystems, are disappearing at an alarming rate. More than 50 million acres have been lost in the past decade alone. Industrial agriculture, invasive species, and unchecked development are pushing the biome toward collapse, with implications for biodiversity, climate resilience, and food systems across the continent. As temperatures rise and rainfall patterns shift, the need for grounded, Indigenous-led stewardship becomes not only urgent but instructive.[133] The Central Grasslands Roadmap, a bioregional initiative uniting stakeholders across Mexico, Canada, the United States, and Indigenous nations, aims to rethink this landscape as a whole.[134] Its shared vision includes healthy soils and native species, resilient working lands, sustainable agriculture and energy systems, and thriving Indigenous and rural communities.[135] But the roadmap acknowledges that technical coordination is not enough. "We must think, collaborate, and act differently."[136] This includes reexamining the foundations of American land use: boundaries drawn without consent, private property regimes, the dominance of the cattle industry (often subsidized to graze on public lands),[137] and the assumption that domesticated cattle can simply substitute for buffalo in regenerative agriculture narratives. To regenerate the Great Plains is not to return to an imagined past—it is to reckon with the present. It is to listen to those who never left, who have carried the memory of buffalo through every attempted erasure. It is to recognize that what we call "land management" must become something more: a living practice of repair.

NEW GOVERNANCE FOR THE NEW PARADIGM

The regeneration of landscapes calls for more than shifts in agricultural practice. It asks us to fundamentally reimagine the systems that shape how we relate to land, food, governance, and law. Moving from transactional to relational paradigms means not only transforming economies and institutions but also reweaving our legal frameworks to reflect the interdependence of all life.[138] An essential part of this paradigm shift is the recognition that nature is not a backdrop or a resource to be managed but a living, rights-bearing community of which humans are a part. Earth jurisprudence, sometimes called ecocentric or wild law, emerged as a response to this call, articulated by cultural historian Thomas Berry and rooted in many Indigenous legal and cosmological traditions. It proposes that legal systems should no longer serve only human interests but instead uphold the health, integrity, and continuity of Earth's living systems.[139] Earth jurisprudence offers a framework for regenerative law: legal practices that reflect our embeddedness within planetary cycles.[140] Drawing from Indigenous understandings that view land and water as relatives rather than commodities, it insists that harm to the land is harm to the people. From this perspective, law is not a tool of control but a practice of reciprocal responsibility. Because ecosystems and cultures vary,

Earth jurisprudence must also be bioregionally specific, shaped by local ecologies, histories, and traditions of governance.[141]

Among the most powerful expressions of this approach is the growing movement for the Rights of Nature, which affirms that rivers, mountains, forests, and other ecosystems possess inherent rights to exist, flourish, regenerate, and evolve—independent of their utility to humans.[142] These rights enable ecosystems to be represented in court not as property but as living legal persons. Ecuador was the first nation to enshrine the Rights of Nature in its constitution in 2008, recognizing Pachamama, Mother Earth, as a sacred being with the right to maintain and regenerate its vital cycles.[143] In 2017, the Whanganui River in Aotearoa (New Zealand) was granted legal personhood through the Te Awa Tupua Act, formalizing Māori relational worldviews in national law.[144]

What began as a visionary shift is now one of the fastest-growing movements in environmental governance. As of 2024, over thirty countries, including Australia, Canada, Bolivia, Panama, India, Colombia, and Nigeria, have adopted some form of Rights of Nature legislation. In the United States, more than three dozen cities, townships, and counties have passed ordinances recognizing legal rights of ecosystems. Increasingly, Indigenous nations are leading the way. Six US tribes have passed their own versions of Rights of Nature, often grounded in traditional ecological knowledge and treaty obligations.[145] In 2018, for example, the White Earth Band of Ojibwe, part of the Minnesota Chippewa Tribe, adopted the Rights of Manoomin, a legal protection for wild rice and the freshwater systems it depends on. The tribal resolution emphasized the necessity of protecting manoomin as a primary treaty food and cultural relative, affirming its right to thrive for future generations.[146] These laws do more than create new legal tools; they affirm a worldview in which rivers, seeds, and soil are not "managed" but related with. They invite communities to rethink what restoration really means: not a return to a pristine past but a renewal of relationship.

This shift is already influencing restoration practices. In the Pacific Northwest, the removal of obsolete dams along the Elwha[147] and Klamath Rivers has reopened ancestral fish runs for species such as lamprey, salmon, and herring.[148] These efforts, led by the Yurok Tribe and other Indigenous communities, are grounded in the recognition of rivers as living beings. In 2019, the Yurok passed a resolution declaring the Klamath River a legal

person under tribal law, with rights to exist, flourish, and be free from pollution and climate harm.[149] The Lake Erie Bill of Rights, passed by the citizens
of Toledo, Ohio, in 2019, attempted to codify similar protections for Lake
Erie, challenging agricultural runoff and toxic pollution. Though overturned
in federal court, the initiative galvanized broader conversations about community rights, ecological governance, and democratic accountability.[150]
Unlike conventional environmental law, which tends to regulate harm
rather than prevent it, Earth jurisprudence is grounded in a relational ethic.
It asks not how much harm is allowable but how we might live in ways that
uphold the dignity and vitality of all beings. It is less about domination or
extraction and more about belonging, reciprocity, and co-flourishing.
Ultimately, Earth jurisprudence does not seek to replace one legal code with
another. It seeks to root our systems of governance in the web of life itself,
acknowledging that the rivers, forests, wetlands, seeds, and animals are not
silent objects but animate relations. Regenerating our landscapes, then, requires not only restoring ecosystems but also restoring our role as caretak-
ers within them.

REGENERATIVE FUTURES

Alberto Acosta, who helped shape Ecuador's groundbreaking constitution,
once observed: "[O]nly by imagining other worlds will this one be changed."[151]
These words invite us not into escapism but into radical remembering of
what it means to be human *with* the Earth rather than apart from it. So, imagine this: You step into a time machine and travel two decades ahead. The
year is 2046. You find yourself back in your town. You step outside and begin to walk.[152]

What do you see?

Are the seeds of action sown in 2026 now roots and branches? Has a
culture of shared stewardship taken hold in response to the ecological unraveling that once felt so overwhelming? Do you see neighbors cultivating
their own food: balconies blooming with herbs, schoolyards transformed
into orchards, old warehouses repurposed for community canning, drying,
and fermenting? Is your neighborhood now home to edible commons and
pollinator corridors where monocultures once stood? And what of water? As
you pass a local river, creek, or lake, can you smell the life in it? Hear frogs

and kingfishers? See children swimming, elders wading, fish returning to ancestral spawning grounds? You are invited to a shared meal, a ritual of nourishment and storytelling. Around the table, neighbors recount what has transpired since 2026: how land was revitalized, gardens replanted, relationships reimagined. You reflect on your own role, not as hero but as one of many who helped weave something beautiful and enduring. Before returning to the present, you take one last look. What does it feel like to live in a place where the economy is measured in relationships, not just transactions? Where care for the land is inseparable from care for each other? Now, return to this moment. Right now.

Ask yourself: What are the seeds I can plant today to bring that world closer? Food remains one of our most potent relational teachers. It connects us to seasons, soils, ancestors, and neighbors. It carries memory, offers medicine, and invites reciprocity. Across the globe, food is anchoring a new—yet ancient—story about belonging, one that is rooted not in domination or escape, but in interconnection, mutual care, and regeneration. By participating in regenerative food systems, as growers, eaters, organizers, storytellers, investors, and kin, we deepen our commitment to place and to each other. We begin to re-member what it means to live well, not as isolated individuals but as part of a larger, breathing whole.

This is not a utopian project. It is already underway.

The question is: *How will you join it?*

Let food be our gathering point. Let land be our teacher. Let our daily acts of care be the building blocks of a future in which the Earth and all its peoples may once again flourish, together.

Acknowledgments

This project would not have been possible without the love and support of Adam Rick, Nell Casey, Eric Maisel, Katherine Golub, Chara Armon, Karryn Olson, Mary Negowetti, and Teodor Negowetti. Thank you also to my generous teachers and mentors who have courageously shared their work and wisdom with the world.

Notes

INTRODUCTION

1. Chrissy Sexton, "Humans Are Now Using 1.7 Times the Amount of Earth's Resources," Earth.com, July 25, 2018, https://www.earth.com/news/humans-using -earths-resources/.

2. "Planetary Boundaries," Stockholm Resilience Center, accessed August 15, 2024, https://www.stockholmresilience.org/research/planetary-boundaries.html.

3. Katherine Richardson et al., "Earth Beyond Six of Nine Planetary Boundaries," *Science Advances* 9, no. 37 (September 15, 2023): eadh2458, https://doi.org/10.1126 /sciadv.adh2458.

4. William J. Ripple et al., "The 2023 State of the Climate Report: Entering Uncharted Territory," *BioScience* 73, no. 12 (December 29, 2023): 841–50, https:// doi.org/10.1093/biosci/biad080.

5. Walter Willett et al., "Food in the Anthropocene: The EAT–Lancet Commission on Healthy Diets from Sustainable Food Systems," *The Lancet* 393, no. 10170 (February 2019): 447–92, https://doi.org/10.1016/S0140-6736(18)31788-4, 449.

6. "Planetary Health," Planetary Health Alliance, accessed August 19, 2024, https://www.planetaryhealthalliance.org/planetary-health.

7. Woody Tasch, "Will the Real Food Movement Please Stand Up?," *Grist*, May 3, 2011, https://grist.org/sustainable-food/2011-05-02-will-the-real-food-movement -please-stand-up/.

8. "Global Stocktake," United Nations, Climate Change, accessed August 15, 2024, https://unfccc.int/topics/global-stocktake.

9. "Biodiversity Protections Off-Track to Meet Sustainable Development Goals— IUCN Report," IUCN, September 18, 2023, https://iucn.org/news/202309 /biodiversity-protections-track-meet-sustainable-development-goals-iucn-report.

10. "Halfway to 2030, World 'Nowhere Near' Reaching Global Goals, UN Warns," *UN News*, July 17, 2023, https://news.un.org/en/story/2023/07/1138777.

11. FAO, IFAD, UNICEF, WFP, and WHO, "The State of Food Security and Nutrition in the World 2024" (Rome, Italy, July 23, 2024), https://doi.org/10.4060/ cd1254en.

12. Food and Agriculture Organization of the United Nations (FAO), International Fund for Agricultural Development (IFAD), United Nations Children's Fund

(UNICEF), World Food Programme (WFP), and World Health Organization (WHO). "The State of Food Security and Nutrition in the World 2024." Rome, Italy, July 23, 2024. https://doi.org/10.4060/cd1254en.

13. Patrick Webb et al., "The Urgency of Food System Transformation Is Now Irrefutable," *Nature Food* 1, no. 10 (September 28, 2020): 584–85, https://doi.org/10 .1038/s43016-020-00161-0.

14. Mackenzie Battle et al., "2023 State of the Industry Report: Cultivated Meat and Seafood," Good Food Institute, 2024, https://gfi.org/resource/cultivated-meat -and-seafood-state-of-the-industry-report/.

15. Natalie Cargill, "How Exactly Clean Meat Is Created & the Advances Needed to Get It in Every Supermarket," 80,000, accessed August 15, 2024, https:// 80000hours.org/podcast/episodes/marie-gibbons-clean-meat/.

16. "Cellular Agriculture and Food Systems Priorities," *Nature Food* 3, no. 10 (October 18, 2022): 781–781, https://doi.org/10.1038/s43016-022-00628-2.

17. Michael Lawrence et al., "Global Polycrisis: The Causal Mechanisms of Crisis Entanglement," *Global Sustainability* 7 (2024): e6, https://doi.org/10.1017/sus.2024.1, 2.

18. Lawrence et al., 2.

19. David Bohm, *Wholeness and the Implicate Order* (London: Routledge, 2002).

20. Jamie Bristow et al., "The System Within: Addressing the Inner Dimensions of Sustainability and Systems Change" (The Club of Rome, 2024), https://www .clubofrome.org/wp-content/uploads/2024/05/Earth4All_Deep_Dive_Jamie _Bristow.pdf, 7.

21. Donald Trent Jacobs and Darcia Narvaez, *Restoring the Kinship Worldview: Indigenous Voices Introduce 28 Precepts for Rebalancing Life on Planet Earth* (Berkeley, CA: North Atlantic Books, 2022), 2, 8.

22. Thomas S. Kuhn, *The Structure of Scientific Revolutions*, 2nd ed., International Encyclopedia of Unified Science, vol 2 (Chicago: University of Chicago Press, 1994), 2.

23. Isabel Rimanoczy and Ana Maria Llamazares, "Twelve Principles to Guide a Long-Overdue Paradigm Shift," *Journal of Management, Spirituality & Religion* 18, no. 6 (December 1, 2021): 54–76, https://doi.org/10.51327/JKKI4753.

24. Carol Sanford, *The Regenerative Life: Transform Any Organization, Our Society, and Your Destiny* (Boston: Nicholas Brealey Publishing, 2020).

25. "Why Regenerative Agriculture?" *Regeneration International* (blog), accessed May 5, 2024, https://regenerationinternational.org/why-regenerative-agriculture/.

26. Bryony Sands et al., "Moving Towards an Anti-Colonial Definition for Regenerative Agriculture," *Agriculture and Human Values* 40, no. 4 (December 2023): 1697–716, https://doi.org/10.1007/s10460-023-10429-3.

27. Sam J. Buckton et al., "The Regenerative Lens: A Conceptual Framework for Regenerative Social-Ecological Systems," *One Earth* 6, no. 7 (July 2023): 824–42, https://doi.org/10.1016/j.oneear.2023.06.006.

28. Ruchi Shroff and Carla Ramos Cortés, "The Biodiversity Paradigm: Building Resilience for Human and Environmental Health," *Development* 63, no. 2–4 (December 2020): 172–80, https://doi.org/10.1057/s41301-020-00260-2, 173–74.

29. Carol Sanford, "The Regenerative Economic Shaper Perspective Paper—Part 2," *The Regenerative Economy Collaborative* (blog), June 24, 2020, https://medium .com/the-regenerative-economy-collaborative/the-regenerative-economic-shaper -perspective-paper-part-2-418f35369ded.

30. Joanna Macy and Chris Johnstone, *Active Hope*, Revised ed. (New World Library, 2022), 4–5.

31. Brian D. Fath et al., "Measuring Regenerative Economics: 10 Principles and Measures Undergirding Systemic Economic Health," *Global Transitions* 1 (2019): 15–27, https://doi.org/10.1016/j.glt.2019.02.002, 18.

CHAPTER 1: THE STORY OF SEPARATION

1. Bethany Brookshire, "Explainer: How Photosynthesis Works," *Science News Explores*, October 28, 2020, https://www.snexplores.org/article/explainer-how -photosynthesis-works.

2. Fritjof Capra, "The New Facts of Life," ecoliteracy.org, June 29, 2009, https:// www.ecoliteracy.org/article/new-facts-life.

3. Garry W. McDonald and Murray G. Patterson, "Bridging the Divide in Urban Sustainability: From Human Exceptionalism to the New Ecological Paradigm," *Urban Ecosystems* 10, no. 2 (June 1, 2007): 169–92, https://doi.org/10.1007/s11252 -006-0017-0. 0519, 171.

4. Johan Rockström et al., "Planetary Boundaries: Exploring the Safe Operating Space for Humanity," *Ecology and Society* 14, no. 2 (2009), http://www.jstor.org/stable /26268316.

5. Jeremy Lent, *The Web of Meaning: Integrating Science and Traditional Wisdom to Find Our Place in the Universe* (Gabriola Island: New Society Publishers, 2021).

6. Fritjof Capra, *The Web of Life: A New Scientific Understanding of Living Systems* (New York: Anchor Books, 1997); Peter M. Senge, *The Fifth Discipline: The Art and Practice of the Learning Organization*, Rev. and updated ed. (New York: Currency Doubleday, 2006).

7. "Nature," *Cambridge Dictionary*, July 10, 2024, https://dictionary.cambridge .org/dictionary/english/nature.

8. Catherine A. Lozupone et al., "Diversity, Stability and Resilience of the Human Gut Microbiota," *Nature* 489, no. 7415 (September 2012): 220–30, https://doi.org/10 .1038/nature11550.

9. "Animal Definition & Meaning—Black's Law Dictionary," *The Law Dictionary*, November 4, 2011, https://thelawdictionary.org/animal/.

10. Egleé L. Zent, "Unfurling Western Notions of Nature and Amerindian Alternatives," *Ethics in Science and Environmental Politics* 15, no. 2 (October 2, 2015): 105–23, https://doi.org/10.3354/esep00159, 114–15.

11. Egleé Zent and Stanford Zent, "Love Sustains Life: *Jkyo Jkwainï* and Allied Strategies in Caring for the Earth," *Journal of Ethnobiology* 42, no. 1 (March 2022): 86–104, https://doi.org/10.2993/0278-0771-42.1.86.

12. "What You Should Know About Genetically Engineered Salmon," *Earthjustice*, November 5, 2020, https://earthjustice.org/feature/what-you-should-know-about-ge -salmon.

13. Rachel Carson, *Silent Spring*, 40th anniversary ed. (Boston: Houghton Mifflin, 2002).

14. "The Story of Silent Spring," August 13, 2015, https://www.nrdc.org/stories /story-silent-spring.

15. "The 2022 Living Planet Report," accessed July 15, 2024, https://livingplanet .panda.org/en-US/.

16. Jonathan Rushton et al., "A Food System Paradigm Shift: From Cheap Food at Any Cost to Food within a One Health Framework," *NAM Perspectives* 11 (November 22, 2021), https://doi.org/10.31478/202111b.

17. "Purpose of the Dietary Guidelines," Dietary Guidelines for Americans, accessed August 21, 2024, https://www.dietaryguidelines.gov/about-dietary-guidelines /purpose-dietary-guidelines.

18. Andy Bellatti, "The Woman Fighting to Make Sustainability Part of the American Diet," *Civil Eats*, February 1, 2016, https://civileats.com/2016/02/01 /miriam-nelson-sustainability-dietary-guidelines/.

19. Dietary Guidelines Advisory Committee, "Scientific Report of the 2015 Dietary Guidelines Advisory Committee: Advisory Report to the Secretary of Health and Human Services and the Secretary of Agriculture" (Washington, DC: US Department of Agriculture, Agricultural Research Service, 2015), 283, https://www .dietaryguidelines.gov/sites/default/files/2019-05/Scientific-Report-of-the-2015 -Dietary-Guidelines-Advisory-Committee.pdf.

20. "2015 Dietary Guidelines: Giving You the Tools You Need to Make Healthy Choices," US Department of Agriculture, accessed August 21, 2024, https://www .usda.gov/media/blog/2015/10/06/2015-dietary-guidelines-giving-you-tools-you -need-make-healthy-choices.

21. "Consolidated Appropriations Act, 2016," Pub. L. No. 114–113, § 734, 129 Stat 2242 (2015), 2280.

22. "USDA Charter for the 2020 Dietary Guidelines Advisory Committee," October 5, 2018, https://www.dietaryguidelines.gov/sites/default/files/2019-03/Die taryGuidelinesAdvisoryCommitteeCharter-10-05-18.pdf.

23. Gyorgy Scrinis, *Nutritionism: The Science and Politics of Dietary Advice* (New York: Columbia University Press, 2013).

24. "Use of the Term Healthy on Food Labeling," US Food & Drug Administration, March 28, 2024, https://www.fda.gov/food/food-labeling-nutrition/use-term-healthy-food-labeling.

25. Jennifer Clapp and Gyorgy Scrinis, "Big Food, Nutritionism and Corporate Power," *Globalizations* 14, no. 4 (2017): 578–95, http://www.tandfonline.com/doi/abs/10.1080/14747731.2016.1239806.

26. "Stop Counting Calories," *Harvard Health*, October 1, 2020, https://www.health.harvard.edu/staying-healthy/stop-counting-calories.

27. Eamonn M. M. Quigley and Prianka Gajula, "Recent Advances in Modulating the Microbiome," *F1000Research* 9 (January 27, 2020): 46, https://doi.org/10.12688/f1000research.20204.1.

28. "Personalized Nutrition," *Health & Nutrition Letter* (Tufts University Friedman School of Nutrition Science and Policy, July 9, 2020), https://www.nutritionletter.tufts.edu/healthy-eating/personalized-nutrition/.

29. Ann Gibbons, "The Evolution of Diet," *National Geographic*, accessed March 7, 2024, http://www.nationalgeographic.com/foodfeatures/evolution-of-diet/.

30. Carlos Augusto Monteiro et al., "Increasing Consumption of Ultra-Processed Foods and Likely Impact on Human Health: Evidence from Brazil," *Public Health Nutrition* 14, no. 1 (December 20, 2010): 5–13, https://doi.org/10.1017/S1368980010003241.

31. Carlos A. Monteiro et al., "Ultra-Processed Foods: What They Are and How to Identify Them," *Public Health Nutrition* 22, no. 5 (April 2019): 936–41, https://doi.org/10.1017/S1368980018003762, 937.

32. Kevin D. Hall et al., "Ultra-Processed Diets Cause Excess Calorie Intake and Weight Gain: An Inpatient Randomized Controlled Trial of Ad Libitum Food Intake," *Cell Metabolism* 30, no. 1 (July 2019): 67–77.e3, https://doi.org/10.1016/j.cmet.2019.05.008.

33. Filippa Juul et al., "Ultra-Processed Food Consumption among US Adults from 2001 to 2018," *The American Journal of Clinical Nutrition* 115, no. 1 (January 2022): 211–21, https://doi.org/10.1093/ajcn/nqab305.

34. Abigail S. Baldridge et al., "The Healthfulness of the US Packaged Food and Beverage Supply: A Cross-Sectional Study," *Nutrients* 11, no. 8 (July 24, 2019): 1704, https://doi.org/10.3390/nu11081704.

35. Ashley N. Gearhardt and Erica M. Schulte, "Is Food Addictive? A Review of the Science," *Annual Review of Nutrition* 41, no. 1 (October 11, 2021): 387–410, https://doi.org/10.1146/annurev-nutr-110420-111710, 388.

36. Bernard Srour et al., "Ultraprocessed Food Consumption and Risk of Type 2 Diabetes Among Participants of the NutriNet-Santé Prospective Cohort," *JAMA Internal Medicine* 180, no. 2 (February 1, 2020): 283, https://doi.org/10.1001/jamainternmed.2019.5942; Anaïs Rico-Campà et al., "Association

between Consumption of Ultra-Processed Foods and All Cause Mortality: SUN Prospective Cohort Study," *BMJ*, May 29, 2019, l1949, https://doi.org/10.1136/bmj.l1949.

37. Elizabeth K. Dunford, Donna R. Miles, and Barry Popkin, "Food Additives in Ultra-Processed Packaged Foods: An Examination of US Household Grocery Store Purchases," *Journal of the Academy of Nutrition and Dietetics* 123, no. 6 (June 2023): 889–901, https://doi.org/10.1016/j.jand.2022.11.007.

38. Meghan B. Azad et al., "Nonnutritive Sweeteners and Cardiometabolic Health: A Systematic Review and Meta-Analysis of Randomized Controlled Trials and Prospective Cohort Studies," *Canadian Medical Association Journal* 189, no. 28 (July 17, 2017): E929–39, https://doi.org/10.1503/cmaj.161390.

39. Azad et al.

40. Dunford et al., "Food Additives in Ultra-Processed Packaged Foods."

41. "Understanding How the FDA Regulates Food Additives and GRAS Ingredients," U.S. Food & Drug Administration, June 6, 2024, https://www.fda.gov/food/food-additives-and-gras-ingredients-information-consumers/understanding-how-fda-regulates-food-additives-and-gras-ingredients.

42. "Food Safety: FDA Should Strengthen Its Oversight of Food Ingredients Determined to Be Generally Recognized as Safe (GRAS)" (US Government Accountability Office, March 5, 2010), https://www.gao.gov/products/gao-10-246.

43. "Food Additives," 21 U.S.C. § 348(c)(5) (2018), https://www.law.cornell.edu/uscode/text/21/348.

44. Maricel V. Maffini and Thomas Neltor, "How the FDA Ignores the Law when Approving New Chemical Additives to Food," *Environmental Health News*, December 23, 2020, https://www.ehn.org/health-issues-associated-with-food-additives-2649620272.html.

45. A. C. Gore et al., "EDC-2: The Endocrine Society's Second Scientific Statement on Endocrine-Disrupting Chemicals," *Endocrine Reviews* 36, no. 6 (December 1, 2015): E1–150, https://doi.org/10.1210/er.2015-1010.

46. Leonardo Trasande et al., "Food Additives and Child Health," *Pediatrics* 142, no. 2 (August 2018): e20181408, https://doi.org/10.1542/peds.2018-1408.

47. Claude Monneret, "What Is an Endocrine Disruptor?" *Comptes Rendus Biologies* 340, no. 9–10 (September 1, 2017): 403–5, https://doi.org/10.1016/j.crvi.2017.07.004.

48. Henrieta Hlisníková et al., "Effects and Mechanisms of Phthalates' Action on Reproductive Processes and Reproductive Health: A Literature Review," *International Journal of Environmental Research and Public Health* 17, no. 18 (September 18, 2020): 6811, https://doi.org/10.3390/ijerph17186811; Carla Giovana Basso, Anderson Tadeu De Araújo-Ramos, and Anderson Joel Martino-Andrade, "Exposure to Phthalates and Female Reproductive Health: A Literature Review,"

Reproductive Toxicology 109 (April 2022): 61–79, https://doi.org/10.1016/j.reprotox
.2022.02.006.

49. Antonia M. Calafat et al., "Exposure of the U.S. Population to Bisphenol A
and 4-*Tertiary*-Octylphenol: 2003–2004," *Environmental Health Perspectives* 116, no. 1
(January 2008): 39–44, https://doi.org/10.1289/ehp.10753.

50. "Bisphenol A (BPA)," National Institute of Environmental Health Sciences,
August 31, 2023, https://www.niehs.nih.gov/health/topics/agents/sya-bpa.

51. Aleksandra Konieczna, Aleksandra Rutkowska, and Dominik Rachoń,
"Health Risk of Exposure to Bisphenol A (BPA)," *Roczniki Panstwowego Zakladu
Higieny* 66, no. 1 (2015): 5–11.

52. Mahiba Shoeib et al., "Survey of Polyfluorinated Chemicals (PFCs) in the
Atmosphere over the Northeast Atlantic Ocean," *Atmospheric Environment* 44, no. 24
(August 2010): 2887–93, https://doi.org/10.1016/j.atmosenv.2010.04.056.

53. James Armitage et al., "Modeling Global-Scale Fate and Transport of
Perfluorooctanoate Emitted from Direct Sources," *Environmental Science & Technology*
40, no. 22 (November 1, 2006): 6969–75, https://doi.org/10.1021/es0614870.

54. Keegan Rankin et al., "A North American and Global Survey of Perfluoroalkyl
Substances in Surface Soils: Distribution Patterns and Mode of Occurrence,"
Chemosphere 161 (October 2016): 333–41, https://doi.org/10.1016/j.chemosphere.2016
.06.109.

55. "PFAS in the US Population," Agency for Toxic Substances and Disease
Registry, January 18, 2024, https://www.atsdr.cdc.gov/pfas/health-effects/us
-population.html.

56. "Our Current Understanding of the Human Health and Environmental
Risks of PFAS," Overviews and Factsheets, US Environmental Protection Agency,
May 16, 2024, https://www.epa.gov/pfas/our-current-understanding-human-health
-and-environmental-risks-pfas. "PFAS in the US Population," Agency for Toxic
Substances and Disease Registry, January 18, 2024, https://www.atsdr.cdc.gov/pfas
/health-effects/us-population.html.

57. Xindi C. Hu et al., "Detection of Poly- and Perfluoroalkyl Substances (PFASs)
in U.S. Drinking Water Linked to Industrial Sites, Military Fire Training Areas, and
Wastewater Treatment Plants," *Environmental Science & Technology Letters* 3, no. 10
(October 11, 2016): 344–50, https://doi.org/10.1021/acs.estlett.6b00260.

58. Clare Death et al., "Per- and Polyfluoroalkyl Substances (PFAS) in Livestock
and Game Species: A Review," *Science of The Total Environment* 774 (June 2021):
144795, https://doi.org/10.1016/j.scitotenv.2020.144795; Amila O. De Silva et al.,
"PFAS Exposure Pathways for Humans and Wildlife: A Synthesis of Current
Knowledge and Key Gaps in Understanding," *Environmental Toxicology and Chemistry*
40, no. 3 (March 2021): 631–57, https://doi.org/10.1002/etc.4935; José L. Domingo
and Martí Nadal, "Per- and Polyfluoroalkyl Substances (PFASs) in Food and Human
Dietary Intake: A Review of the Recent Scientific Literature," *Journal of Agricultural*

and Food Chemistry 65, no. 3 (January 25, 2017): 533–43, https://doi.org/10.1021/acs
.jafc.6b04683.

59. Andrea C. Blaine et al., "Uptake of Perfluoroalkyl Acids into Edible
Crops via Land Applied Biosolids: Field and Greenhouse Studies," *Environmental
Science & Technology* 47, no. 24 (December 17, 2013): 14062–69, https://doi.org
/10.1021/es403094q; Committee on the Guidance on PFAS Testing and Health
Outcomes et al., *Guidance on PFAS Exposure, Testing, and Clinical Follow-Up*
(Washington, DC: National Academies Press, 2022), https://doi.org/10.17226
/26156; Deanna P. Scher et al., "Occurrence of Perfluoroalkyl Substances (PFAS)
in Garden Produce at Homes with a History of PFAS-Contaminated Drinking
Water," *Chemosphere* 196 (April 2018): 548–55, https://doi.org/10.1016/j.chemosphere
.2017.12.179.

60. Tom Perkins, "'I Don't Know How We'll Survive': The Farmers Facing Ruin
in America's 'Forever Chemicals' Crisis," *The Guardian*, March 22, 2022, https://
www.theguardian.com/environment/2022/mar/22/i-dont-know-how-well-survive
-the-farmers-facing-ruin-in-americas-forever-chemicals-crisis.

61. "Biden-Harris Administration Launches Plan to Combat PFAS Pollution,"
The White House, October 18, 2021, https://www.whitehouse.gov/briefing-room
/statements-releases/2021/10/18/fact-sheet-biden-harris-administration-launches
-plan-to-combat-pfas-pollution/.

62. "Biden-Harris Administration Takes New Action to Protect Communities
from PFAS Pollution," The White House, March 14, 2023, https://www.whitehouse
.gov/briefing-room/statements-releases/2023/03/14/fact-sheet-biden-harris
-administration-takes-new-action-to-protect-communities-from-pfas-pollution/.

63. Consumer Reports, "Dangerous PFAS Chemicals Are in Your Food
Packaging," March 24, 2022, https://www.consumerreports.org/health/food
-contaminants/dangerous-pfas-chemicals-are-in-your-food-packaging
-a3786252074/.

64. Claudia Campanale et al., "A Detailed Review Study on Potential Effects of
Microplastics and Additives of Concern on Human Health," *International Journal of
Environmental Research and Public Health* 17, no. 4 (January 2020): 1212, https://doi
.org/10.3390/ijerph17041212.

65. Kristian Syberg et al., "Regulation of Plastic from a Circular Economy
Perspective," *Current Opinion in Green and Sustainable Chemistry* 29 (June 2021):
100462, https://doi.org/10.1016/j.cogsc.2021.100462.

66. Alonzo Alfaro-Núñez et al., "Microplastic Pollution in Seawater and Marine
Organisms across the Tropical Eastern Pacific and Galápagos," *Scientific Reports* 11,
no. 1 (March 19, 2021): 6424, https://doi.org/10.1038/s41598-021-85939-3.

67. "The New Plastics Economy: Rethinking the Future of Plastics" (World
Economic Forum, January 19, 2016), 7. https://www.weforum.org/publications/the
-new-plastics-economy-rethinking-the-future-of-plastics/.

68. Silvia Lomartire, João C. Marques, and Ana M. M. Gonçalves, "The Key Role of Zooplankton in Ecosystem Services: A Perspective of Interaction between Zooplankton and Fish Recruitment," *Ecological Indicators* 129 (October 2021): 107867, https://doi.org/10.1016/j.ecolind.2021.107867.

69. K. Kvale et al., "Zooplankton Grazing of Microplastic Can Accelerate Global Loss of Ocean Oxygen," *Nature Communications* 12, no. 1 (April 21, 2021): 2358, https://doi.org/10.1038/s41467-021-22554-w.

70. "Ocean Deoxygenation," IUCN, December 2019, https://iucn.org/resources/issues-brief/ocean-deoxygenation.

71. "Global Plastic Waste Set to Almost Triple by 2060, Says OECD," *OECD*, June 3, 2022, https://www.oecd.org/en/about/news/press-releases/2022/06/global-plastic-waste-set-to-almost-triple-by-2060.html.

72. "The Real Truth about the U.S. Plastics Recycling Rate," *Beyond Plastics*, Bennington College, May 4, 2022, https://static1.squarespace.com/static/5eda91260bbb7e7a4bf528d8/t/62b2238152acae761414d698/1655841666913/The-Real-Truth-about-the-US-Plastic-Recycling-Rate-2021-Facts-and-Figures-_5-4-22.pdf.

73. "Report Reveals That U.S. Plastics Recycling Rate Has Fallen to <6%," *Beyond Plastics*, May 2022, https://www.beyondplastics.org/plastics-recycling-rates.

74. "About Genetically Engineered Foods," Center for Food Safety, accessed March 11, 2024, https://www.centerforfoodsafety.org/issues/311/ge-foods/about-ge-foods.

75. Committee on Genetically Engineered Crops: Past Experience and Future Prospects et al., *Genetically Engineered Crops: Experiences and Prospects* (Washington, DC: National Academies Press, 2016), https://doi.org/10.17226/23395.

76. "Food, Genetically Modified," World Health Organization, May 1, 2014, https://www.who.int/news-room/questions-and-answers/item/food-genetically-modified.

77. Charles M. Benbrook, "Trends in Glyphosate Herbicide Use in the United States and Globally," *Environmental Sciences Europe* 28, no. 1 (December 2016): 3, https://doi.org/10.1186/s12302-016-0070-0.

78. "Glyphosate," Environmental Working Group, July 14, 2020, https://www.ewg.org/areas-focus/toxic-chemicals/glyphosate.

79. Carey Gillam, "Not Just for Corn and Soy: A Look at Glyphosate Use in Food Crops," US Right to Know, May 4, 2016, https://usrtk.org/pesticides/glyphosate-use-in-food-crops/.

80. Kate Vaiknoras, "U.S. Soybean Production Expands Since 2002 as Farmers Adopt New Practices, Technologies," USDA Economic Research Service, July 26, 2023, https://www.ers.usda.gov/amber-waves/2023/july/u-s-soybean-production-expands-since-2002-as-farmers-adopt-new-practices-technologies/.

81. "Recent Trends in GE Adoption," USDA Economic Research Service, July 26, 2024, https://www.ers.usda.gov/data-products/adoption-of-genetically -engineered-crops-in-the-u-s/recent-trends-in-ge-adoption/.

82. Andre Schütze et al., "Quantification of Glyphosate and Other Organophosphorus Compounds in Human Urine via Ion Chromatography Isotope Dilution Tandem Mass Spectrometry," *Chemosphere* 274 (July 2021): 129427, https://doi.org/10.1016/j.chemosphere.2020.129427.

83. "Backgrounder: History of Monsanto's Glyphosate Herbicides" (June 2005), Wayback Machine, April 25, 2014. https://web.archive.org/web/20140425045757 /http://www.monsanto.com/products/documents/glyphosate-background -materials/back_history.pdf.

84. *Monsanto RoundUp 90s Commercial,* 1996. https://www.youtube.com /watch?v=VRsolvgEQy8.

85. Robin Mesnage and Michael N. Antoniou, "Computational Modelling Provides Insight into the Effects of Glyphosate on the Shikimate Pathway in the Human Gut Microbiome," *Current Research in Toxicology* 1 (June 2020): 25–33, https://doi.org/10.1016/j.crtox.2020.04.001.

86. "IARC Monograph on Glyphosate," World Health Organization, International Agency for Research on Cancer, July 19, 2018, https://www.iarc.who .int/featured-news/media-centre-iarc-news-glyphosate.

87. "Glyphosate," Overviews and Factsheets, US Environmental Protection Agency, September 11, 2023, https://www.epa.gov/ingredients-used-pesticide -products/glyphosate.

88. Charles Benbrook, Robin Mesnage, and William Sawyer, "Genotoxicity Assays Published since 2016 Shed New Light on the Oncogenic Potential of Glyphosate-Based Herbicides," *Agrochemicals* 2, no. 1 (January 16, 2023): 47–68, https://doi.org/10.3390/agrochemicals2010005.

89. Robin Mesnage et al., "Use of Shotgun Metagenomics and Metabolomics to Evaluate the Impact of Glyphosate or Roundup MON 52276 on the Gut Microbiota and Serum Metabolome of Sprague-Dawley Rats," *Environmental Health Perspectives* 129, no. 1 (January 2021): 017005, https://doi.org/10.1289/EHP6990.

90. John Peterson Myers et al., "Concerns over Use of Glyphosate-Based Herbicides and Risks Associated with Exposures: A Consensus Statement," *Environmental Health* 15, no. 1 (December 2016): 19, https://doi.org/10.1186/s12940 -016-0117-0.

91. Brenda Eskenazi et al., "Association of Lifetime Exposure to Glyphosate and Aminomethylphosphonic Acid (AMPA) with Liver Inflammation and Metabolic Syndrome at Young Adulthood: Findings from the CHAMACOS Study," *Environmental Health Perspectives* 131, no. 3 (March 2023): 037001, https://doi.org/10.1289/EHP11721.

92. Sheila Kaplan, "Childhood Exposure to Common Herbicide May Increase the Risk of Disease in Young Adulthood," Berkeley Public Health, March 1, 2023,

https://publichealth.berkeley.edu/news-media/research-highlights/childhood
-exposure-to-common-herbicide-may-increase-the-risk-of-disease-in-young-adulthood.

93. Matthew J. Maenner et al., "Prevalence and Characteristics of Autism
Spectrum Disorder Among Children Aged 8 Years—Autism and Developmental
Disabilities Monitoring Network, 11 Sites, United States, 2020," *MMWR Surveillance
Summaries* 72, no. 2 (March 24, 2023): 1–14, https://doi.org/10.15585/mmwr
.ss7202a1.

94. Stephanie Seneff, *Toxic Legacy: How the Weedkiller Glyphosate Is Destroying
Our Health and the Environment* (White River Junction, VT: Chelsea Green
Publishing, 2021), 42–43.

95. Seneff, *Toxic Legacy*, 90.

96. Aiyong Cui et al., "Prevalence, Trend, and Predictor Analyses of Vitamin D
Deficiency in the US Population, 2001–2018," *Frontiers in Nutrition* 9 (October 3,
2022): 965376, https://doi.org/10.3389/fnut.2022.965376.

97. G. S. Johal and D. M. Huber, "Glyphosate Effects on Diseases of Plants,"
European Journal of Agronomy 31, no. 3 (October 2009): 144–52, https://doi.org/10
.1016/j.eja.2009.04.004.

98. Selim Eker et al., "Foliar-Applied Glyphosate Substantially Reduced Uptake
and Transport of Iron and Manganese in Sunflower (*Helianthus Annuus* L.) Plants,"
Journal of Agricultural and Food Chemistry 54, no. 26 (December 1, 2006): 10019–
25, https://doi.org/10.1021/jf0625196.

99. Ramdas Kanissery et al., "Glyphosate: Its Environmental Persistence and
Impact on Crop Health and Nutrition," *Plants* 8, no. 11 (November 13, 2019): 499,
https://doi.org/10.3390/plants8110499.

100. Erick V. S. Motta, Kasie Raymann, and Nancy A. Moran, "Glyphosate
Perturbs the Gut Microbiota of Honey Bees," *Proceedings of the National Academy of
Sciences* 115, no. 41 (October 9, 2018): 10305–10, https://doi.org/10.1073/pnas
.1803880115.

101. James Crall, "Glyphosate Impairs Bee Thermoregulation," *Science* 376,
no. 6597 (June 3, 2022): 1051–52, https://doi.org/10.1126/science.abq5554.

102. Jennifer Sass, "Neonic Pesticides: Potential Risks to Brain and Sperm,"
NRDC (blog), January 6, 2021, https://www.nrdc.org/bio/jennifer-sass/neonic
-pesticides-potential-risks-brain-and-sperm.

103. Julie M. Miwa, Robert Freedman, and Henry A. Lester, "Neural Systems
Governed by Nicotinic Acetylcholine Receptors: Emerging Hypotheses," *Neuron* 70,
no. 1 (April 2011): 20–33, https://doi.org/10.1016/j.neuron.2011.03.014.

104. Duo Zhang and Shaoyou Lu, "Human Exposure to Neonicotinoids and the
Associated Health Risks: A Review," *Environment International* 163 (May 2022):
107201, https://doi.org/10.1016/j.envint.2022.107201.

105. Maria Ospina et al., "Exposure to Neonicotinoid Insecticides in the U.S.
General Population: Data from the 2015–2016 National Health and Nutrition

Examination Survey," *Environmental Research* 176 (September 2019): 108555, https://doi.org/10.1016/j.envres.2019.108555.

106. Angelico Mendy and Susan M. Pinney, "Exposure to Neonicotinoids and Serum Testosterone in Men, Women, and Children," *Environmental Toxicology* 37, no. 6 (June 2022): 1521–28, https://doi.org/10.1002/tox.23503.

107. Ann M. Vuong, Cai Zhang, and Aimin Chen, "Associations of Neonicotinoids with Insulin and Glucose Homeostasis Parameters in US Adults: NHANES 2015–2016," *Chemosphere* 286 (January 2022): 131642, https://doi.org/10.1016/j.chemosphere.2021.131642.

108. Amruta M. Godbole et al., "Exploratory Analysis of the Associations between Neonicotinoids and Measures of Adiposity among US Adults: NHANES 2015–2016," *Chemosphere* 300 (August 2022): 134450, https://doi.org/10.1016/j.chemosphere.2022.134450.

109. Jennifer Sass, "Neonic Pesticide May Become More Toxic in Tap Water," *NRDC* (blog), February 4, 2019, https://www.nrdc.org/bio/jennifer-sass/neonic-pesticide-may-become-more-toxic-tap-water.

110. Courtney Lindwall, "Neonicotinoids 101: The Effects on Humans and Bees," *NRDC*, May 25, 2022, https://www.nrdc.org/stories/neonicotinoids-101-effects-humans-and-bees.

111. "The Importance of Pollinators," US Department of Agriculture, accessed March 13, 2024, https://www.usda.gov/peoples-garden/pollinators.

112. Jeffery S. Pettis et al., "Pesticide Exposure in Honey Bees Results in Increased Levels of the Gut Pathogen Nosema," *Naturwissenschaften* 99, no. 2 (February 2012): 153–58, https://doi.org/10.1007/s00114-011-0881-1.

113. Lucas Rhoads, "The Rusty Patched Bumble Bee Needs Another Legal Win," *NRDC* (blog), March 23, 2021, https://www.nrdc.org/bio/lucas-rhoads/rusty-patched-bumble-bee-needs-another-legal-win.

114. J. R. Reilly et al., "Crop Production in the USA Is Frequently Limited by a Lack of Pollinators," *Proceedings of the Royal Society B: Biological Sciences* 287, no. 1931 (July 29, 2020): 20200922, https://doi.org/10.1098/rspb.2020.0922.

115. Lennard Pisa et al., "An Update of the Worldwide Integrated Assessment (WIA) on Systemic Insecticides. Part 2: Impacts on Organisms and Ecosystems," *Environmental Science and Pollution Research* 28, no. 10 (March 2021): 11749–97, https://doi.org/10.1007/s11356-017-0341-3.

116. Daniel Cressey, "Largest-Ever Study of Controversial Pesticides Finds Harm to Bees," *Nature*, June 29, 2017, nature.2017.22229, https://doi.org/10.1038/nature.2017.22229.

117. Marek Cuhra, Thomas Bøhn, and Petr Cuhra, "Glyphosate: Too Much of a Good Thing?," *Frontiers in Environmental Science* 4 (April 28, 2016), https://doi.org/10.3389/fenvs.2016.00028.

118. Katrine Banke Nørgaard and Nina Cedergreen, "Pesticide Cocktails Can Interact Synergistically on Aquatic Crustaceans," *Environmental Science and Pollution Research* 17, no. 4 (May 2010): 957–67, https://doi.org/10.1007/s11356-009 -0284-4.

119. European Food Safety Authority, "The 2012 European Union Report on Pesticide Residues in Food," *EFSA Journal* 12, no. 12 (December 2014), https://doi .org/10.2903/j.efsa.2014.3942.

120. "Pesticide Data Program Annual Summary, Calendar Year 2022" (USDA Agricultural Marketing Service, January 2024), https://www.ams.usda.gov/sites /default/files/media/2022PDPAnnualSummary.pdf, Appendix J, 2, 4.

121. Helena Sandoval-Insausti et al., "Intake of Fruits and Vegetables According to Pesticide Residue Status in Relation to All-Cause and Disease-Specific Mortality: Results from Three Prospective Cohort Studies," *Environment International* 159 (January 2022): 107024, https://doi.org/10.1016/j.envint.2021.107024.

CHAPTER 2: THE CULT OF PRODUCTIVITY

1. Philip Lowe et al., "Regulating the New Rural Spaces: The Uneven Development of Land," *Journal of Rural Studies* 9, no. 3 (July 1993): 205–22 at 221, https://doi.org /10.1016/0743-0167(93)90067-T.

2. Philip McMichael, "A Food Regime Genealogy," *The Journal of Peasant Studies* 36, no. 1 (January 2009): 139–69, https://doi.org/10.1080/03066150902820354.

3. Ethan Gordon, Federico Davila, and Chris Riedy, "Transforming Landscapes and Mindscapes Through Regenerative Agriculture," *Agriculture and Human Values* 39, no. 2 (June 2022): 809–26, https://doi.org/10.1007/s10460-021-10276-0.

4. Raj Patel, "The Long Green Revolution," *Journal of Peasant Studies* 40, no. 1 (January 2013): 1–63 at 5, https://doi.org/10.1080/03066150.2012.719224.

5. Glenn Davis Stone, "Commentary: New Histories of the Indian Green Revolution," *The Geographical Journal* 185, no. 2 (June 2019): 243–50, https://doi.org /10.1111/geoj.12297.

6. Nick Cullather, *The Hungry World: America's Cold War Battle Against Poverty in Asia* (Harvard University Press, 2011), 45, https://doi.org/10.2307/j.ctvjnrv65.

7. R. E. Evenson and D. Gollin, "Assessing the Impact of the Green Revolution, 1960 to 2000," *Science* 300, no. 5620 (May 2, 2003): 758–62, https://doi.org/10.1126 /science.1078710.

8. Norman E. Borlaug, *Norman E. Borlaug: The Green Revolution, Peace and Humanity: Speech Delivered upon Receipt of the 1970 Nobel Prize, Oslo Norway, 1970* (CIMMYT, 1972), https://repository.cimmyt.org/handle/10883/19284.

9. Amir Kassam and Laila Kassam, "Paradigms of Agriculture," in *Rethinking Food and Agriculture* (Elsevier, 2021), 181–218 at 184, https://doi.org/10.1016/B978-0 -12-816410-5.00010-4.

10. "The Green Revolution: Norman Borlaug and the Race to Fight Global Hunger," PBS, *American Experience*, April 3, 2020, https://www.pbs.org/wgbh /americanexperience/features/green-revolution-norman-borlaug-race-to-fight -global-hunger/.

11. "Caught Up in the War on Communism: Norman Borlaug and the 'Green Revolution,'" PBS, *American Experience*, April 3, 2020, https://www.pbs.org/wgbh /americanexperience/features/caught-war-on-communism-norman-borlaug-and -green-revolution/.

12. Tim G. Benton and Rob Bailey, "The Paradox of Productivity: Agricultural Productivity Promotes Food System Inefficiency," *Global Sustainability* 2 (2019): e6, 3, https://doi.org/10.1017/sus.2019.3.

13. "Staple Foods: What Do People Eat?," in *Dimensions of Need: An Atlas of Food and Agriculture* (Food and Agriculture Organization of the United Nations, 1995), https://www.fao.org/3/u8480e/u8480e07.htm.

14. "Staple Foods: What Do People Eat?"

15. OECD, *Agricultural Policy Monitoring and Evaluation 2023: Adapting Agriculture to Climate Change*, Agricultural Policy Monitoring and Evaluation (OECD, 2023), https://doi.org/10.1787/b14de474-en.

16. Susan Schneider, "Climate Change, Food Security, and the Myth of Unlimited Abundance," *Journal of Food Law & Policy* 19, no. 1 (October 30, 2023), 50, https://scholarworks.uark.edu/jflp/vol19/iss1/7.

17. Susan Schneider, "A Reconsideration of Agricultural Law: A Call for the Law of Food, Farming, and Sustainability," *William & Mary Environmental Law and Policy Review* 34, no. 3 (April 1, 2010): 935, 949.

18. "Feed Grains Sector at a Glance," USDA Economic Research Service, December 21, 2023, https://www.ers.usda.gov/topics/crops/corn-and-other-feed -grains/feed-grains-sector-at-a-glance/.

19. "US Corn Farm Price Received Monthly Trends: USDA Farm Price Received," YCharts, accessed August 24, 2024, https://ycharts.com/indicators/us_corn_price; "Commodity Costs and Returns," USDA Economic Research Service, June 20, 2024, https://www.ers.usda.gov/data-products/commodity-costs-and-returns/commodity -costs-and-returns/#Historical%20Costs%20and%20Returns:%20Corn.

20. Ruixue Wang, Roderick M. Rejesus, and Serkan Aglasan, "Warming Temperatures, Yield Risk and Crop Insurance Participation," *European Review of Agricultural Economics* 48, no. 5 (November 2, 2021): 1109–31, https://doi.org/10.1093 /erae/jbab034.

21. Noah S. Diffenbaugh, Frances V. Davenport, and Marshall Burke, "Historical Warming Has Increased U.S. Crop Insurance Losses," *Environmental Research Letters* 16, no. 8 (August 1, 2021): 084025, https://doi.org/10.1088/1748-9326/ac1223.

22. "Good Farming Practice Determination Standards Handbook 2024 and Succeeding Years" (USDA Federal Crop Insurance Corporation, November 20, 2023),

https://www.rma.usda.gov/sites/default/files/press-release/2024-14060-Good-Farming-Practice-Determination-Standards.pdf.

23. "USDA Improves Crop Insurance to Better Support Conservation, Climate-Smart Practices," USDA Risk Management Agency, December 6, 2023, https://www.rma.usda.gov/news-events/news/2023/washington-dc/usda-improves-crop-insurance-better-support-conservation.

24. Tyler J. Lark et al., "Environmental Outcomes of the US Renewable Fuel Standard," *Proceedings of the National Academy of Sciences* 119, no. 9 (March 2022): e2101084119, https://doi.org/10.1073/pnas.2101084119.

25. "Farms and Farmland," 2022 Census of Agriculture Highlights (USDA National Agricultural Statistics Service, March 2024), https://www.nass.usda.gov/Publications/Highlights/2024/Census22_HL_FarmsFarmland.pdf.

26. Mario Loyola, "Stop the Ethanol Madness," *The Atlantic*, November 23, 2019, https://www.theatlantic.com/ideas/archive/2019/11/ethanol-has-forsaken-us/602191/.

27. Megan Horst and Amy Marion, "Racial, Ethnic and Gender Inequities in Farmland Ownership and Farming in the U.S.," *Agriculture and Human Values* 36, no. 1 (March 2019): 1–16, https://doi.org/10.1007/s10460-018-9883-3.

28. Ken Drexler, "Research Guides: Indian Removal Act: Primary Documents in American History: Introduction," Library of Congress, May 14, 2019, https://guides.loc.gov/indian-removal-act/introduction.

29. J. Weston Phippen, "Kill Every Buffalo You Can! Every Buffalo Dead Is an Indian Gone," *The Atlantic*, May 13, 2016, https://www.theatlantic.com/national/archive/2016/05/the-buffalo-killers/482349/.

30. Ken Drexler, "Research Guides: Homestead Act: Primary Documents in American History: Introduction," Library of Congress, December 11, 2020, https://guides.loc.gov/homestead-act/introduction.

31. Brian Barth, "Black Land Matters," *Modern Farmer*, August 27, 2018, https://modernfarmer.com/2018/08/black-land-matters/.

32. "Food Justice," *FoodPrint*, February 28, 2024, https://foodprint.org/issues/food-justice/.

33. Nina Lakhani, "'America Is a Factory Farming Nation': Key Takeaways from US Agriculture Census," *The Guardian*, February 15, 2024, https://www.theguardian.com/environment/2024/feb/15/us-agriculture-census-farming.

34. Lakhani.

35. "Farms and Farmland."

36. Lakhani.

37. "Fisheries Subsidies Agreement: What's the Big Deal?," Pew, May 10, 2023, https://pew.org/3proEft.

38. Enric Sala et al., "The Economics of Fishing the High Seas," *Science Advances* 4, no. 6 (June 2018): eaat2504, https://doi.org/10.1126/sciadv.aat2504.

39. U. Rashid Sumaila et al., "Winners and Losers in a World Where the High Seas Is Closed to Fishing," *Scientific Reports* 5, no. 1 (February 12, 2015): 8481, https://doi.org/10.1038/srep08481.

40. "Collapsing Fisheries—Have We Reached the End of the Line?," *Oceaneos*, accessed March 21, 2024, https://www.oceaneos.org/state-of-our-oceans/collapsing-fisheries-examples-of-different-species/.

41. "Collapsing Fisheries—Have We Reached the End of the Line?"

42. "High Seas Treaty Must Reflect Critical Role of Fish in Marine Ecosystems," Pew, March 14, 2022, https://pew.org/3J6l7uY.

43. *The State of World Fisheries and Aquaculture 2022* (FAO, 2022), 46. https://doi.org/10.4060/cc0461en.

44. "High Seas Treaty Must Reflect Critical Role of Fish in Marine Ecosystems," Pew, March 14, 2022, https://pew.org/3J6l7uY.

45. "High Seas Treaty Ratification: Track Progress," High Seas Alliance Treaty Ratification, accessed March 21, 2024, https://highseasalliance.org/treaty-ratification/track-progress/.

46. "High Seas Treaty Ratification."

47. "Lecture: Focus Panel 3—Beyond Growth Beyond Europe: What Policies and Partnerships?," Beyond Growth 2023 Conference, May 15, 2023, https://www.beyond-growth-2023.eu/lecture/focus-panel-3/.

48. Audrey Denvir et al., "Ecological and Human Dimensions of Avocado Expansion in México: Towards Supply-Chain Sustainability," *Ambio* 51, no. 1 (January 2022): 152–66, https://doi.org/10.1007/s13280-021-01538-6.

49. Denvir et al.

50. "Unholy Guacamole: Deforestation, Water Capture, and Violence Behind Mexico's Avocado Exports to the U.S. and Other Major Markets" (Climate Rights International, November 2023), https://cri.org/reports/unholy-guacamole/.

51. "Unholy Guacamole."

52. Denvir et al.

53. Denvir, Audrey, Eugenio Y. Arima, Antonio González-Rodríguez, and Kenneth R. Young. "Ecological and Human Dimensions of Avocado Expansion in México: Towards Supply-Chain Sustainability." *Ambio* 51, no. 1 (January 2022): 152–66. https://doi.org/10.1007/s13280-021-01538-6.

54. "Unholy Guacamole."

55. "International Agricultural Productivity."

56. Benton and Bailey.

57. Global Alliance for the Future of Food, "Power Shift: Why We Need to Wean Industrial Food Systems Off Fossil Fuels" (Global Alliance for the Future of Food, 2023), https://futureoffood.org/wp-content/uploads/2023/10/ga_food-energy-nexus_report.pdf, 2.

58. Francesco Tubiello et al., "Food Systems Emissions Shares, 1990–2019" (Zenodo, October 29, 2021), https://doi.org/10.5281/ZENODO.5615082; M. Crippa et al., "Food Systems Are Responsible for a Third of Global Anthropogenic GHG Emissions," *Nature Food* 2, no. 3 (March 8, 2021): 198–209, https://doi.org/10.1038/s43016-021-00225-9.

59. IPES-Food, "From Uniformity to Diversity: A Paradigm Shift from Industrial Agriculture to Diversified Agroecological Systems" (International Panel of Experts on Sustainable Food Systems, 2016), https://ipes-food.org/report/from-uniformity-to-diversity/.

60. Max Roser, Hannah Ritchie, and Pablo Rosado, "Food Supply," *Our World in Data*, January 2, 2024, https://ourworldindata.org/food-supply.

61. "Calories on the Nutrition Facts Label," U.S. Food & Drug Administration, March 5, 2024, https://www.fda.gov/food/nutrition-facts-label/calories-nutrition-facts-label.

62. Roser, Ritchie, and Rosado.

63. FAO et al., *Urbanization, Agrifood Systems Transformation and Healthy Diets Across the Rural-Urban Continuum*, The State of Food Security and Nutrition in the World 2023 (Rome: FAO, 2023), 8, https://doi.org/10.4060/cc3017en.

64. FAO et al., 18.

65. FAO et al., vii.

66. "Key Statistics & Graphics," USDA Economic Research Service, October 25, 2023, https://www.ers.usda.gov/topics/food-nutrition-assistance/food-security-in-the-u-s/key-statistics-graphics/#foodsecure.

67. "Agriculture; Plantations; Other Rural Sectors," International Labor Organization, accessed March 28, 2024, https://www.ilo.org/global/industries-and-sectors/agriculture-plantations-other-rural-sectors/lang--en/index.htm.

68. Philip H. Howard and Mary Hendrickson, "Op-Ed: Monopolies in the Food System Make Food More Expensive and Less Accessible," *Civil Eats*, February 17, 2021, https://civileats.com/2021/02/17/op-ed-monopolies-in-the-food-system-make-food-more-expensive-and-less-accessible/.

69. Philip H. Howard, *Concentration and Power in the Food System: Who Controls What We Eat?*, Revised edition, Contemporary Food Studies: Economy, Culture and Politics (London: Bloomsbury Academic, 2021).

70. James M. MacDonald, Xiao Dong, and Keith Owen Fuglie, "Concentration and Competition in U.S. Agribusiness" (Washington, DC: Economic Research Service, U.S. Department of Agriculture, June 2023), https://doi.org/10.32747/2023.8054022.ers.

71. Nina Lakhani, Aliya Uteuova, and Alvin Chang, "Revealed: The True Extent of America's Food Monopolies, and Who Pays the Price," *The Guardian*, July 14, 2021, http://www.theguardian.com/environment/ng-interactive/2021/jul/14/food-monopoly-meals-profits-data-investigation.

72. Lakhani, Uteuova, and Chang.

73. Howard and Hendrickson.

74. Kristina Kiki Hubbard, "The Sobering Details Behind the Latest Seed Monopoly Chart," *Civil Eats*, January 11, 2019, https://civileats.com/2019/01/11/the -sobering-details-behind-the-latest-seed-monopoly-chart/.

75. Claire Kelloway and Sarah Miller, "Food and Power: Addressing Monopolization in America's Food System" (Open Markets Institute, March 2019), https://static1.squarespace.com/static/5e449c8c3ef68d752f3e70dc/t /614a2ebebf7d510debfd53f3/1632251583273/200921_MonopolyFoodReport _endnote_v3.pdf.

76. Deese, Brian, Sameera Fazili, and Bharat Ramamurti. "Recent Data Show Dominant Meat Processing Companies Are Taking Advantage of Market Power to Raise Prices and Grow Profit Margins," The White House, December 10, 2021, https://www.whitehouse.gov/briefing-room/blog/2021/12/10/recent-data-show -dominant-meat-processing-companies-are-taking-advantage-of-market-power-to -raise-prices-and-grow-profit-margins/

77. Deese, Fazili, and Ramamurti.

78. Christopher Leonard, *The Meat Racket: The Secret Takeover of America's Food Business*, 1st ed. (New York, NY: Simon & Schuster, 2014).

79. "Fees Paid to Growers for Raising Broiler Chickens Varied Widely in 2020," USDA Economic Research Service, September 8, 2022, http://www.ers.usda.gov /data-products/chart-gallery/gallery/chart-detail/?chartId=104642.

80. "RAFI Praises USDA's Proposed Rule on Poultry Tournament Systems," RAFI, June 3, 2024, https://www.rafiusa.org/blog/rafi-praises-usdas-proposed-rule -on-poultry-tournament-systems/.

81. James M. MacDonald, "Financial Risks and Incomes in Contract Broiler Production," Amber Waves, USDA Economic Research Service, August 4, 2014, https://www.ers.usda.gov/amber-waves/2014/august/financial-risks-and-incomes -in-contract-broiler-production/.

82. "Understanding Contract Agriculture," Rural Advancement Foundation International-USA, accessed March 29, 2024, https://www.rafiusa.org/programs /contract-agriculture-reform/understanding-contract-agriculture/.

83. "Fees Paid to Growers for Raising Broiler Chickens Varied Widely in 2020."

84. "RAFI Praises USDA's Proposed Rule on Poultry Tournament Systems."

85. "Food Dollar Application," USDA Economic Research Service, November 15, 2023, https://data.ers.usda.gov/reports.aspx?ID=17885.

86. Phoebe Galt, "As Biden Looks to Tackle 'Shrinkflation,' New Analysis Highlights Extreme Stickiness of High Food Prices," Food & Water Watch, March 5, 2024, https://www.foodandwaterwatch.org/2024/03/05/as-biden-looks-to-tackle -shrinkflation-new-analysis-highlights-extreme-stickiness-of-high-food-prices/.

87. Howard and Hendrickson.

88. "Agribusiness Sector Total," *OpenSecrets*, accessed March 28, 2024, https://www.opensecrets.org/industries/totals?cycle=2020&ind=A.

89. "Agribusiness Sector Total."

90. Rob Bailey and Laura Wellesley, *Chokepoints and Vulnerabilities in Global Food Trade* (Toronto, ON, CA: Chatham House, 2017).

91. Whitehead, Dalton, and Yuan H. Brad Kim. "The Impact of COVID 19 on the Meat Supply Chain in the USA: A Review," *Food Science of Animal Resources* 42, no. 5 (September 2022): 762–74. https://doi.org/10.5851/kosfa.2022.e39.

92. Leah Douglas, "Chicken Company to Cull Birds as Processing Capacity Plummets," Food and Environment Reporting Network, April 12, 2020, https://thefern.org/ag_insider/chicken-company-to-cull-birds-as-processing-capacity-plummets/; Sophie Kevany, "Millions of US Farm Animals to Be Culled by Suffocation, Drowning and Shooting," *The Guardian*, May 19, 2020, https://www.theguardian.com/environment/2020/may/19/millions-of-us-farm-animals-to-be-culled-by-suffocation-drowning-and-shooting-coronavirus.

93. Ruchi Shroff and Carla Ramos Cortés, "The Biodiversity Paradigm: Building Resilience for Human and Environmental Health," *Development* 63, no. 2–4 (December 2020): 172–80, https://doi.org/10.1057/s41301-020-00260-2.

94. "The Problem of Food Waste," *FoodPrint*, February 28, 2024, https://foodprint.org/issues/the-problem-of-food-waste/.

95. "Food Waste Index Report 2024. Think Eat Save: Tracking Progress to Halve Global Food Waste" (Nairobi, Kenya: United Nations Environment Programme, March 21, 2024), 2, https://wedocs.unep.org/xmlui/handle/20.500.11822/45230.

96. FAO, "Discards in the World's Marine Fisheries: An Update," accessed March 23, 2024. https://www.fao.org/3/y5936e/y5936e09.htm.

97. "Food Waste Problem," ReFED, accessed March 23, 2024, https://refed.org/food-waste/the-problem/.

98. Kirsten Jaglo, Shannon Kenny, and Jenny Stephenson, "From Farm to Kitchen: The Environmental Impacts of U.S. Food Waste," Reports and Assessments (U.S. Environmental Protection Agency Office of Research and Development, November 2021), https://www.epa.gov/land-research/farm-kitchen-environmental-impacts-us-food-waste, 1.

99. Jaglo, Kenny, and Stephenson, ii.

100. "Reduced Food Waste," Project Drawdown, February 6, 2020, https://drawdown.org/solutions/reduced-food-waste.

101. *Food Waste Is the World's Dumbest Problem*, Climate Lab, 2017, https://www.youtube.com/watch?v=6RlxySFrkIM.

102. Andy Fisher, *Big Hunger: The Unholy Alliance between Corporate America and Anti-Hunger Groups*, Food, Health, and the Environment (Cambridge, MA: The MIT Press, 2017).

103. Andy Fisher, "The Way Cheap Food Feeds Big Hunger and Inequality," *Sustainable America* (blog), March 16, 2018, https://sustainableamerica.org/blog/the-way-cheap-food-feeds-big-hunger-and-inequality/.

104. Peter B. R. Hazell, "Green Revolution: Curse or Blessing?" (International Food Policy Research Institute [IFPRI], 2002), https://ebrary.ifpri.org/utils/getfile/collection/p15738coll2/id/64639/filename/64640.cpd.

105. Hannah Ritchie, "Yields vs. Land Use: How the Green Revolution Enabled Us to Feed a Growing Population," *Our World in Data*, February 1, 2024, https://ourworldindata.org/yields-vs-land-use-how-has-the-world-produced-enough-food-for-a-growing-population.

106. Philip K. Thornton, "Livestock Production: Recent Trends, Future Prospects," *Philosophical Transactions of the Royal Society B: Biological Sciences* 365, no. 1554 (September 27, 2010): 2853–67, https://doi.org/10.1098/rstb.2010.0134.

107. Patricio Grassini, Kent M. Eskridge, and Kenneth G. Cassman, "Distinguishing Between Yield Advances and Yield Plateaus in Historical Crop Production Trends," *Nature Communications* 4, no. 1 (December 17, 2013): 2918, https://doi.org/10.1038/ncomms3918.

108. Deepak K. Ray et al., "Recent Patterns of Crop Yield Growth and Stagnation," *Nature Communications* 3, no. 1 (December 18, 2012): 1293, https://doi.org/10.1038/ncomms2296.

109. IPES-Food, 15.

110. "NASA Analysis Confirms 2023 as Warmest Year on Record," NASA, January 12, 2024, https://www.nasa.gov/news-release/nasa-analysis-confirms-2023-as-warmest-year-on-record/.

111. "NASA Clocks July 2023 as Hottest Month on Record Ever Since 1880," NASA Global Climate Change: Vital Signs of the Planet, August 14, 2023, https://climate.nasa.gov/news/3279/nasa-clocks-july-2023-as-hottest-month-on-record-ever-since-1880.

112. Bob Henson and Jeff Masters, "NOAA: July 2024 Was Earth's Hottest Month on Record," Yale Climate Connections, August 15, 2024, http://yaleclimateconnections.org/2024/08/noaa-july-2024-was-earths-hottest-month-on-record/.

113. "Climate Change Indicators: Length of Growing Season," US Environmental Protection Agency, June 2024, https://www.epa.gov/climate-indicators/climate-change-indicators-length-growing-season.

114. Chuang Zhao et al., "Temperature Increase Reduces Global Yields of Major Crops in Four Independent Estimates," *Proceedings of the National Academy of Sciences* 114, no. 35 (August 29, 2017): 9326–31, https://doi.org/10.1073/pnas.1701762114.

115. "What the Western US Megadrought Tells Us About Climate Change," World Economic Forum, March 9, 2022, https://www.weforum.org/agenda/2022/03/western-us-megadrought-climate-change/.

116. "Climate Change Leads to More Extreme Weather, but Early Warnings Save Lives," United Nations Climate Change, September 1, 2021, https://unfccc.int/news/climate-change-leads-to-more-extreme-weather-but-early-warnings-save-lives.

117. Josie Garthwaite, "Climate of Chaos: Why Warming Makes Weather Less Predictable," Stanford Doerr School of Sustainability, December 14, 2021, https://sustainability.stanford.edu/news/climate-chaos-why-warming-makes-weather-less-predictable.

118. Adam B. Smith, "U.S. Billion-Dollar Weather and Climate Disasters, 1980—Present" (NOAA National Centers for Environmental Information, 2020), https://doi.org/10.25921/STKW-7W73.

119. Renee Cho, "How Climate Change Will Affect Plants," *State of the Planet* (blog), January 27, 2022, https://news.climate.columbia.edu/2022/01/27/how-climate-change-will-affect-plants/.

120. Schneider, 949.

121. Kenny Torrella, "Why American Farms Often Get a Free Pass on Critical Environmental and Labor Laws," *Vox*, August 31, 2023, https://www.vox.com/future-perfect/2023/8/31/23852325/farming-myths-agricultural-exceptionalism-pollution-labor-animal-welfare-laws.

122. "What Is Ag-Gag Legislation?," ASPCA, accessed March 29, 2024, https://www.aspca.org/improving-laws-animals/public-policy/what-ag-gag-legislation.

123. "Right-to-Farm," National Agricultural Law Center, accessed March 29, 2024, https://nationalaglawcenter.org/state-compilations/right-to-farm/.

124. "Agriculture & Food," TEEB, The Economics of Ecosystems and Biodiversity, accessed March 25, 2024, https://teebweb.org/our-work/agrifood/.

125. "The Evaluation Framework," TEEB, The Economics of Ecosystems and Biodiversity, accessed April 1, 2024, https://teebweb.org/our-work/agrifood/understanding-teebagrifood/evaluation-framework/.

126. "True Cost Accounting: Implementation Guidance and Inventory," Global Alliance for the Future of Food, October 8, 2020, https://futureoffood.org/insights/tca-implementation-inventory/.

127. CONABIO, "Ecosystems and Agro-Biodiversity Across Small and Large-Scale Maize Production Systems, Feeder Study to the 'TEEB for Agriculture and Food,'" 2017, https://www.teebweb.org/wp-content/uploads/2018/01/Final-Maize-TEEB-report_290817.pdf, 1.

128. "True Cost of Food: Measuring What Matters to Transform the U.S. Food System" (Rockefeller Foundation, July 2021), 16, https://www.rockefellerfoundation.org/wp-content/uploads/2021/07/True-Cost-of-Food-Full-Report-Final.pdf.

129. "True Cost of Food," 24.

130. "True Cost of Food," 20.

131. Gerardo Ceballos, Paul R. Ehrlich, and Rodolfo Dirzo, "Biological Annihilation via the Ongoing Sixth Mass Extinction Signaled by Vertebrate

Population Losses and Declines," *Proceedings of the National Academy of Sciences* 114, no. 30 (July 25, 2017), https://doi.org/10.1073/pnas.1704949114.

132. Damian Carrington, "Earth's Sixth Mass Extinction Event Under Way, Scientists Warn," *The Guardian*, July 10, 2017, https://www.theguardian.com /environment/2017/jul/10/earths-sixth-mass-extinction-event-already-underway -scientists-warn.

133. Mark A. Anthony, S. Franz Bender, and Marcel G. A. Van Der Heijden, "Enumerating Soil Biodiversity," *Proceedings of the National Academy of Sciences* 120, no. 33 (August 15, 2023): e2304663120, https://doi.org/10.1073/pnas.2304663120.

134. Ronald Vargas, "Soils, Where Food Begins," United Nations (United Nations, December 2, 2022), https://www.un.org/en/un-chronicle/soils-where -food-begins.

135. FAO and ITPS, "Status of the World's Soil Resources (SWSR)—Main Report" (Rome, Italy: Food and Agriculture Organization of the United Nations and Intergovernmental Technical Panel on Soils, 2015), 4, https://www.fao.org/3/bc590e /bc590e.pdf.

136. David R. Montgomery and Anne Biklé, *The Hidden Half of Nature: The Microbial Roots of Life and Health*, First edition (New York: W.W. Norton & Company, 2016).

137. Vargas, "Soils, Where Food Begins."

138. Winfried E. H. Blum, Sophie Zechmeister-Boltenstern, and Katharina M. Keiblinger, "Does Soil Contribute to the Human Gut Microbiome?," *Microorganisms* 7, no. 9 (August 23, 2019): 287, https://doi.org/10.3390/microorganisms7090287.

139. Vargas.

140. Blum, Zechmeister-Boltenstern, and Keiblinger, "Does Soil Contribute to the Human Gut Microbiome?"

141. Vargas.

142. Vargas.

143. Michael Via, "The Malnutrition of Obesity: Micronutrient Deficiencies That Promote Diabetes," *ISRN Endocrinology* 2012 (March 15, 2012): 1–8, https://doi.org /10.5402/2012/103472.

144. Brian Halweil, "Still No Free Lunch: Nutrient Levels in U.S. Food Supply Eroded by Pursuit of High Yields," The Organic Center (September 1, 2007), https:// www.organic-center.org/still-no-free-lunch-nutrient-levels-us-food-supply-eroded -pursuit-high-yields.

145. Donald R. Davis, Melvin D. Epp, and Hugh D. Riordan, "Changes in USDA Food Composition Data for 43 Garden Crops, 1950 to 1999," *Journal of the American College of Nutrition* 23, no. 6 (December 2004): 669–82, https://doi.org/10.1080 /07315724.2004.10719409.

146. David Thomas, "The Mineral Depletion of Foods Available to US as A Nation (1940–2002)—A Review of the 6th Edition of McCance and Widdowson,"

Nutrition and Health 19, no. 1–2 (July 2007): 21–55, https://doi.org/10.1177 /026010600701900205.

147. Davis, Epp, and Riordan, "Changes in USDA Food Composition Data for 43 Garden Crops, 1950 to 1999."

148. Rachel Lovell, "Why Modern Food Lost Its Nutrients," *BBC*, accessed March 25, 2024, https://www.bbc.com/future/bespoke/follow-the-food/why-modern -food-lost-its-nutrients/?ct=t.

149. Heribert Hirt, "Healthy Soils for Healthy Plants for Healthy Humans: How Beneficial Microbes in the Soil, Food and Gut Are Interconnected and How Agriculture Can Contribute to Human Health," *EMBO Reports* 21, no. 8 (August 5, 2020): e51069, https://doi.org/10.15252/embr.202051069.

150. Winfried E. H. Blum, Sophie Zechmeister-Boltenstern, and Katharina M. Keiblinger, "Does Soil Contribute to the Human Gut Microbiome?," *Microorganisms* 7, no. 9 (August 23, 2019): 287, https://doi.org/10.3390/microorganisms7090287.

151. The World Counts, "CO_2 Concentration," accessed April 1, 2024, https:// www.theworldcounts.com/challenges/global-warming/CO2-concentration.

152. "Carbon Dioxide Concentration," NASA Climate Change: Vital Signs of the Planet, July 2024, https://climate.nasa.gov/vital-signs/carbon-dioxide?intent=121.

153. Renee Cho, "How Climate Change Will Affect Plants," *State of the Planet* (blog), January 27, 2022, https://news.climate.columbia.edu/2022/01/27/how -climate-change-will-affect-plants/.

154. Cho.

155. Helena Bottemiller Evich, "The Great Nutrient Collapse," *The Agenda*, September 13, 2017, http://politi.co/2zACS5k.

156. Danielle E. Medek, Joel Schwartz, and Samuel S. Myers, "Estimated Effects of Future Atmospheric CO_2 Concentrations on Protein Intake and the Risk of Protein Deficiency by Country and Region," *Environmental Health Perspectives* 125, no. 8 (August 16, 2017): 087002, https://doi.org/10.1289/EHP41.

157. Daniel R. Taub, Brian Miller, and Holly Allen, "Effects of Elevated CO_2 on the Protein Concentration of Food Crops: A Meta-analysis," *Global Change Biology* 14, no. 3 (March 2008): 565–75, https://doi.org/10.1111/j.1365-2486.2007.01511.x.

158. Taub, Miller, and Allen.

159. Samuel S. Myers et al., "Effect of Increased Concentrations of Atmospheric Carbon Dioxide on the Global Threat of Zinc Deficiency: A Modelling Study," *The Lancet Global Health* 3, no. 10 (October 2015): e639–45, https://doi.org/10.1016/S2214 -109X(15)00093-5.

160. Matthew R. Smith and Samuel S. Myers, "Impact of Anthropogenic CO_2 Emissions on Global Human Nutrition," *Nature Climate Change* 8, no. 9 (September 2018): 834–39, https://doi.org/10.1038/s41558-018-0253-3.

161. Lewis H. Ziska et al., "Rising Atmospheric CO_2 Is Reducing the Protein Concentration of a Floral Pollen Source Essential for North American Bees,"

Proceedings of the Royal Society B: Biological Sciences 283, no. 1828 (April 13, 2016): 20160414, https://doi.org/10.1098/rspb.2016.0414.

162. "Water and Sanitation," UN Department of Economic and Social Affairs, accessed August 24, 2024, https://sdgs.un.org/topics/water-and-sanitation.

163. Scott Jasechko et al., "Rapid Groundwater Decline and Some Cases of Recovery in Aquifers Globally," *Nature* 625, no. 7996 (January 25, 2024): 715–21, https://doi.org/10.1038/s41586-023-06879-8.

164. Lifeng Li, "Water Scarcity, the Climate Crisis and Global Food Security: A Call for Collaborative Action," United Nations (United Nations, October 12, 2023), https://www.un.org/en/un-chronicle/water-scarcity-climate-crisis-and-global-food-security-call-collaborative-action.

165. Rachel Becker, "Colorado River Water Cut Back—Except for California," *CalMatters*, August 16, 2022, http://calmatters.org/environment/2022/08/colorado-river-water-california/.

166. California Department of Water Resources, "Critically Overdrafted Basins," accessed April 3, 2024, https://water.ca.gov/Programs/Groundwater-Management/Bulletin-118/Critically-Overdrafted-Basins.

167. Gregg M. Garfin et al., "Chapter 25: Southwest. Impacts, Risks, and Adaptation in the United States: The Fourth National Climate Assessment, Volume II" (U.S. Global Change Research Program, 2018), https://doi.org/10.7930/NCA4.2018.CH25.

168. Ashley Kerna, Dari Duval, and George Frisvold, "Arizona Leafy Greens: Economic Contributions of the Industry Cluster," Agricultural and Resource Economics, The University of Arizona, August 30, 2017, https://economics.arizona.edu/arizona-leafy-greens-economic-contributions-industry-cluster.

169. "California Agricultural Statistics Review 2021–22," California Department of Food & Agriculture, 2022, https://www.cdfa.ca.gov/Statistics/PDFs/2022_Ag_Stats_Review.pdf.

170. Twilight Greenaway, "California Dairy Uses Lots of Water. Here's Why It Matters," *Civil Eats*, June 30, 2022, https://civileats.com/2022/06/30/california-dairy-water-uses-climate-change-drought-pollution/.

171. "Big Ag, Big Oil and California's Big Water Problem," Food & Water Watch, 2021, 6, https://www.foodandwaterwatch.org/wp-content/uploads/2021/10/CA-Water-White-Paper.pdf.

172. Greenaway.

173. Associated Press, "Saudi Land Purchases in California and Arizona Fuel Debate over Water Rights," *Los Angeles Times*, March 29, 2016, https://www.latimes.com/business/la-fi-saudi-arabia-alfalfa-20160329-story.html.

174. Lauren Markham, "Who Keeps Buying California's Scarce Water? Saudi Arabia," *The Guardian*, March 25, 2019, https://www.theguardian.com/us-news/2019/mar/25/california-water-drought-scarce-saudi-arabia.

175. Markham.

176. Associated Press.

177. Garfin et al.

178. Ritchie.

179. Damian Carrington, "Humans Just 0.01% of All Life but Have Destroyed 83% of Wild Mammals—Study," *The Guardian*, May 21, 2018, https://www .theguardian.com/environment/2018/may/21/human-race-just-001-of-all-life-but -has-destroyed-over-80-of-wild-mammals-study.

180. "Overview of U.S. Livestock, Poultry, and Aquaculture Production in 2017," USDA Animal and Plant Health Inspection Service (April 2018), https://www.aphis .usda.gov/sites/default/files/demographics2017.pdf; "Animals on Factory Farms," ASPCA, accessed August 24, 2024, https://www.aspca.org/protecting-farm-animals /animals-factory-farms.

181. Alberto Cesarani and Giuseppe Pulina, "Farm Animals Are Long Away from Natural Behavior: Open Questions and Operative Consequences on Animal Welfare," *Animals* 11, no. 3 (March 6, 2021): 724, https://doi.org/10.3390/ani11030724.

182. Shibing You et al., "African Swine Fever Outbreaks in China Led to Gross Domestic Product and Economic Losses," *Nature Food* 2, no. 10 (September 27, 2021): 802–8, https://doi.org/10.1038/s43016-021-00362-1.

183. "U.S. Approaches Record Number of Avian Influenza Outbreaks," Centers for Disease Control and Prevention, November 3, 2022, https://www.cdc.gov/flu /avianflu/spotlights/2022-2023/nearing-record-number-avian-influenza.htm.

184. Josh Funk, "Egg, Poultry Prices Recover, but Bird Flu Outbreak Isn't Over," *AP News*, November 14, 2023, https://apnews.com/article/bird-flu-outbreak-turkey -egg-chicken-prices-afca48e861a54091d0e224a5fcbed17b?lctg=237284891.

185. "U.S. Approaches Record Number of Avian Influenza Outbreaks."

186. Funk, "Egg, Poultry Prices Recover, but Bird Flu Outbreak Isn't Over."

187. Benji Jones, "Yes, Bird Flu Is a Threat. It's Time to Take It Seriously," *Vox*, July 17, 2023, https://www.vox.com/science/23793697/bird-flu-avian-influenza -explained-pandemic-potential-wildlife?lctg=237284891.

188. Emily Anthes and Apoorva Mandavilli, "What to Know About the Bird Flu Outbreak in Dairy Cows," *The New York Times*, April 9, 2024, https://www.nytimes .com/article/bird-flu-cattle-human.html.

189. Renee Cho, "Making Fish Farming More Sustainable," *State of the Planet* (blog), April 13, 2016, https://news.climate.columbia.edu/2016/04/13/making-fish -farming-more-sustainable/.

190. U.S. Soy Staff Writer, "Feeding Sustainable Animal Protein Growth," *U.S. Soy*, January 18, 2023, https://ussoy.org/feeding-sustainable-animal-protein-growth/.

191. Hannah Ritchie and Max Roser, "Land Use," *Our World in Data*, February 16, 2024, https://ourworldindata.org/land-use.

192. Shepon et al.

193. Alon Shepon et al., "The Opportunity Cost of Animal Based Diets Exceeds All Food Losses," *Proceedings of the National Academy of Sciences* 115, no. 15 (April 10, 2018): 3804–9, https://doi.org/10.1073/pnas.1713820115.

194. Shepon et al.

195. Nina Lakhani, "'America Is a Factory Farming Nation': Key Takeaways from US Agriculture Census," *The Guardian*, February 15, 2024, https://www.theguardian.com/environment/2024/feb/15/us-agriculture-census-farming.

196. "Pew Commission Says Industrial Scale Farm Animal Production Poses 'Unacceptable' Risks to Public Health, Environment," Pew, April 29, 2008, http://pew.org/1Q6ZCrz.

197. "Lake Erie," State of the Great Lakes, June 20, 2022, https://stateofgreatlakes.net/lake-assessments/lake-erie/.

198. Tony Briscoe, "Lake Erie Provides Drinking Water for More People than Any Other, but Algae Blooms Are Making It Toxic," Phys.org, November 20, 2019, https://phys.org/news/2019-11-lake-erie-people-algae-blooms.html.

199. Briscoe.

200. "Lake Erie."

201. "China's 26-Storey Pig Skyscraper Ready to Slaughter 1 Million Pigs a Year," *The Guardian*, November 25, 2022, https://www.theguardian.com/environment/2022/nov/25/chinas-26-storey-pig-skyscraper-ready-to-produce-1-million-pigs-a-year.

CHAPTER 3: THE PROGRESS TRAP

1. Charles C. Mann, *The Wizard and the Prophet: Two Remarkable Scientists and Their Dueling Visions to Shape Tomorrow's World* (New York: Alfred A. Knopf, 2018).

2. Georgia Jiang, "Vertical Farming—No Longer a Futuristic Concept," USDA Agricultural Research Service, July 12, 2023, https://www.ars.usda.gov/oc/utm/vertical-farming-no-longer-a-futuristic-concept/.

3. Laurens Klerkx and Pablo Villalobos, "Are AgriFoodTech Start-Ups the New Drivers of Food Systems Transformation? An Overview of the State of the Art and a Research Agenda," *Global Food Security* 40 (March 2024): 100726, https://doi.org/10.1016/j.gfs.2023.100726.

4. Mario Herrero et al., "Innovation Can Accelerate the Transition Towards a Sustainable Food System," *Nature Food* 1, no. 5 (May 19, 2020): 266–72, https://doi.org/10.1038/s43016-020-0074-1.

5. David Christian Rose and Jason Chilvers, "Agriculture 4.0: Broadening Responsible Innovation in an Era of Smart Farming," *Frontiers in Sustainable Food Systems* 2 (December 21, 2018): 87, https://doi.org/10.3389/fsufs.2018.00087.

6. An Ecomodernist Manifesto, "An Ecomodernist Manifesto," accessed April 8, 2024. http://www.ecomodernism.org/manifesto-english.

7. Helen Breewood and Tara Garnett, "What Is Ecomodernism?" *TABLE* (June 9, 2022), 8, https://doi.org/10.56661/041dba86.

8. Linus Blomqvist, Ted Nordhaus, and Michael Shellenberger, "Nature Unbound: Decoupling for Conservation," The Breakthrough Institute, September 2015, 69, https://thebreakthrough.org/articles/nature-unbound.

9. Linus Blomqvist, "Raising Agricultural Yields Spares Land," The Breakthrough Institute, March 13, 2024, https://thebreakthrough.org/issues/food -agriculture-environment/raising-agricultural-yields-spares-land.

10. George Monbiot, *Regenesis: Feeding the World Without Devouring the Planet* (Great Britain: Penguin Books, 2022), 230–1.

11. "Price Nature or Make Nature Priceless?" The Breakthrough Institute, July 14, 2015, https://thebreakthrough.org/articles/price-nature-or-make-nature-priceless.

12. Bruno Latour, "Fifty Shades of Green," *Environmental Humanities* 7, no. 1 (May 1, 2016): 219–25, https://doi.org/10.1215/22011919-3616416.

13. Maywa Montenegro De Wit, "Can Agroecology and CRISPR Mix? The Politics of Complementarity and Moving Toward Technology Sovereignty," *Agriculture and Human Values* 39, no. 2 (June 2022): 733–55, https://doi.org/10.1007/s10460-021-10284-0.

14. An Ecomodernist Manifesto.

15. Erle C. Ellis et al., "People Have Shaped Most of Terrestrial Nature for at Least 12,000 Years," *Proceedings of the National Academy of Sciences* 118, no. 17 (April 27, 2021): e2023483118, https://doi.org/10.1073/pnas.2023483118.

16. An Mo Notenbaert et al., "Policies in Support of Pastoralism and Biodiversity in the Heterogeneous Drylands of East Africa," *Pastoralism: Research, Policy and Practice* 2, no. 1 (2012): 14, https://doi.org/10.1186/2041-7136-2-14.

17. Bryony Sands et al., "Moving Towards an Anti-Colonial Definition for Regenerative Agriculture," *Agriculture and Human Values* 40, no. 4 (December 2023): 1697–716, https://doi.org/10.1007/s10460-023-10429-3.

18. Amir Kassam and Laila Kassam, "Paradigms of Agriculture," in A. Kassam & L. Kassam (eds.), *Rethinking Food and Agriculture* (Elsevier, 2021), 181–218 at 188, https://doi.org/10.1016/B978-0-12-816410-5.00010-4.

19. Richard Heinberg, "Something Wicked This Way Comes," *Resilience*, December 7, 2023, https://www.resilience.org/stories/2023-12-07/something-wicked -this-way-comes/.

20. Latour.

21. Robert Wood, Radhika Nagpal, and Gu-Yeon Wei, "The Robobee Project Is Building Flying Robots the Size of Insects," *Scientific American*, March 1, 2013, https://www.scientificamerican.com/article/robobee-project-building-flying -robots-insect-size/.

22. "RoboBees: Autonomous Flying Microrobots," Wyss Institute, August 5, 2016, https://wyss.harvard.edu/technology/robobees-autonomous-flying-microrobots/.

23. Serena Solomon, "In the Face of Climate Change and Food Insecurity, New Zealand Considers Lab-Grown Fruit," *The Guardian*, September 7, 2023, https:// www.theguardian.com/world/2023/sep/07/new-zealand-lab-grown-fruit.

24. Gaëtan Vanloqueren and Philippe V. Baret, "How Agricultural Research Systems Shape a Technological Regime That Develops Genetic Engineering but Locks out Agroecological Innovations," *Research Policy* 38, no. 6 (July 2009): 971–83, https://doi.org/10.1016/j.respol.2009.02.008.

25. Adrian Dubock, "Golden Rice: To Combat Vitamin A Deficiency for Public Health," in *Vitamin A*, eds. Leila Queiroz Zepka, Veridiana Vera De Rosso, and Eduardo Jacob-Lopes (IntechOpen, 2019), https://doi.org/10.5772/intechopen.84445.

26. "Philippines Becomes First Country to Approve Nutrient-Enriched 'Golden Rice' for Planting," International Rice Research Institute, July 23, 2021, https://www.irri.org/news-and-events/news/philippines-becomes-first-country-approve-nutrient-enriched-golden-rice.

27. Nick Aspinwall, "Why Is the Philippines Blocking 'Miracle Crops'?," *Foreign Policy*, June 28, 2024, https://foreignpolicy.com/2024/06/13/philippines-gmos-golden-rice-bt-eggplant/.

28. Felicia Wu et al., "Allow Golden Rice to Save Lives," *Proceedings of the National Academy of Sciences* 118, no. 51 (December 21, 2021): e2112090118, https://doi.org/10.1073/pnas.2112090118.

29. Ruchi Shroff and Carla Ramos Cortés, "The Biodiversity Paradigm: Building Resilience for Human and Environmental Health," *Development* 63, no. 2–4 (December 2020): 172–80, https://doi.org/10.1057/s41301-020-00260-2.

30. Allison K. Wilson, "Will Gene-Edited and Other GM Crops Fail Sustainable Food Systems?," in A. Kassam & L. Kassam (eds.), *Rethinking Food and Agriculture* (Elsiever, 2021), 262.

31. Katherine Dentzman, Ryan Gunderson, and Raymond Jussaume, "Techno-Optimism as a Barrier to Overcoming Herbicide Resistance: Comparing Farmer Perceptions of the Future Potential of Herbicides," *Journal of Rural Studies* 48 (December 2016): 22–32 at 23, https://doi.org/10.1016/j.jrurstud.2016.09.006.

32. Stephen O. Duke, "Perspectives on Transgenic, Herbicide-Resistant Crops in the United States Almost 20 Years After Introduction," *Pest Management Science* 71, no. 5 (May 2015): 652–57, https://doi.org/10.1002/ps.3863.

33. Glenn Davis Stone and Andrew Flachs, "The Ox Fall Down: Path-Breaking and Technology Treadmills in Indian Cotton Agriculture," *The Journal of Peasant Studies* 45, no. 7 (November 10, 2018): 1272–96 at 1276, https://doi.org/10.1080/03066150.2017.1291505.

34. Stone and Flachs, 1277.

35. Margaret R. Douglas and John F. Tooker, "Large-Scale Deployment of Seed Treatments Has Driven Rapid Increase in Use of Neonicotinoid Insecticides and Preemptive Pest Management in U.S. Field Crops," *Environmental Science & Technology* 49, no. 8 (April 21, 2015): 5088–97, https://doi.org/10.1021/es506141g.

36. Fred Gould, Zachary S. Brown, and Jennifer Kuzma, "Wicked Evolution: Can We Address the Sociobiological Dilemma of Pesticide Resistance?," *Science* 360, no. 6390 (May 18, 2018): 728–32, https://doi.org/10.1126/science.aar3780.

37. Nicholas G. Karavolias et al., "Application of Gene Editing for Climate Change in Agriculture," *Frontiers in Sustainable Food Systems* 5 (September 7, 2021): 685801, https://doi.org/10.3389/fsufs.2021.685801.

38. Joseph Opoku Gakpo, "Scientific Group Says Gene Editing Key Tool for Transforming Global Food Systems," Alliance for Science, September 24, 2021, https://allianceforscience.org/blog/2021/09/scientific-group-says-gene-editing-key -tool-for-transforming-global-food-systems/.

39. Hongge Jia et al., "Genome Editing of the Disease Susceptibility Gene in Citrus Confers Resistance to Citrus Canker," *Plant Biotechnology Journal* 15, no. 7 (July 2017): 817–23, https://doi.org/10.1111/pbi.12677.

40. Christine Tait-Burkard et al., "Livestock 2.0—Genome Editing for Fitter, Healthier, and More Productive Farmed Animals," *Genome Biology* 19, no. 1 (December 2018): 204, https://doi.org/10.1186/s13059-018-1583-1.

41. Alison Louise Van Eenennaam, "Genetic Modification of Food Animals," *Current Opinion in Biotechnology* 44 (April 2017): 27–34, https://doi.org/10.1016/j .copbio.2016.10.007.

42. Nicholas G. Karavolias et al., "Application of Gene Editing for Climate Change in Agriculture," *Frontiers in Sustainable Food Systems* 5 (September 7, 2021): 685801, https://doi.org/10.3389/fsufs.2021.685801.

43. "The Nobel Prize in Chemistry 2020," NobelPrize.org, accessed April 22, 2024, https://www.nobelprize.org/prizes/chemistry/2020/summary/.

44. Emily Waltz, "CRISPR-Edited Crops Free to Enter Market, Skip Regulation," *Nature Biotechnology* 34, no. 6 (June 2016): 582–2, https://doi.org/10.1038/nbt0616-582.

45. Andrew Kessel, "Calyxt Debuts Premium Soybean Cooking Oil Calyno," Proactive Investors, April 30, 2020, https://www.proactiveinvestors.com/companies /news/918560/calyxt-debuts-premium-soybean-cooking-oil-calyno-918560.html.

46. Emily Waltz, "GABA-Enriched Tomato Is First CRISPR-Edited Food to Enter Market," *Nature Biotechnology* 40, no. 1 (December 14, 2021): 9–11, https://doi.org/10 .1038/d41587-021-00026-2.

47. Yubing He and Yunde Zhao, "Technological Breakthroughs in Generating Transgene-Free and Genetically Stable CRISPR-Edited Plants," *aBIOTECH* 1, no. 1 (January 2020): 88–96, https://doi.org/10.1007/s42994-019-00013-x.

48. Rupert Sheldrake, "Setting Innovation Free in Agriculture," in *Rethinking Food and Agriculture* (Elsevier, 2021), 1–29 at, https://doi.org/10.1016/B978-0-12 -816410-5.00001-3.

49. P. B. Thompson, "Why Using Genetics to Address Welfare May Not Be a Good Idea," *Poultry Science* 89, no. 4 (April 2010): 814–21, https://doi.org/10.3382/ps .2009-00307.

50. Pauline Fricot, "Here Is Ultraphotosynthesis!," *RIPE: Realizing Increased Photosynthetic Efficiency for Sustainable Increases in Crop Yield*, November 10, 2020, https://ripe.illinois.edu/news/here-ultraphotosynthesis.

51. Willian Batista-Silva et al., "Engineering Improved Photosynthesis in the Era of Synthetic Biology," *Plant Communications* 1, no. 2 (March 2020): 100032, https://doi.org/10.1016/j.xplc.2020.100032.

52. Susan Milius, "Tweaking How Plants Manage a Crisis Boosts Photosynthesis," *Science News*, November 17, 2016, https://www.sciencenews.org/article/tweaking-how-plants-manage-crisis-boosts-photosynthesis.

53. Denise Chow, "Plant Scientists Have Found a Way to 'Hack' Photosynthesis. Here's Why That's a Big Deal," *NBC News*, January 10, 2019, https://www.nbcnews.com/mach/science/plant-scientists-have-found-way-hack-photosynthesis-here-s-why-ncna956706.

54. Maria Temming, "A New Way to Genetically Tweak Photosynthesis Boosts Plant Growth," *Science News*, January 3, 2019, https://www.sciencenews.org/article/new-way-genetically-tweak-photosynthesis-boosts-plant-growth.

55. Temming.

56. Fricot.

57. "Home | RIPE," *RIPE*, accessed April 7, 2024, https://ripe.illinois.edu/.

58. Willian Batista-Silva et al., "Engineering Improved Photosynthesis in the Era of Synthetic Biology," *Plant Communications* 1, no. 2 (March 2020): 100032, https://doi.org/10.1016/j.xplc.2020.100032.

59. Thomas R. Sinclair, Thomas W. Rufty, and Ramsey S. Lewis, "Increasing Photosynthesis: Unlikely Solution for World Food Problem," *Trends in Plant Science* 24, no. 11 (November 2019): 1032–39, https://doi.org/10.1016/j.tplants.2019.07.008.

60. Chow.

61. Klerkx and Rose.

62. David Nally, "Against Food Security: On Forms of Care and Fields of Violence," *Global Society* 30, no. 4 (October 2016): 558–82, https://doi.org/10.1080/13600826.2016.1158700.

63. Benjamin Nowak, "Precision Agriculture: Where Do We Stand? A Review of the Adoption of Precision Agriculture Technologies on Field Crops Farms in Developed Countries," *Agricultural Research* 10, no. 4 (December 2021): 515–22, https://doi.org/10.1007/s40003-021-00539-x.

64. "Robot Uses Machine Learning to Harvest Lettuce," University of Cambridge, July 8, 2019, https://www.cam.ac.uk/research/news/robot-uses-machine-learning-to-harvest-lettuce.

65. Simon Birrell et al., "A Field-Tested Robotic Harvesting System for Iceberg Lettuce," *Journal of Field Robotics* 37, no. 2 (March 2020): 225–45, https://doi.org/10.1002/rob.21888.

66. Mina Mesbahi, "The Future of Agriculture: Adopting More Advanced Robots for Next-Gen Farming," *Wevolver*, January 18, 2021, https://www.wevolver.com/article/the-future-of-agriculture-adopting-more-advanced-robots-for-next-gen-farming.

67. Jess Lowenberg-DeBoer, "The Precision Agriculture Revolution," *Foreign Affairs*, April 20, 2015, https://www.foreignaffairs.com/articles/united-states/2015-04-20/precision-agriculture-revolution.

68. Sarah Rotz et al., "Automated Pastures and the Digital Divide: How Agricultural Technologies Are Shaping Labour and Rural Communities," *Journal of Rural Studies* 68 (May 2019): 112–22, https://doi.org/10.1016/j.jrurstud.2019.01.023.

69. Mario Herrero et al., "Articulating the Effect of Food Systems Innovation on the Sustainable Development Goals," *The Lancet Planetary Health* 5, no. 1 (January 2021): e50–62, https://doi.org/10.1016/S2542-5196(20)30277-1.

70. Donya-Faye Wix, "(Infographic) The U.S. Farm Labor Shortage," *AgAmerica*, February 26, 2020, https://agamerica.com/blog/the-impact-of-the-farm-labor-shortage/.

71. Isabelle M. Carbonell, "The Ethics of Big Data in Big Agriculture," *Internet Policy Review* 5, no. 1 (March 31, 2016), https://doi.org/10.14763/2016.1.405.

72. Sabina Wex, "The Farming Industry Is Struggling with Labor Shortages and the Climate Crisis. But Agtech like Digital Cow Collars and AI-Quality-Control Tools Are Solving Problems in Unexpected Ways.," *Business Insider*, September 13, 2023, https://www.businessinsider.com/agtech-agriculture-industry-farming-methods-climate-change-labor-solutions-2023-9.

73. Steven A. Wolf and Spencer D. Wood, "Precision Farming: Environmental Legitimation, Commodification of Information, and Industrial Coordination," *Rural Sociology* 62, no. 2 (June 1997): 180–206, https://doi.org/10.1111/j.1549-0831.1997.tb00650.x.

74. Sally Brooks, "Configuring the Digital Farmer: A Nudge World in the Making?," *Economy and Society* 50, no. 3 (July 3, 2021): 374–96, https://doi.org/10.1080/03085147.2021.1876984.

75. Kelly Bronson, "Looking Through a Responsible Innovation Lens at Uneven Engagements with Digital Farming," *NJAS: Wageningen Journal of Life Sciences* 90–91, no. 1 (December 1, 2019): 1–6 at 4, https://doi.org/10.1016/j.njas.2019.03.001.

76. Joseph Poore and Thomas Nemecek, "Reducing Food's Environmental Impacts Through Producers and Consumers," *Science* 360, no. 6392 (2018): 987.

77. National Chicken Council, "Per Capita Consumption of Poultry and Livestock, 1965 to Forecast 2022, in Pounds," National Chicken Council, accessed April 13, 2024, https://www.nationalchickencouncil.org/about-the-industry/statistics/per-capita-consumption-of-poultry-and-livestock-1965-to-estimated-2012-in-pounds/.

78. Jelle Bruinsma, "Livestock Production," in *World Agriculture: Towards 2015/2030 an FAO Perspective* (London: Earthscan Publications, 2003).

79. Seth Millstein, "How Many Animals Are Killed for Food Every Day?," *Sentient*, July 30, 2024, https://sentientmedia.org/how-many-animals-are-killed-for -food-every-day/.

80. Walter Willett et al., "Food in the Anthropocene: The EAT–Lancet Commission on Healthy Diets from Sustainable Food Systems," *The Lancet* 393, no. 10170 (February 2019): 447–92, https://doi.org/10.1016/S0140-6736(18)31788-4.

81. Jennie I. Macdiarmid, Flora Douglas, and Jonina Campbell, "Eating Like There's No Tomorrow: Public Awareness of the Environmental Impact of Food and Reluctance to Eat Less Meat as Part of a Sustainable Diet," *Appetite* 96 (January 2016): 487–93, https://doi.org/10.1016/j.appet.2015.10.011.

82. Brock Bastian and Steve Loughnan, "Resolving the Meat-Paradox: A Motivational Account of Morally Troublesome Behavior and Its Maintenance," *Personality and Social Psychology Review* 21, no. 3 (August 2017): 278–99, https://doi .org/10.1177/1088868316647562.

83. Marleen C. Onwezen and Cor N. Van Der Weele, "When Indifference Is Ambivalence: Strategic Ignorance About Meat Consumption," *Food Quality and Preference* 52 (September 2016): 96–105, https://doi.org/10.1016/j.foodqual.2016.04 .001.

84. Bastian and Loughnan.

85. Fiona Harvey, "Americans Can Eat Meat While Cutting Global Heating, Says Agriculture Secretary," *The Guardian*, November 6, 2021, https://www.theguardian .com/environment/2021/nov/06/americans-can-eat-meat-while-cutting-global -heating-says-agriculture-secretary.

86. "Joint US-EU Press Release on the Global Methane Pledge," The White House, September 18, 2021, https://www.whitehouse.gov/briefing-room/statements -releases/2021/09/18/joint-us-eu-press-release-on-the-global-methane-pledge/.

87. Emma Bryce, "Kowbucha, Seaweed, Vaccines: The Race to Reduce Cows' Methane Emissions," *The Guardian*, September 30, 2021, https://www.theguardian .com/environment/2021/sep/30/cow-methane-emissions-reduce-seaweed-kowbucha.

88. "AgSTAR: Biogas Recovery in the Agriculture Sector," Data and Tools, U.S. Environmental Protection Agency, August 5, 2024, https://www.epa.gov/agstar.

89. Ruthie Lazenby, "Rethinking Manure Biogas: Policy Considerations to Promote Equity and Protect the Climate and Environment," Center for Agriculture & Food Systems, August 2022, https:// www.vermontlaw.edu/sites/default/files /2022-08/ Rethinking_Manure_Biogas.pdf.

90. Derek Harrison, "Is California Overstating the Climate Benefit of Dairy Manure Methane Digesters?," *Inside Climate News* (blog), December 30, 2023, https:// insideclimatenews.org/news/30122023/milking-it-california-overstating-climate -benefit-dairy-manure-methane-digesters/.

91. "AgSTAR."

92. Chloe Waterman and Molly Armus, "Biogas or Bull****? The Deceptive Promise of Manure Biogas as a Methane Solution," Friends of the Earth, 2024, 15, https://foe.org/wp-content/uploads/2024/03/Factory-Farm-Gas-Brief_final-0312.pdf.

93. Kaya Laterman, "This California Dairy Farm's Secret Ingredient for Clean Electricity: Cow Poop," *Yahoo!News*, January 21, 2022, https://www.yahoo.com/news/california-dairy-farm-secret-ingredient-100314801.html.

94. Michael McCully, "Energy Revenue Could Be a Game Changer for Dairy Farms," *Hoard's Dairyman*, September 23, 2021, https://hoards.com/article-30925-energy-revenue-could-be-a-game-changer-for-dairy-farms.html.

95. "Is Anaerobic Digestion Right for Your Farm?," US Environmental Protection Agency, December 2023, https://www.epa.gov/agstar/anaerobic-digestion-right-your-farm.

96. Waterman and Armus, 23.

97. "N.C. Residents Living Near Large Hog Farms Have Elevated Disease, Death Risks," *Duke Health*, September 18, 2018, https://corporate.dukehealth.org/news/nc-residents-living-near-large-hog-farms-have-elevated-disease-death-risks.

98. "N.C. Residents Living Near Large Hog Farms Have Elevated Disease, Death Risks."

99. Katelyn Weisbrod, "North Carolina's New Farm Bill Speeds the Way for Smithfield's Massive Biogas Plan for Hog Farms," *Inside Climate News* (blog), August 31, 2021, https://insideclimatenews.org/news/31082021/north-carolina-farm-bill-biogas-smithfield/.

100. Julie Guthman and Charlotte Biltekoff, "Magical Disruption? Alternative Protein and the Promise of De-materialization," *Environment and Planning E: Nature and Space* 4, no. 4 (December 2021): 1583–1600, https://doi.org/10.1177/2514848620963125.

101. Catherine Tubb and Tony Seba, "Rethinking Food & Agriculture," *Rethinkx*, September 2019, 6, https://www.rethinkx.com/publications/rethinkingfoodandagriculture2019.en.

102. "Beyond Meat: Vegan Meat, Plant Based Meat Substitutes," Beyond Meat, accessed April 25, 2024, https://www.beyondmeat.com/en-US/.

103. "Heme + The Science Behind Impossible," Impossible Foods, accessed April 25, 2024, https://impossiblefoods.com/heme.

104. Elaine Watson, "Kraft Heinz Creates Joint Venture with AI-Powered Foodtech Startup NotCo," *FoodNavigator-USA*, February 22, 2022, https://www.foodnavigator-usa.com/Article/2022/02/22/Kraft-Heinz-creates-joint-venture-with-AI-powered-foodtech-startup-NotCo.

105. "What Is Cellular Agriculture?," *New Harvest*, accessed April 6, 2024, https://new-harvest.org/what-is-cellular-agriculture/.

106. Nicola Jones, "Lab-Grown Meat: The Science of Turning Cells into Steaks and Nuggets," *Nature* 619, no. 7968 (July 4, 2023): 22–24, https://doi.org/10.1038/d41586-023-02095-6.

107. "Human Food Made with Cultured Animal Cells," US Food & Drug Administration, March 21, 2023, https://www.fda.gov/food/food-ingredients-packaging/human-food-made-cultured-animal-cells.

108. "Reboot Food," accessed April 11, 2024, https://www.rebootfood.org.

109. Michael Eisen, "How GMOs Can Save Civilization (and Probably Already Have)," Impossible Foods, March 16, 2018, https://impossiblefoods.com/blog/how-gmos-can-save-civilization-and-probably-already-have.

110. "Perfect Day's ProFerm—Animal Free Milk Protein," Perfect Day, accessed April 11, 2024, https://perfectday.com/proferm/.

111. "EVERY EggWhite," *Every*, accessed August 26, 2024, https://www.every.com/every-eggwhite.

112. "Precision Fermentation," Reboot Food, accessed April 25, 2024, https://www.rebootfood.org/precision-fermentation.

113. "Food Out of Thin Air," Solar Foods, accessed April 25, 2024, https://solarfoods.com/.

114. "Reboot Food."

115. Laurel Oldach, "Biochemistry of a Burger," ASBMB Today, October 1, 2019, https://www.asbmb.org/asbmb-today/industry/100119/biochemistry-of-a-burger.

116. Katherine Dentzman, Ryan Gunderson, and Raymond Jussaume, "Techno-Optimism as a Barrier to Overcoming Herbicide Resistance: Comparing Farmer Perceptions of the Future Potential of Herbicides," *Journal of Rural Studies* 48 (December 2016): 22–32, https://doi.org/10.1016/j.jrurstud.2016.09.006, 24, quoting Alvin M. Weinberg, "Can Technology Replace Social Engineering?" in *Technology and Man's Future*, ed. Albert H. Teich (New York: St. Martin's Press, 1966/1981), 29–39.

117. "Technology Is Not Values Neutral: Ending the Reign of Nihilistic Design," The Consilience Project, June 26, 2022, https://consilienceproject.org/technology-is-not-values-neutral-ending-the-reign-of-nihilistic-design-2/.

118. "State of Global Policy 2022: Public Investment in Alternative Proteins to Feed a Growing World," Good Food Institute, 2023, 3, https://gfi.org/wp-content/uploads/2023/01/State-of-Global-Policy-Report_2022.pdf.

119. "How Singapore Is Using Alternative Proteins to Boost Food Security," EDB Singapore, July 4, 2022, https://www.edb.gov.sg/en/business-insights/insights/how-singapore-is-using-alternative-proteins-to-boost-food-security.html.

120. "WePlanet | Who We Are," WePlanet, accessed April 13, 2024, https://www.weplanet.org/whoweare.

121. "Open Letter," FundFutureFood, accessed April 13, 2024, https://www.fundfuturefood.org/openletter.

122. "Open Letter."

123. Million Belay, "Harvesting Hypocrisy: The Global Food Fight and Africa's Forgotten Farmers," *Nation*, November 6, 2023, https://nation.africa/kenya/blogs -opinion/blogs/-the-global-food-fight-and-africa-s-forgotten-farmers -4426380#google_vignette.

124. IPES-Food, "The Politics of Protein: Examining Claims About Livestock, Fish, 'Alternative Proteins' and Sustainability," 2022, 53–54, https://ipes-food.org /report/the-politics-of-protein/.

125. "Food Out of Thin Air."

126. IPES-Food, 23.

127. Katri Behm et al., "Comparison of Carbon Footprint and Water Scarcity Footprint of Milk Protein Produced by Cellular Agriculture and the Dairy Industry," *The International Journal of Life Cycle Assessment* 27, no. 8 (August 2022): 1017–34, https://doi.org/10.1007/s11367-022-02087-0.

128. Olivier Hamant, "Plant Scientists Can't Ignore Jevons Paradox Anymore," *Nature Plants* 6, no. 7 (July 14, 2020): 720–22, https://doi.org/10.1038/s41477-020 -0722-3.

129. Christopher L. Magee and Tessaleno C. Devezas, "A Simple Extension of Dematerialization Theory: Incorporation of Technical Progress and the Rebound Effect," *Technological Forecasting and Social Change* 117 (April 2017): 196–205, https:// doi.org/10.1016/j.techfore.2016.12.001.

130. Sam Bliss, "Resources for a Better Future: Jevons Paradox," *Resilience*, June 17, 2020, https://www.resilience.org/stories/2020-06-17/jevons-paradox/.

131. John Bellamy Foster, Brett Clark, and Richard York, "Capitalism and the Curse of Energy Efficiency," *Monthly Review* 62, no. 6 (November 1, 2010), 11, month- lyreview.org/2010/11/01/capitalism-and-the-curse-of-energy-efficiency.

132. Louis Sears et al., "Jevons' Paradox and Efficient Irrigation Technology," *Sustainability* 10, no. 5 (May 16, 2018): 1590, https://doi.org/10.3390/su10051590.

133. Hamant.

134. Hannah Ritchie, "Palm Oil," *Our World in Data*, January 15, 2024, https:// ourworldindata.org/palm-oil.

135. Hamant.

136. Thomas K. Rudel et al., "Agricultural Intensification and Changes in Cultivated Areas, 1970–2005," *Proceedings of the National Academy of Sciences* 106, no. 49 (December 8, 2009): 20675–80, https://doi.org/10.1073/pnas .0812540106.

137. Pedro Pellegrini and Roberto J. Fernández, "Crop Intensification, Land Use, and on-Farm Energy-Use Efficiency during the Worldwide Spread of the Green Revolution," *Proceedings of the National Academy of Sciences* 115, no. 10 (March 6, 2018): 2335–40, https://doi.org/10.1073/pnas.1717072115.

138. Foster, Clark, and York, 13.

139. Kate Vaiknoras, "U.S. Soybean Production Expands Since 2002 as Farmers Adopt New Practices, Technologies," USDA Economic Research Service, July 26, 2023, https://www.ers.usda.gov/amber-waves/2023/july/u-s-soybean-production -expands-since-2002-as-farmers-adopt-new-practices-technologies/.

140. Richard York, "Do Alternative Energy Sources Displace Fossil Fuels?," *Nature Climate Change* 2, no. 6 (June 2012): 441–43, https://doi.org/10.1038/ nclimate1451.

141. Giampiero Grossi et al., "Livestock and Climate Change: Impact of Livestock on Climate and Mitigation Strategies," *Animal Frontiers* 9, no. 1 (January 3, 2019): 69–76, https://doi.org/10.1093/af/vfy034.

142. Seth Millstein, "How Many Animals Are Killed for Food Every Day?," *Sentient*, July 30, 2024, https://sentientmedia.org/how-many-animals-are-killed-for -food-every-day/.

143. Richard York, "Poultry and Fish and Aquatic Invertebrates Have Not Displaced Other Meat Sources," *Nature Sustainability* 4, no. 9 (April 26, 2021): 766– 68, https://doi.org/10.1038/s41893-021-00714-6.

144. Stefano B. Longo et al., "Aquaculture and the Displacement of Fisheries Captures," *Conservation Biology* 33, no. 4 (August 2019): 832–41, https://doi.org/10 .1111/cobi.13295.

145. Emma Ignaszewski, "2023 Outlook: The State of the Plant-Based Meat Category," The Good Food Institute, April 11, 2023, https://gfi.org/blog/2023-outlook -the-state-of-the-plant-based-meat-category/.

146. Emma Alsop, "Plant Based Food Substitutes on the Rise, but Not at the Expense of Meat Sales Says ABS," *Beef Central*, January 1, 2022, https://www .beefcentral.com/news/plant-based-food-substitutes-on-the-rise-but-not-at-the -expense-of-meat-sales-says-abs/.

147. Shuoli Zhao et al., "Meet the Meatless: Demand for New Generation Plant-Based Meat Alternatives," *Applied Economic Perspectives and Policy* 45, no. 1 (March 2023): 4–21, https://doi.org/10.1002/aepp.13232.

148. Kenny Torrella, "Germany's Surprising and Sudden Embrace of Vegan Food," *Vox*, July 22, 2022, https://www.vox.com/future-perfect/23273338/germany -less-meat-plant-based-vegan-vegetarian-flexitarian.

149. Zhao et al.

150. Bob Woods, "'Real' Meat Maker Cargill May Emerge Big Winner in Plant-Based Food as Buzzy Startup Boom Fades," CNBC, June 17, 2023, https://www.cnbc .com/2023/06/17/why-cargill-a-real-meat-maker-may-be-big-winner-in-plant-based -food.html.

151. "Tyson Foods Unveils Alternative Protein Products and New Raised & Rooted Brand," Tyson Foods, June 13, 2019, https://www.tysonfoods.com/news/news -releases/2019/6/tyson-foods-unveils-alternative-protein-products-and-new-raised -rootedr.

152. Julie Guthman et al., "In the Name of Protein," *Nature Food* 3, no. 6 (May 31, 2022): 391–93, https://doi.org/10.1038/s43016-022-00532-9.

153. Hannah Barrett and David Christian Rose, "Perceptions of the Fourth Agricultural Revolution: What's In, What's Out, and What Consequences Are Anticipated?," *Sociologia Ruralis* 62, no. 2 (April 2022): 162–89, https://doi.org/10.1111/soru.12324.

154. "The Limits of a Technological Fix." *Nature Food* 2, no. 4 (April 21, 2021): 211. https://doi.org/10.1038/s43016-021-00275-z.

155. Chris Smaje, "Saying NO to a Farm-Free Future," *Chris Smaje* (blog), April 20, 2023, https://chrissmaje.com/2023/04/saying-no-to-a-farm-free-future/.

CHAPTER 4: FARMING WITH NATURE

1. Otto Scharmer, "4.0 Lab: The Future of Food, Finance, Health, Ed, & Management," *HuffPost*, June 25, 2017, https://www.huffpost.com/entry/40-lab-the-future-of-food-finance-health-ed-management_b_594fa701e4bof078efd98267.

2. Carol Sanford, *The Regenerative Life: Transform Any Organization, Our Society, and Your Destiny* (Boston: Nicholas Brealey Publishing, 2020).

3. Andre Leu, "The Definition of Regenerative Agriculture," *Regeneration International* (blog), December 22, 2023, https://regenerationinternational.org/2023/12/22/the-definition-of-regenerative-agriculture/.

4. Christopher J. Rhodes, "The Imperative for Regenerative Agriculture," *Science Progress* 100, no. 1 (March 2017): 80–129, https://doi.org/10.3184/003685017X14876775256165.

5. Tom Kuhlman and John Farrington, "What Is Sustainability?," *Sustainability* 2, no. 11 (November 1, 2010): 3436–48, https://doi.org/10.3390/su2113436.

6. "THE 17 GOALS | Sustainable Development," United Nations, Department of Economic and Social Affairs, accessed April 29, 2024, https://sdgs.un.org/goals.

7. "THE 17 GOALS | Sustainable Development."

8. "Shifting from 'Sustainability' to Regeneration," *Building Research & Information* 35, no. 6 (November 2007): 674–80, https://doi.org/10.1080/09613210701475753.

9. Tomaso Ferrando, "Commons and Commoning to Build Ecologically Reparatory Food Systems," in *Routledge Handbook of Sustainable and Regenerative Food Systems*, eds. Jessica Duncan, Michael Carolan, and Johannes S. C. Wiskerke, 1st ed. (Routledge, 2020), 262–76 at 272, https://doi.org/10.4324/9780429466823-19.

10. Sam J. Buckton et al., "The Regenerative Lens: A Conceptual Framework for Regenerative Social-Ecological Systems," *One Earth* 6, no. 7 (July 2023): 824–42 at 824, https://doi.org/10.1016/j.oneear.2023.06.006.

11. "Buen Vivir: The Rights of Nature in Bolivia and Ecuador," Rapid Transition Alliance, December 2, 2018, https://rapidtransition.org/stories/the-rights-of-nature -in-bolivia-and-ecuador/.

12. "Lessons from Indigenous Traditions and Innovation / Expert Q&A: Dr. Enrique Salmón," *Biohabitats*, 2018, https://www.biohabitats.com/newsletter /ecology-culture-and-economy-lessons-from-indigenous-traditions-and-innovation /expert-qa-enrique-salmon/.

13. Gwen Grelet et al., "Regenerative Agriculture in Aotearoa New Zealand– Research Pathways to Build Science-Based Evidence and National Narratives," Next Foundation, Landcare Research, Our Land and Water National Science Challenge, February 2021, https://researcharchive.lincoln.ac.nz/server/api/core/bitstreams /ba039547-7452-44de-8986-efb4e5671f8c/content.

14. Buckton et al., 825.

15. Bill Reed, "Shifting from 'Sustainability' to Regeneration," *Building Research & Information* 35, no. 6 (November 2007): 674–80, https://doi.org/10.1080 /09613210701475753.

16. Sandra Waddock, *Catalyzing Transformation: Making System Change Happen* (New York, NY: Business Expert Press, 2024), 34.

17. "What Is Ecology?—The Ecological Society of America," accessed May 5, 2024, https://www.esa.org/about/what-does-ecology-have-to-do-with-me/.

18. Melanie Goodchild, "Relational Systems Thinking: The Dibaajimowin (Story) of Re-Theorizing 'Systems Thinking' and 'Complexity Science,'" *Journal of Awareness-Based Systems Change* 2, no. 1 (May 31, 2022): 53–76, https://doi.org/10.47061/jabsc.v2i1 .2027; Melanie Goodchild, "Relational Systems Thinking: That's How Change Is Going to Come, from Our Earth Mother," *Journal of Awareness-Based Systems Change* 1, no. 1 (February 25, 2021): 75–103, https://doi.org/10.47061/jabsc.v1i1.577.

19. Waddock, 35.

20. "About the PHA," Planetary Health Alliance, accessed May 7, 2024, https:// www.planetaryhealthalliance.org/about-the-pha.

21. "One Health Basics | One Health," CDC, September 28, 2023, https://www .cdc.gov/onehealth/basics/index.html.

22. Emily Anthes and Apoorva Mandavilli, "What to Know About the Bird Flu Outbreak in Dairy Cows," *The New York Times*, April 9, 2024, https://www.nytimes .com/article/bird-flu-cattle-human.html.

23. Jonathan Rushton et al., "A Food System Paradigm Shift: From Cheap Food at Any Cost to Food Within a One Health Framework," *NAM Perspectives* 11 (November 22, 2021), https://doi.org/10.31478/202111b.

24. IPES-Food, "Unravelling the Food–Health Nexus: Addressing Practices, Political Economy, and Power Relations to Build Healthier Food Systems," The Global Alliance for the Future of Food and IPES-Food, 2017, https://futureoffood.org/wp -content/uploads/2021/01/FoodHealthNexus_Full-Report_FINAL.pdf.

25. Heribert Hirt, "Healthy Soils for Healthy Plants for Healthy Humans: How Beneficial Microbes in the Soil, Food and Gut Are Interconnected and How Agriculture Can Contribute to Human Health," *EMBO Reports* 21, no. 8 (August 5, 2020): e51069, https://doi.org/10.15252/embr.202051069.

26. Jared Diamond, "The Worst Mistake in the History of the Human Race," *Discover Magazine*, May 1, 1999, https://www.discovermagazine.com/planet-earth/the-worst-mistake-in-the-history-of-the-human-race.

27. David R. Montgomery, *Dirt: The Erosion of Civilizations* (Berkeley, CA: University of California Press, 2012).

28. Gunnar Rundgren, "Garden Earth—Beyond Sustainability: In Defence of Farming," *Garden Earth—Beyond Sustainability* (blog), June 2, 2022, https://gardenearth.blogspot.com/2022/06/in-defence-of-farming.html.

29. "Traditional Diets," *Oldways*, accessed May 8, 2024, https://oldwayspt.org/traditional-diets.

30. Avik Ray, Rajasri Ray, and E. A. Sreevidya, "How Many Wild Edible Plants Do We Eat—Their Diversity, Use, and Implications for Sustainable Food System: An Exploratory Analysis in India," *Frontiers in Sustainable Food Systems* 4 (June 11, 2020): 56, https://doi.org/10.3389/fsufs.2020.00056.

31. Leena Von Hertzen, Ilkka Hanski, and Tari Haahtela, "Natural Immunity: Biodiversity Loss and Inflammatory Diseases Are Two Global Megatrends That Might Be Related," *EMBO Reports* 12, no. 11 (November 2011): 1089–93, https://doi.org/10.1038/embor.2011.195.

32. *The White/Wiphala Paper on Indigenous Peoples' Food Systems* (FAO, 2021), https://doi.org/10.4060/cb4932en, xiv.

33. Nicole Redvers et al., "Indigenous Natural and First Law in Planetary Health," *Challenges* 11, no. 2 (October 28, 2020): 29, https://doi.org/10.3390/challe11020029, 2.

34. Rene Kuppe, "Protected Areas and Indigenous Territories: A Problematic Coexistence," International Work Group for Indigenous Affairs (IWGIA), October 3, 2023, https://www.iwgia.org/en/news/5270-protected-areas-and-indigenous-territories-a-problematic-coexistence.html?utm_source=substack.

35. Kuppe.

36. Kuppe.

37. "The Amazon in Crisis: Forest Loss Threatens the Region and the Planet," World Wildlife Fund, November 8, 2022, https://www.worldwildlife.org/stories/the-amazon-in-crisis-forest-loss-threatens-the-region-and-the-planet.

38. Alexander Koch et al., "Earth System Impacts of the European Arrival and Great Dying in the Americas After 1492," *Quaternary Science Reviews* 207 (March 2019): 13–36, https://doi.org/10.1016/j.quascirev.2018.12.004.

39. M. W. Palace et al., "Ancient Amazonian Populations Left Lasting Impacts on Forest Structure," *Ecosphere* 8, no. 12 (December 2017): e02035, https://doi.org/10.1002/ecs2.2035.

40. "The Amazon in Crisis."

41. Rodrigo Perez Ortega, "Ancient Amazonians Created Mysterious 'Dark Earth' on Purpose," *Science*, September 20, 2023, https://www.science.org/content /article/ancient-amazonians-created-mysterious-dark-earth-purpose.

42. Kate Evans, "The Nutrient-Rich Legacy in the Amazon's Dark Earths," *Eos*, March 23, 2022, http://eos.org/features/the-nutrient-rich-legacy-in-the-amazons -dark-earths.

43. Victoria Frausin et al., "'God Made the Soil, but We Made It Fertile': Gender, Knowledge, and Practice in the Formation and Use of African Dark Earths in Liberia and Sierra Leone," *Human Ecology* 42, no. 5 (October 2014): 695–710 at 695, https:// doi.org/10.1007/s10745-014-9686-0.

44. "Biochar," Regeneration.org, accessed May 10, 2024, https://regeneration .org/nexus/biochar.

45. "Soil & Water Benefits of Biochar," US Biochar Initiative (USBI), accessed May 10, 2024, https://biochar-us.org/soil-water-benefits-biochar.

46. Lyla June, accessed May 10, 2024, https://www.lylajune.com.

47. Lyla June Johnston, "Architects of Abundance: Indigenous Regenerative Food and Land Management Systems and the Excavation of Hidden History" (University of Alaska Fairbanks, 2022), https://scholarworks.alaska.edu/handle/11122 /13122.

48. Johnston, 283.

49. Johnston, 283.

50. Johnston, 286.

51. Johnston, 286.

52. Vandana Shiva, "Cocreating Responsible Food and Agriculture Systems," in A. Kassam & L. Kassam (eds.), *Rethinking Food and Agriculture* (Elsevier, 2021), 413–18, https://doi.org/10.1016/B978-0-12-816410-5.00019-0, 417.

53. Johnston, 264.

54. Johnston, 264.

55. Johnston, 264.

56. Bryony Sands et al., "Moving Towards an Anti-Colonial Definition for Regenerative Agriculture," *Agriculture and Human Values* 40, no. 4 (December 2023): 1697–716 at 1710, https://doi.org/10.1007/s10460-023-10429-3.

57. IPES-Food, "From Uniformity to Diversity: A Paradigm Shift from Industrial Agriculture to Diversified Agroecological Systems," International Panel of Experts on Sustainable Food Systems, 2016, p. 3, www.ipes-food.org.

58. Rattan Lal, "Regenerative Agriculture for Food and Climate," *Journal of Soil and Water Conservation* 75, no. 5 (2020): 123A–124A at 1A, https://doi.org/10.2489 /jswc.2020.0620A.

59. "Why Regenerative Organic?," Regenerative Organic Certified, accessed August 26, 2024, https://regenorganic.org/why-regenerative-organic/.

60. Andre Leu, "The Definition of Regenerative Agriculture," *Regeneration International* (blog), December 22, 2023, https://regenerationinternational.org/2023/12/22/the-definition-of-regenerative-agriculture/.

61. "America's First Bioethicist: Aldo Leopold," Collaborative for Health & Environment, accessed May 3, 2024, https://www.healthandenvironment.org/environmental-health/social-context/history/americas-first-bioethicist-aldo-leopold.

62. C. Francis et al., "Agroecology: The Ecology of Food Systems," *Journal of Sustainable Agriculture* 22, no. 3 (July 17, 2003): 99–118 at 100, https://doi.org/10.1300/J064v22n03_10.

63. Colin Tudge, "Two Paradigms of Science—and Two Models of Science-Based Agriculture," in A. Kassam & L. Kassam (eds.), *Rethinking Food and Agriculture* (Elsevier, 2021), 165–79 at 170, https://doi.org/10.1016/B978-0-12-816410-5.00009-8.

64. Miguel A. Altieri and Clara I. Nicholls, "Agroecology and the Reconstruction of a Post-COVID-19 Agriculture," *The Journal of Peasant Studies* 47, no. 5 (July 28, 2020): 881–98, https://doi.org/10.1080/03066150.2020.1782891.

65. "Why Regenerative Agriculture?," *Regeneration International* (blog), accessed May 5, 2024, https://regenerationinternational.org/why-regenerative-agriculture/.

66. J. R. Heckman, "A History of Organic Farming: Transitions from Sir Albert Howard's War in the Soil to USDA National Organic Program," *Renewable Agriculture and Food Systems* 21, no. 3 (September 2006): 143–50, https://doi.org/10.1079/RAF2005126.

67. Steve Gliessman, "Defining Agroecology," *Agroecology and Sustainable Food Systems* 42, no. 6 (July 3, 2018): 599–600, https://doi.org/10.1080/21683565.2018.1432329.

68. "Farm like the World Depends on It," Regenerative Organic Certified, accessed May 18, 2024, https://regenorganic.org/.

69. Bobby Gill, "How Do You Define Regenerative Agriculture?," Savory Institute, May 15, 2023, https://savory.global/how-do-you-define-regenerative-agriculture/.

70. "About—SEKEM," accessed May 16, 2024, https://sekem.com/en/about/.

71. "SEKEM Agriculture," *SEKEM* (blog), accessed May 9, 2024, https://sekem.com/en/ecology/sekem-agriculture/.

72. Helmy Abouleish, "SEKEM Vision and Mission 2057," SEKEM, 2018, p. 2, https://www.sekem.com/wp-content/uploads/2018/10/SEKEM-Vision-2057_20180615-3.pdf.

73. Abouleish.

74. Biovision Foundation for Ecological Development and Global Alliance for the Future of Food, "Beacons of Hope: Accelerating Transformations to Sustainable Food Systems," Global Alliance for the Future of Food, 2019, 36, https://futureoffood.org/wp-content/uploads/2021/02/BeaconsOfHope_Report_082019.pdf.

75. "Heliopolis University—SEKEM," accessed May 16, 2024, https://sekem.com /en/cultural-life/heliopolis-university/.

76. Thoraya Seada et al., "The Future of Agriculture in Egypt: Comparative Study of Organic and Conventional Food Production Systems in Egypt," Carbon Footprint Center, 2016, https://www.sekem.com/wp-content/uploads/2016/09/the -future-of-agriculture-in-egypt.pdf.

77. Claudia C. Flores and Santiago J. Sarandón, "Limitations of Neoclassical Economics for Evaluating Sustainability of Agricultural Systems: Comparing Organic and Conventional Systems," *Journal of Sustainable Agriculture* 24, no. 2 (June 28, 2004): 77–91, https://doi.org/10.1300/J064v24n02_08.

78. Andrea Beste, "Comparing Organic, Agroecological and Regenerative Farming Part 1—Organic," *Resilience*, February 10, 2021, https://www.resilience.org /stories/2021-02-10/comparing-organic-agroecological-and-regenerative-farming -part-1-organic/.

79. ETC Group, "Who Will Feed Us? The Peasant Food Web vs. The Industrial Food Chain," 2017, https://www.etcgroup.org/sites/www.etcgroup.org/files/files/etc -whowillfeedus-english-webshare.pdf.

80. Flores and Sarandón.

81. IPES-Food.

82. Xavier Poux and Pierre-Marie Aubert, "An Agroecological Europe in 2050: Multifunctional Agriculture for Healthy Eating. Findings from the Ten Years for Agroecology (TYFA) Modelling Exercise" (Paris, FR: Iddri-AScA, 2018).

83. Giovanni Tamburini et al., "Agricultural Diversification Promotes Multiple Ecosystem Services Without Compromising Yield," *Science Advances* 6, no. 45 (November 6, 2020): eaba1715, https://doi.org/10.1126/sciadv.aba1715.

84. "Farming Systems Trial," *Rodale Institute* (blog), accessed May 13, 2024, https://rodaleinstitute.org/science/farming-systems-trial/.

85. "Farming Systems Trial."

86. "Farming Systems Trial 40-Year Report," Rodale Institute, 2022, https:// rodaleinstitute.org/wp-content/uploads/FST_40YearReport_RodaleInstitute-1.pdf.

87. IPES-Food, 56.

88. Biovision Foundation for Ecological Development and Global Alliance for the Future of Food, "Beacons of Hope: Accelerating Transformations to Sustainable Food Systems," Global Alliance for the Future of Food, 2019, p. 36, https://futureoffood .org/wp-content/uploads/2021/02/BeaconsOfHope_Report_082019.pdf.

89. Pablo Tittonell et al., "Regenerative Agriculture—Agroecology Without Politics?," *Frontiers in Sustainable Food Systems* 6 (August 2, 2022): 844261, 9, https:// doi.org/10.3389/fsufs.2022.844261.

90. Elvira Marin Irigaray, "Restoring the Landscape by Building a Movement," The 4 Returns Community Platform, April 6, 2022, https://4returns.commonland .com/landscapes/revitalizing-land-and-community-in-the-altiplano/.

91. "Reversing Desertification with Regenerative Practices," *Commonland*, accessed August 28, 2024, https://commonland.com/landscapes/reversing -desertification-with-regenerative-practices/.

92. Tittonell et al., 9.

93. Raquel Luján Soto et al., "Restoring Soil Quality of Woody Agroecosystems in Mediterranean Drylands Through Regenerative Agriculture," *Agriculture, Ecosystems & Environment* 306 (February 2021): 107191, https://doi.org/10.1016/j.agee .2020.107191.

94. "The Almond Project—Advancing Soil Health for California Agriculture," The Almond Project, accessed May 16, 2024, https://thealmondproject.com/.

95. Emily Baron Cadloff, "Almonds Are Under Threat. The Key to Saving Them Could Be in the Soil," *Modern Farmer*, July 14, 2023, https://modernfarmer.com/2023 /07/almonds-are-under-threat/.

96. Tommy L. D. Fenster, Patricia Y. Oikawa, and Jonathan G. Lundgren, "Regenerative Almond Production Systems Improve Soil Health, Biodiversity, and Profit," *Frontiers in Sustainable Food Systems* 5 (August 10, 2021): 664359, https://doi .org/10.3389/fsufs.2021.664359.

97. Julian Fulton, Michael Norton, and Fraser Shilling, "Water-Indexed Benefits and Impacts of California Almonds," *Ecological Indicators* 96 (January 2019): 711– 17, https://doi.org/10.1016/j.ecolind.2017.12.063.

98. Cadloff.

99. Cadloff.

100. Fenster, Oikawa, and Lundgren.

101. Fenster, Oikawa, and Lundgren, 18.

102. "Sikkim's State Policy on Organic Farming, India," futurepolicy.org, March 5, 2019, https://www.futurepolicy.org/healthy-ecosystems/sikkims-state -policy-on-organic-farming-and-sikkim-organic-mission-india/.

103. "Sikkim: India's First 100% Organic State Leading the Way in Sustainable Farming," *IndiaFarm*, August 6, 2023, https://indiafarm.org/sikkim-indias-first -organic-farming-state/.

104. FAO, Cirad, and Rythu Sadhikara Samstha, "Re-Thinking Food Systems in Andhra Pradesh, India: How Natural Farming Could Feed the Future," 2023, https://www.fao.org/family-farming/detail/en/c/1679556/.

105. "Community Managed Natural Farming: An Agroecology Movement Takes Root in India's Andhra Pradesh State," *Global Alliance for the Future of Food* (blog), accessed April 26, 2024, https://futureoffood.org/insights/community-managed -natural-farming-india/.

106. FAO, Cirad, and Rythu Sadhikara Samstha.

107. IPES-Food.

108. High Level Panel of Experts on Food Security and Nutrition (HLPE), "Agroecological and Other Innovative Approaches for Sustainable Agriculture and

Food Systems That Enhance Food Security and Nutrition," 2019, https://www
.globalagriculture.org/fileadmin/files/weltagrarbericht/IAASTD-Buch/01Reports
/11FAOAgroecology/HLPEAgroecologyReport.pdf.

109. Rattan Lal, "Regenerative Agriculture for Food and Climate," *Journal of Soil and Water Conservation* 75, no. 5 (2020): 123A–124A at 1A, https://doi.org/10.2489 /jswc.2020.0620A.

110. IPES-Food, "Breaking Away from Industrial Food and Farming Systems: Seven Case Studies of Agroecological Transition," 2018, 5, https://ipes-food.org /report/breaking-away-from-industrial-food-and-farming-systems/.

CHAPTER 5: AN ECOLOGICAL ECONOMY

1. Rootstock Radio, "Eric Holt-Gimenez: Our Food System Isn't Broken (It Just Never Worked Well in the First Place)," *Organic Valley*, February 19, 2018, https:// www.organicvalley.coop/blog/eric-holt-gimenez-food-first-world-hunger/.

2. Emily Kawano and Julie Matthaei, "System Change: A Basic Primer to the Solidarity Economy," *Non Profit News | Nonprofit Quarterly*, July 8, 2020, https:// nonprofitquarterly.org/system-change-a-basic-primer-to-the-solidarity-economy/.

3. Jose Luis Vivero-Pol et al., "Introduction: The Food Commons Are Coming . . . ," in J. L. Vivero-Pol, T. Ferrando, O. De Schutter, & U. Mattei, *Routledge Handbook of Food as a Commons* (London, New York: Earthscan from Routledge, 2019).

4. Mark Fisher, *Capitalist Realism: Is There No Alternative?* (Winchester, UK, Washington, USA: Zero Books, 2009).

5. Ethan Soloviev, "The End of Supply 'Chains,'" *Terra Genesis* (blog), September 19, 2020, https://medium.com/terra-genesis/the-end-of-supply-chains -doadf3c84a40.

6. "Why 'Stakeholder' Is Out of Date," Earthwork Collective, accessed June 24, 2024, https://earthworkcollective.com/why-stakeholder-is-out-of-date/.

7. Sandra Waddock, *Transforming Towards Life-Centered Economies: How Business, Government, and Civil Society Can Build a Better World* [ebook] (Business Expert Press, 2020), 12.

8. Richard M. Ebeling, "In the Beginning: The Mont Pelerin Society, 1947," American Institute for Economic Research, accessed May 30, 2024, https://www.aier .org/article/in-the-beginning-the-mont-pelerin-society-1947/.

9. Waddock, 31–32.

10. Mary Mellor, "Ecofeminist Political Economy and the Politics of Money," in A. Salleh (ed.), *Eco-Sufficiency and Global Justice: Women Write Political Ecology* (London: Pluto Press, 2009), 251–67 at 254.

11. Mario Giampietro and Kozo Mayumi, "Unraveling the Complexity of the Jevons Paradox: The Link Between Innovation, Efficiency, and Sustainability," *Frontiers in Energy Research* 6 (April 4, 2018): 26, 7–8, https://doi.org/10.3389/fenrg .2018.00026.

12. "Starbucks Fiscal 2022 Global Environmental & Social Impact Report," Starbucks, 2023, p. 33, https://stories.starbucks.com/uploads/2023/04/2022 -Starbucks-Global-Environmental-Social-Impact-Report.pdf.

13. Julie Creswell, "For Many Big Food Companies, Emissions Head in the Wrong Direction," *The New York Times*, September 22, 2023, https://www.nytimes .com/2023/09/22/business/food-companies-emissions-climate-pledges.html.

14. Creswell.

15. Joshua Farley, "The Foundations for an Ecological Economy: An Overview," in J. Farley & D. Malghan (eds.), *Beyond Uneconomic Growth: A Festschrift in Honor of Herman Daly*, vol. 2 (Burlington, VT: University of Vermont, 2016), 1–17 at 4.

16. Donella H. Meadows, Club of Rome, and Potomac Associates, eds., *The Limits to Growth: A Report for the Club of Rome's Project on the Predicament of Mankind*, 2nd ed. (New York: Universe Books, 1974), 23–24.

17. William J. Ripple et al., "Corrigendum: World Scientists' Warning of a Climate Emergency," *BioScience* 70, no. 1 (January 1, 2020): 100–100, https://doi.org /10.1093/biosci/biz152.

18. David Burch and Geoffrey Lawrence, "Towards a Third Food Regime: Behind the Transformation," *Agriculture and Human Values* 26, no. 4 (December 2009): 267–79, https://doi.org/10.1007/s10460-009-9219-4.

19. Jennifer Clapp and S. Ryan Isakson, "Risky Returns: The Implications of Financialization in the Food System," *Development and Change* 49, no. 2 (March 2018): 437–60, https://doi.org/10.1111/dech.12376.

20. Marjorie Kelly, "Breaking Up with Capitalism," *YES! Magazine*, May 7, 2024, https://www.yesmagazine.org/opinion/2024/05/07/money-wealth-democracy -capitalism.

21. Kelly.

22. Phoebe Stephens, "Social Finance Investing for a Resilient Food Future," *Sustainability* 13, no. 12 (June 8, 2021): 6512, https://doi.org/10.3390/su13126512.

23. Kristina Karlsson and Lenore Palladino, "Towards Accountable Capitalism: Remaking Corporate Law Through Stakeholder Governance," *The Harvard Law School Forum on Corporate Governance* (blog), February 11, 2019, https://corpgov.law .harvard.edu/2019/02/11/towards-accountable-capitalism-remaking-corporate-law -through-stakeholder-governance/.

24. Attributed to both Fredric Jameson and Slavoj Žižek.

25. Marjorie Kelly, "Can There Be 'Good' Corporations?—YES! Magazine Solutions Journalism," *YES! Magazine*, April 16, 2012, https://www.yesmagazine.org /issue/end-corporate-rule/2012/04/16/can-there-be-201cgood201d-corporations.

26. "Ursula Le Guin National Book Award Acceptance Speech," *American Rhetoric Online Speech Bank*, accessed May 20, 2024, https://www.americanrhetoric.com /speeches/ursulakleguinnationalbookawardsspeech.htm.

27. Rachel Carson, *Silent Spring*, New Edition (London: Penguin Classics, 2000).

28. Brian K. Obach, *Organic Struggle: The Movement for Sustainable Agriculture in the United States* (The MIT Press, 2015), https://doi.org/10.7551/mitpress /9780262029094.001.0001.

29. Philip H. Howard, *Concentration and Power in the Food System: Who Controls What We Eat?*, Revised edition, (London: Bloomsbury Academic, 2021), 133–34.

30. Howard, 133.

31. Daniel Jaffee and Philip H. Howard, "Corporate Cooptation of Organic and Fair Trade Standards," *Agriculture and Human Values* 27, no. 4 (December 2010): 387–99 at 389 https://doi.org/10.1007/s10460-009-9231-8.

32. Jaffee and Howard, 397.

33. Pablo Tittonell et al., "Regenerative Agriculture—Agroecology Without Politics?," *Frontiers in Sustainable Food Systems* 6 (August 2, 2022): 844261, 11, https://doi.org/10.3389/fsufs.2022.844261.

34. FAIRR, "The Four Labours of Regenerative Agriculture: Paving the Way towards Meaningful Commitments," September 2023, https://go.fairr.org/FAIRR _Report_The_Four_Labours_of_Regenerative_Agriculture_2023.

35. FAIRR.

36. IPES-Food, "From Uniformity to Diversity: A Paradigm Shift from Industrial Agriculture to Diversified Agroecological Systems," International Panel of Experts on Sustainable Food Systems, 2016, 58, http://www.ipes-food.org.

37. Alberto Alonso-Fradejas et al., "'Junk Agroecology': The Corporate Capture of Agroecology for a Partial Ecological Transition Without Social Justice" (Friends of the Earth International, Transnational Institute, and Crocevia, 2020), https://www .foei.org/publication/junk-agroecology/.

38. Ruchi Shroff and Carla Ramos Cortés, "The Biodiversity Paradigm: Building Resilience for Human and Environmental Health," *Development* 63, no. 2–4 (December 2020): 172–80 at 178 https://doi.org/10.1057/s41301-020-00260-2.

39. Haley Zaremba et al., "Toward a Feminist Agroecology," *Sustainability* 13, no. 20 (October 12, 2021): 11244, https://doi.org/10.3390/su132011244.

40. "Declaration of Nyéléni," Nyéléni: International Movement for Food Sovereignty, February 27, 2007, https://nyeleni.org/en/declaration-of-nyeleni/.

41. "Declaration of Nyéléni."

42. Alexander Zaitchik, "The U.S.-Mexico Tortilla War," Food and Environment Reporting Network, May 28, 2024, https://thefern.org/2024/05/the-u-s-mexico -tortilla-war/.

43. Zaitchik.

44. "Food Sovereignty, a Manifesto for the Future of Our Planet," La Via Campesina—EN, October 13, 2021, https://viacampesina.org/en/food-sovereignty-a -manifesto-for-the-future-of-our-planet-la-via-campesina/.

45. UN Human Rights Council (39th sess., 2018), "United Nations Declaration on the Rights of Peasants and Other People Working in Rural Areas: Resolution /

Adopted by the Human Rights Council on 28 September 2018," October 8, 2018, https://digitallibrary.un.org/record/1650694.

46. Alexander Zaitchik, "The U.S.-Mexico Tortilla War," Food and Environment Reporting Network, May 28, 2024, https://thefern.org/2024/05/the-u-s-mexico -tortilla-war/.

47. Gordon Conway and Samrat Singh, "Food Sovereignty in Practice: Developing Climate Resilient Food Systems," Imperial College London, Centre for Environmental Policy, 2021.

48. Haley Zaremba et al., "Toward a Feminist Agroecology," *Sustainability* 13, no. 20 (October 12, 2021): 11244, https://doi.org/10.3390/su132011244.

49. "New Economics," Global Assessment for a New Economics, University of York, accessed June 3, 2024, https://www.york.ac.uk/new-economics/new-economics/.

50. Carol Sanford, "The Regenerative Economic Shaper Perspective Paper—Part 5," *The Regenerative Economy Collaborative* (blog), June 24, 2020, https://medium .com/the-regenerative-economy-collaborative/the-regenerative-economic-shaper -perspective-paper-part-5-a739af12d9bb.

51. Sandra Waddock, *Catalyzing Transformation: Making System Change Happen* (New York, NY: Business Expert Press, 2024), 48.

52. Waddock, 48.

53. "OECD Better Life Index," accessed June 10, 2024, https://www .oecdbetterlifeindex.org/#/11111111111.

54. "Genuine Progress Indicator," Gross National Happiness USA, accessed June 10, 2024, https://gnhusa.org/genuine-progress-indicator/.

55. "Human Development Index," UNDP Human Development Reports, accessed June 10, 2024, https://hdr.undp.org/data-center/human-development -index.

56. "Gross National Happiness Index," United Nations Department of Economic and Social Affairs, accessed June 10, 2024, https://sdgs.un.org/partnerships/gross -national-happiness-index.

57. Lisa Hough-Stewart and Nick Meynen, "GDP Is a Useless Measurement. But What Should Replace It?," openDemocracy, August 23, 2022, https://www .opendemocracy.net/en/oureconomy/gdp-measurement-new-zealand-beyond -economic-growth/.

58. "Well-Being of Future Generations (Wales) Act 2015," The Future Generations Commissioner for Wales, accessed June 6, 2024, https://www.futuregenerations .wales/about-us/future-generations-act/.

59. "Well-Being of Future Generations (Wales) Act 2015," 2015 anaw 2 (2015), https://www.futuregenerations.wales/about-us/future-generations-act/.

60. "Well-being of Future Generations (Wales) Act 2015."

61. "Wales—Coordination & Alignment of Implementation Towards Wellbeing Goals," Wellbeing Economy Alliance, accessed June 4, 2024, https://weall.org

/resource/wales-coordination-alignment-of-implementation-towards-wellbeing -goals.

62. "Agriculture (Wales) Act 2023," 2023 asc 4 (2023), https://www.legislation .gov.uk/asc/2023/4/section/1.

63. Ethan Roland and Gregory Landua, "Regenerative Enterprise Part 2: The Eight Forms of Capital—Regenerative Enterprise," *Regenerative Enterprise*, accessed June 10, 2024, https://www.regenterprise.com/regenerative-enterprise-chapter-2-the -eight-forms-of-capital/.

64. "True Cost Accounting: Implementation Guidance and Inventory," *Global Alliance for the Future of Food* (blog), accessed March 31, 2024, https://futureoffood .org/insights/tca-implementation-inventory/.

65. Steven R. McGreevy et al., "Sustainable Agrifood Systems for a Post-Growth World," *Nature Sustainability* 5, no. 12 (August 4, 2022): 1011–17, https://doi.org/10 .1038/s41893-022-00933-5.

66. "Global Report on Food Crises (GRFC) 2024 | World Food Programme," April 23, 2024, https://www.wfp.org/publications/global-report-food-crises-grfc -2024. FAO, IFAD, UNICEF, WFP, WHO, "The State of Food Security and Nutrition in the World 2024," Rome, Italy, July 23, 2024, https://doi.org/10.4060/cd1254en.

67. "The Problem of Food Waste," *FoodPrint*, February 28, 2024, https:// foodprint.org/issues/the-problem-of-food-waste/.

68. Wendell Berry, "The Agrarian Standard," *Orion Magazine*, accessed June 4, 2024, https://orionmagazine.org/article/the-agrarian-standard/.

69. Berry.

70. Steven R. McGreevy et al., "Sustainable Agrifood Systems for a Post-Growth World," *Nature Sustainability* 5, no. 12 (August 4, 2022): 1011–17 at 1012, https://doi .org/10.1038/s41893-022-00933-5.

71. "The Mission," Nature Governance, accessed August 29, 2024, https://www .naturegovernance.org/about.

72. "The Mission."

73. "Earth Law Center | Aligning Our Laws with Nature's Laws," Earth Law Center, accessed June 20, 2024, https://www.earthlawcenter.org/apply-for-the -dandelion-fellowship.

74. "Faith in Nature: Nature on the Board, an Open Source Guide," Lawyers for Nature and Earth Law Center, 2022, 2, https://ecojurisprudence.org/wp-content /uploads/2022/11/Faith-In-Nature_NOTB_GUIDE.pdf.

75. "Faith in Nature," 5.

76. "Faith in Nature," 2.

77. "Faith in Nature," 9.

78. "Faith in Nature," 8.

79. "Nature on the Board," Earth Law Center, accessed May 2, 2024, https:// www.earthlawcenter.org/notb.

80. Carol Sanford, "The Regenerative Economic Shaper Perspective Paper—Part 1," *The Regenerative Economy Collaborative* (blog), September 1, 2023, https://medium .com/the-regenerative-economy-collaborative/the-regenerative-economic-shaper -perspective-paper-part-1-8cd56d77f4b0.

81. "Lessons from Indigenous Traditions and Innovation / Expert Q&A: Dr. Ronald Trosper," *Biohabitats*, 2018, https://www.biohabitats.com/newsletter /ecology-culture-and-economy-lessons-from-indigenous-traditions-and-innovation /expert-qa-ronald-trosper/.

82. "Buen Vivir: The Rights of Nature in Bolivia and Ecuador," Rapid Transition Alliance, December 2, 2018, https://rapidtransition.org/stories/the-rights-of-nature -in-bolivia-and-ecuador/.

83. Patrick Greenfield, "Plans to Mine Ecuador Forest Violate Rights of Nature, Court Rules," *The Guardian*, December 2, 2021, https://www.theguardian.com /environment/2021/dec/02/plan-to-mine-in-ecuador-forest-violate-rights-of-nature -court-rules-aoe.

84. "Buen Vivir."

85. Emily Kawano, "Imaginal Cells of the Solidarity Economy," *Nonprofit Quarterly*, September 8, 2021, https://nonprofitquarterly.org/imaginal-cells-of-the -solidarity-economy/.

86. Ferris Jabr, "How Does a Caterpillar Turn into a Butterfly?," *Scientific American*, August 10, 2012, https://www.scientificamerican.com/article/caterpillar -butterfly-metamorphosis-explainer/.

87. Kawano.

88. "Solidarity Economy Map & Directory," US Solidarity Economy Network, accessed June 4, 2024, https://ussen.org/resources/solidarity-economy -map-directory/.

89. Penn Loh and Julian Agyeman, "Urban Food Sharing and the Emerging Boston Food Solidarity Economy," *Geoforum* 99 (February 2019): 213–22, https://doi .org/10.1016/j.geoforum.2018.08.017.

90. Emily Kawano and Julie Matthaei, "System Change: A Basic Primer to the Solidarity Economy," *Non Profit News / Nonprofit Quarterly*, July 8, 2020, https://nonprofitquarterly.org/system-change-a-basic-primer-to-the-solidarity -economy/.

91. "The Solidarity Economy," New Economy Coalition, accessed June 4, 2024, https://neweconomy.net/solidarity-economy/.

92. "The People's Network for Land & Liberation," ArcGIS StoryMaps, May 3, 2024, https://storymaps.arcgis.com/stories/65a2231d95d641e9b1586063894ca09f.

93. Kawano.

94. Kelly.

95. Sanford.

96. Kelly.

97. Rachel Keidan, "Up in the County: From Spuds to Grains," Maine Farmland Trust, August 27, 2018, https://www.mainefarmlandtrust.org/blogs/up -in-the-county-from-spuds-to-grains.

98. "Family Owned and Family Grown!," Aurora Mills and Farm, accessed June 8, 2024, https://auroramillsandfarm.com/the-farm.

99. Laurie Schreiber, "Going with the Grain: Central Maine's Grain Economy Is Expanding Across the State," *Mainebiz*, July 12, 2021, https://www.mainebiz.biz /article/going-with-the-grain-central-maines-grain-economy-is-expanding-across -the-state.

100. Catie Joyce-Bulay, "A Grainshed Rises in the Northeast," *Modern Farmer*, February 28, 2021, https://modernfarmer.com/2021/02/a-grainshed-rises-in-the -northeast/.

101. Kelly Brownell, "A Call to Invest in Agriculture's Missing Middle," *The Leading Voices in Food*, accessed June 6, 2024, https://wfpc.sanford.duke.edu /podcasts/e178-invest-in-agricultures-missing-middle/.

102. "Visioning a Resilient Food-Grade Grainshed in the Midwest Characteristics, Barriers, and Opportunities," Artisan Grain Collaborative, November 2023, 13, https://cdn.prod.website-files.com/63e3d9a0146ea26d60fe357f /665dfa15ff7a6eb343dc5383_Visioning_Resilient%20Food-Grade%20Grainshed _Nov2023.pdf.

103. "About Us," Maine Grains, June 3, 2019, https://mainegrains.com/about-us/.

104. "Maine Grains," *CEI*, accessed June 19, 2024, https://www.ceimaine.org /about/cei-stories/maine-grains-2/.

105. "About Us."

106. Phoebe Stephens, "Social Finance Investing for a Resilient Food Future," *Sustainability* 13, no. 12 (June 8, 2021): 6512, https://doi.org/10.3390/su13126512.

107. "Resilient Food Systems Infrastructure Program," USDA Agricultural Marketing Service, accessed June 17, 2024, https://www.ams.usda.gov/services /grants/rfsi.

108. Global Alliance for the Future of Food, "Cultivating Change: A Collaborative Philanthropic Initiative to Accelerate and Scale Agroecology and Regenerative Approaches," 2024, https://futureoffood.org/wp-content/uploads /2024/05/GA_CultivatingChange_Report_052124.pdf.

109. Stephens.

110. Doug Petry et al., "Cultivating Farmer Prosperity: Investing in Regenerative Agriculture," Boston Consulting Group, OP2B, WBCSD, May 2023, https://www .wbcsd.org/contentwbc/download/16321/233420/1.

111. Brownell.

112. "Maine Grains."

113. "Beetcoin and the Slow Money Movement," *Beetcoin*, accessed May 20, 2024, https://beetcoin.org/movement.

114. "Local Groups," Slow Money Institute, accessed June 20, 2024, https:// slowmoney.org/local-groups/.

115. "Slow Money."

116. "Beetcoin—The World's First Non-Crypto, Non-Currency," *Beetcoin*, accessed June 12, 2024, https://beetcoin.org/.

117. "Beetcoin and the Slow Money Movement."

118. "SOIL Boulder," accessed June 20, 2024, https://soilboulder.org/ about.

119. Catie Joyce-Bulay, "A Grainshed Rises in the Northeast," *Modern Farmer*, February 28, 2021, https://modernfarmer.com/2021/02/a-grainshed-rises-in-the -northeast/.

120. "Artisan Grain Collaborative," accessed June 8, 2024, https://www .graincollaborative.com/about-us.

121. "About the Regional Grain Movement in the Northwest," Cascadia Grains, accessed June 17, 2024, https://www.cascadiagrains.com/about-us.

122. "Grains in the Midwest," Artisan Grain Collaborative, accessed June 8, 2024, https://www.graincollaborative.com/learn.

123. "Grains in the Midwest."

124. Laurie Schreiber, "Going with the Grain: Central Maine's Grain Economy Is Expanding Across the State," *Mainebiz*, July 12, 2021, https://www.mainebiz.biz /article/going-with-the-grain-central-maines-grain-economy-is-expanding-across -the-state.

125. Catie Joyce-Bulay, "A Grainshed Rises in the Northeast," *Modern Farmer*, February 28, 2021, https://modernfarmer.com/2021/02/a-grainshed-rises-in-the -northeast/.

126. Dawn Thilmany et al., "Local Food Supply Chain Dynamics and Resilience During COVID-19," *Applied Economic Perspectives and Policy* 43, no. 1 (March 2021): 86–104, https://doi.org/10.1002/aepp.13121.

127. "Maine Grains."

128. Audre Lorde, "The Master's Tools Will Never Dismantle the Master's House," *Genius*, accessed June 18, 2024, https://genius.com/Audre-lorde-the-masters -tools-will-never-dismantle-the-masters-house-annotated.

129. John Fullerton, "Reimagining Finance: The Path Toward Genuine Financial Reform," Capital Institute, n.d., 18, https://d11n7da8rpqbjy.cloudfront.net/nrhythm /31073624_1718222528538Finance_For_A_Regenerative_World_Whitepaper_3_1 -compressed.pdf?lid=12205.

130. "Vision," *RSF Social Finance*, accessed June 19, 2024, https://rsfsocialfinance .org/vision/.

131. "How Regenerative Finance Can Build Community Wealth," *RSF Social Finance*, April 24, 2024, https://rsfsocialfinance.org/2024/04/24/how-regenerative -finance-can-build-community-wealth/.

132. "Integrated Capital," *RSF Social Finance*, accessed May 20, 2024, https://rsfsocialfinance.org/vision/how-we-work/integrated-capital/.

133. "How Regenerative Finance Can Build Community Wealth."

134. "Manifesto," Mad Capital, accessed June 19, 2024, https://madcapital.com/about/manifesto.

135. "The EFOD Fund," EFOD, June 12, 2021, https://efod.org/fund/.

136. "Detroit Black Community Food Sovereignty Network," DBCFSN, accessed June 21, 2024, https://www.dbcfsn.org.

137. "Black Food Sovereignty Coalition," Black Food Sovereignty Coalition, accessed June 21, 2024, https://blackfoodnw.org.

138. "Boston Farms Community Land Trust," Boston Farms, accessed June 21, 2024, https://www.bostonfarms.org.

139. "Oakland Bloom—Empowering Refugee and Immigrant Chefs Through Food Entrepreneurship," *Oakland Bloom*, accessed June 21, 2024, https://oaklandbloom.org/.

140. "Celebrating the First Round of EFOD Fund Grantees," EFOD, October 19, 2021, https://efod.org/2021/10/19/celebrating-the-first-round-of-efod-fund-grantees/.

141. Kelly.

142. "Freedom Farms JXN," Freedom Farms Urban Farming Cooperative, September 15, 2022, https://www.freedomfarmsjxn.com/.

143. "Cooperation Jackson," accessed June 10, 2024, https://cooperationjackson.org/.

144. Kelly.

145. "Organically Grown Company Pioneers Groundbreaking Ownership Structure to Maintain Mission & Independence in Perpetuity," *RSF Social Finance*, July 9, 2018, https://rsfsocialfinance.org/2018/07/09/organically-grown-company-pioneers-groundbreaking-ownership-structure-to-maintain-mission-independence-in-perpetuity/.

146. Sarah Joannides, "Organically Grown Company's Journey to Trust Ownership—and Beyond," Provender Alliance, December 19, 2019, https://provender.org/organically-grown-companys-journey-to-trust-ownership-and-beyond/.

147. Joannides.

148. "Organically Grown Company's Groundbreaking Move to New Ownership Structure," *RSF Social Finance*, October 31, 2018, http://rsfsocialfinance.org/2018/10/31/organically-grown-companys-groundbreaking-move-to-new-ownership-structure/.

149. "Organically Grown Company Pioneers Groundbreaking Ownership Structure to Maintain Mission & Independence in Perpetuity."

150. "Purpose Trust," Organically Grown Company, accessed May 23, 2024, https://www.organicgrown.com/purpose-trust.

151. "Organically Grown Company Pioneers Groundbreaking Ownership Structure to Maintain Mission & Independence in Perpetuity."

152. Joannides.

153. "Our Story & Mission," Hummingbird Wholesale, accessed June 4, 2024, https://hummingbirdwholesale.com/pages/our-story-mission.

154. Dania Francis et al., "How the Government Helped White Americans Steal Black Farmland," *The New Republic*, May 5, 2022, https://newrepublic.com/article /166276/black-farm-land-lost-20th-century-billions.

155. Holly Rippon-Butler, "Land Policy: Towards a More Equitable Farming Future," National Young Farmers Association, 2020, https://www.youngfarmers.org /land/wp-content/uploads/2020/11/LandPolicyReport.pdf.

156. Sophie Ackoff et al., "Building a Future with Farmers 2022: Results and Recommendations from the National Young Farmer Survey," National Young Farmers Coalition, 2022, 10, https://www.youngfarmers.org/wp-content/uploads /2022/11/2022nationalsurveyreport.pdf.

157. Ackoff et al., 10.

158. Mitch Hunter et al., "Farms Under Threat 2040: Choosing an Abundant Future," Washington, DC: American Farmland Trust, 2022, p. 3, https:// farmlandinfo.org/wp-content/uploads/sites/2/2022/08/AFT_FUT_Abundant -Future-7_29_22-WEB.pdf.

159. Hunter et al., 43.

160. Hunter et al., ii.

161. Hunter et al., iii.

162. Hunter et al., 3.

163. "The Sogorea Te Land Trust: Led by Urban Indigenous Women," The Sogorea Te Land Trust, March 13, 2024, https://sogoreate-landtrust.org/.

164. Abril Castro and Caius Z. Willingham, "Progressive Governance Can Turn the Tide for Black Farmers," Center for American Progress, April 3, 2019, https:// www.americanprogress.org/article/progressive-governance-can-turn-tide-black -farmers/.

165. Holly Rippon-Butler, "Land Policy: Towards a More Equitable Farming Future," National Young Farmers Association, 2020, p. 24, https://www .youngfarmers.org/land/wp-content/uploads/2020/11/LandPolicyReport.pdf.

166. Donna Bransford and Jocelyn Wong, "How Returning Land Can Build Power and Advance Healing Justice," *Non Profit News | Nonprofit Quarterly*, December 12, 2023, https://nonprofitquarterly.org/how-returning-land-can-build -power-and-advance-healing-justice/.

167. "Jubilee Justice," *Jubilee Justice*, accessed June 23, 2024, https://www .jubileejustice.org.

168. Trisha Gopal, "On a Former Cotton Plantation in Louisiana, a Black Farmers' Cooperative Is Reclaiming the Land and Money Their Ancestors Lost," *Business Insider*, October 7, 2023, https://www.businessinsider.com/black-farmers -jubilee-justice-louisiana-regenerative-farming-2023-10.

169. "The Jubilee Justice Black Farmers' Rice Project," *Jubilee Justice*, accessed June 10, 2024, https://www.jubileejustice.org/sri-rice.

170. Laura Flanders, "Jubilee Justice Regenerative Farming: Tackling Racism with Rice," *The Nation*, July 17, 2023, https://www.thenation.com/article/society /black-farmers-jubilee-justice-rice/.

171. Trisha Gopal, "On a Former Cotton Plantation in Louisiana, a Black Farmers' Cooperative Is Reclaiming the Land and Money Their Ancestors Lost," *Business Insider*, October 7, 2023, https://www.businessinsider.com/black-farmers-jubilee -justice-louisiana-regenerative-farming-2023-10.

172. Bransford and Wong.

173. "About Agrarian Trust," *Agrarian Trust* (blog), accessed June 10, 2024, https://www.agrariantrust.org/about/.

174. "Little Jubba Central Maine Agrarian Commons," The Somali Bantu Community Association, accessed June 6, 2024, https://somalibantumaine.org/little -jubba-central-maine-agrarian-commons/.

175. "Liberation Farms," The Somali Bantu Community Association, accessed June 22, 2024, https://somalibantumaine.org/liberation-farms/.

176. "Liberation Farms."

177. "Somali Bantu Community Association Finds Dream Farm—Little Jubba Central Maine AC," Agrarian Trust, June 23, 2020, https://www.agrariantrust.org /little-jubba-central-maine-ac-dream-farm/.

CHAPTER 6: A NEW LENS ON TECHNOLOGY, SCIENCE, AND INNOVATION

1. Robin Wall Kimmerer, "Corn Tastes Better on the Honor System," *Emergence Magazine*, accessed June 24, 2024, https://emergencemagazine.org/feature/corn -tastes-better/.

2. Kimmerer.

3. "2019 Tufts Food Systems Symposium Program—Food[at]Tufts," *Food[at]Tufts* (blog), 2019, https://sites.tufts.edu/foodattufts/tufts-food-system-symposium/2019 -tufts-food-systems-symposium/program/.

4. Pamela C. Ronald and Raoul W. Adamchak, *Tomorrow's Table: Organic Farming, Genetics, and the Future of Food* (New York, NY: Oxford University Press, 2008), 167.

5. The Consilience Project, "Technology Is Not Values Neutral: Ending the Reign of Nihilistic Design," Civilization Research Institute, June 26, 2022, https:// consilienceproject.org/technology-is-not-values-neutral-ending-the-reign-of -nihilistic-design-2/.

6. Lewis Mumford, *Technics and Civilization* (Hardcott, 1934); Langdon Winner, *Autonomous Technology* (MIT Press, 1978).

7. The Consilience Project.

8. David Christian Rose and Jason Chilvers, "Agriculture 4.0: Broadening Responsible Innovation in an Era of Smart Farming," *Frontiers in Sustainable Food Systems* 2 (December 21, 2018): 87, https://doi.org/10.3389/fsufs.2018.00087.

9. Patrick Baur and Alastair Iles, "Replacing Humans with Machines: A Historical Look at Technology Politics in California Agriculture," *Agriculture and Human Values* 40, no. 1 (March 2023): 113–40, https://doi.org/10.1007/s10460-022 -10341-2.

10. Simone Van Der Burg, Marc-Jeroen Bogaardt, and Sjaak Wolfert, "Ethics of Smart Farming: Current Questions and Directions for Responsible Innovation towards the Future," *NJAS: Wageningen Journal of Life Sciences* 90–91, no. 1 (December 1, 2019): 1–10, https://doi.org/10.1016/j.njas.2019.01.001.

11. Clemens Driessen and Leonie F. M. Heutinck, "Cows Desiring to Be Milked? Milking Robots and the Co-Evolution of Ethics and Technology on Dutch Dairy Farms," *Agriculture and Human Values* 32, no. 1 (March 2015): 3–20, https://doi.org /10.1007/s10460-014-9515-5.

12. Driessen and Heutinck.

13. Louisa Dahmani and Véronique D. Bohbot, "Habitual Use of GPS Negatively Impacts Spatial Memory During Self-Guided Navigation," *Scientific Reports* 10, no. 1 (April 14, 2020): 6310, https://doi.org/10.1038/s41598-020-62877-0.

14. A Growing Culture and ETC Group, "Politics of Technology," 2023, accessed March 18, 2025, https://www.etcgroup.org/sites/www.etcgroup.org/files/files/politics _of_technology_en_v2.pdf.

15. IPES-Food, "From Uniformity to Diversity: A Paradigm Shift from Industrial Agriculture to Diversified Agroecological Systems," International Panel of Experts on Sustainable Food Systems, 2016, https://ipes-food.org/report/from-uniformity-to -diversity/.

16. Maywa Montenegro De Wit, "Can Agroecology and CRISPR Mix? The Politics of Complementarity and Moving Toward Technology Sovereignty," *Agriculture and Human Values* 39, no. 2 (June 2022): 733–55 at 740, https://doi.org /10.1007/s10460-021-10284-0.

17. "UK Government Approves First-Ever Genetically Edited Wheat Field Trial," Rothamsted Research, 2021, https://www.rothamsted.ac.uk/news/genome-edited -wheat-field-trial-gets-go-ahead-uk-government.

18. Tanya Y. Curtis et al., "Contrasting Gene Expression Patterns in Grain of High and Low Asparagine Wheat Genotypes in Response to Sulphur Supply," *BMC Genomics* 20, no. 1 (December 2019): 628, https://doi.org/10.1186/s12864-019 -5991-8.

19. Joseph Oddy et al., "Stress, Nutrients and Genotype: Understanding and Managing Asparagine Accumulation in Wheat Grain," *CABI Agriculture and Bioscience* 1, no. 1 (December 2020): 10, https://doi.org/10.1186/s43170-020-00010-x.

20. Rebecca Mackelprang, "Organic Farming with Gene Editing: An Oxymoron or a Tool for Sustainable Agriculture?," *The Conversation*, October 10, 2018, http://theconversation.com/organic-farming-with-gene-editing-an-oxymoron-or-a-tool-for-sustainable-agriculture-101585.

21. Montenegro De Wit, 740.

22. "Politics of Technology."

23. The Consilience Project.

24. Jack Stilgoe, Richard Owen, and Phil Macnaghten, "Developing a Framework for Responsible Innovation," *Research Policy* 42, no. 9 (November 2013): 1568–80, https://doi.org/10.1016/j.respol.2013.05.008.

25. Stilgoe, Owen, and Macnaghten.

26. Maaz Gardezi et al., "In Pursuit of Responsible Innovation for Precision Agriculture Technologies," *Journal of Responsible Innovation* 9, no. 2 (May 4, 2022): 224–47, https://doi.org/10.1080/23299460.2022.2071668.

27. Stilgoe, Owen, and Macnaghten, Table 1.

28. Stilgoe, Owen, and Macnaghten.

29. Philip Loring et al., "Solving the Great Food Puzzle: Right Innovation, Right Impact, Right Place" (Gland, Switzerland: WWF, 2023), 28, https://wwfint.awsassets.panda.org/downloads/solving-the-great-food-puzzle-right-innovation--right-impact--right-place.pdf.

30. Loring et al., 28.

31. The Consilience Project.

32. Klerkx and Rose, 5.

33. Loring et al., 21.

34. Baur and Iles, 116.

35. David Christian Rose et al., "The Old, the New, or the Old Made New? Everyday Counter-Narratives of the So-Called Fourth Agricultural Revolution," *Agriculture and Human Values* 40, no. 2 (June 2023): 423–39, https://doi.org/10.1007/s10460-022-10374-7.

36. Michel P. Pimbert and Boukary Barry, "Let the People Decide: Citizen Deliberation on the Role of GMOs in Mali's Agriculture," *Agriculture and Human Values* 38, no. 4 (December 2021): 1097–1122, https://doi.org/10.1007/s10460-021-10221-1.

37. "Living Laboratories Initiative," Agriculture and Agri-Food Canada, March 15, 2019, https://agriculture.canada.ca/en/science/living-laboratories-initiative.

38. "Government of Canada Launches Nine New Living Labs: Collaborative on-Farm Solutions to Combat Climate Change in Agriculture," News releases, Agriculture and Agri-Food Canada, July 14, 2022, https://www.canada.ca/en/agriculture-agri-food/news/2022/07/government-of-canada-launches-nine-new-living-labs-collaborative-on-farm-solutions-to-combat-climate-change-in-agriculture.html.

39. "Diversity by Design: Emergent Agricultural Technologies for Small-Scale Farming," Science and Society Collective, accessed June 27, 2024, https://scienceandsocietycollective.com/diversity-by-design-emergent-agricultural-technologies-for-small-scale-farming/.

40. "Semillero de Ideas | Nursery of Ideas," Semillero de Ideas | Nursery of Ideas, accessed June 28, 2024, https://www.semilleroideas.org/

41. Erik Nicholson and Alexia Estrada, "Op-Ed: Want an Agtech Revolution? Center Farmworkers' Expertise," *Civil Eats*, January 23, 2023, https://civileats.com/2023/01/23/op-ed-want-an-agtech-revolution-center-farmworkers-expertise/.

42. Nicholson and Estrada.

43. Nicholson and Estrada.

44. Robin Wall Kimmerer, "A Letter from Indigenous Scientists in Support of the March for Science," *Milkweed*, April 21, 2017, https://milkweed.org/blog/a-letter-from-indigenous-scientists-in-support-of-the-march-for-science.

45. Fikret Berkes, *Sacred Ecology*, 4th ed. (Routledge, 2017), 8, https://doi.org/10.4324/9781315114644.

46. Harriet V. Kuhnlein et al., eds., *Indigenous Peoples' Food Systems: The Many Dimensions of Culture, Diversity and Environment for Nutrition and Health* (Rome: Food and Agriculture Organization of the United Nations, 2009).

47. Michael Wassegijig Price, "Wild Rice and the Anishinaabe Scientist," *Tribal College Journal of American Indian Higher Education* (blog), February 25, 2013, https://tribalcollegejournal.org/wild-rice-anishinaabe-scientist/.

48. Montenegro De Wit, "Can Agroecology and CRISPR Mix? The Politics of Complementarity and Moving Toward Technology Sovereignty," 737.

49. Jessica Milgroom, "Wild Rice and the Ojibwe," *MNopedia*, July 20, 2020, https://www.mnopedia.org/thing/wild-rice-and-ojibwe.

50. Laura Matson et al., "Transforming Research and Relationships Through Collaborative Tribal-University Partnerships on Manoomin (Wild Rice)," *Environmental Science & Policy* 115 (January 2021): 108–15, https://doi.org/10.1016/j.envsci.2020.10.010.

51. Nancy Averett, "The Future of Wild Rice May Depend on an Unlikely Alliance," Food and Environment Reporting Network, February 15, 2023, https://thefern.org/2023/02/the-future-of-wild-rice-may-depend-on-an-unlikely-alliance/.

52. Aurelien Bouayad, "Wild Rice Protectors: An Ojibwe Odyssey," *Environmental Law Review* 22, no. 1 (March 2020): 25–42 at 32, https://doi.org/10.1177/1461452920912909.

53. Winona LaDuke, "Wild Rice and Ethics," *Cultural Survival*, May 7, 2010, https://www.culturalsurvival.org/publications/cultural-survival-quarterly/wild-rice-and-ethics.

54. Winona LaDuke, "The Political Economy of Wild Rice Indigenous Heritage and University Research," *The Multinational Monitor* 25, no. 4 (April 2004), https://www.multinationalmonitor.org/mm2004/04012004/april04corp4.html.

55. Bouayad.

56. Ken Foster, "Hybrid Wild Rice Production Utilizing Cytoplasmic-Genetic Male Sterility System," United States US5773680A, filed April 28, 1995, and issued June 30, 1998, https://patents.google.com/patent/US5773680A/en.

57. Amanda Raster and Christina Gish Hill, "The Dispute over Wild Rice: An Investigation of Treaty Agreements and Ojibwe Food Sovereignty," *Agriculture and Human Values* 34, no. 2 (June 2017): 267–81, https://doi.org/10.1007/s10460-016-9703-6.

58. LaDuke, "The Political Economy of Wild Rice Indigenous Heritage and University Research."

59. Gracie Stockton, "Q&A with David Biesboer, Wild Rice Researcher Who's Fatalistic About Its Future," *Minnesota Reformer* (blog), April 2, 2021, https://minnesotareformer.com/2021/04/02/qa-with-david-biesboer-wild-rice-researcher-whos-fatalistic-about-its-future/.

60. Matson et al.

61. Nancy Averett, "The Future of Wild Rice May Depend on an Unlikely Alliance," Food and Environment Reporting Network, February 15, 2023, https://thefern.org/2023/02/the-future-of-wild-rice-may-depend-on-an-unlikely-alliance/.

62. Benedict E. Singleton et al., "Toward Productive Complicity: Applying 'Traditional Ecological Knowledge' in Environmental Science," *The Anthropocene Review* 10, no. 2 (August 2023): 393–414, https://doi.org/10.1177/20530196211057026.

63. Averett.

64. Cheryl Bartlett, Murdena Marshall, and Albert Marshall, "Two-Eyed Seeing and Other Lessons Learned Within a Co-Learning Journey of Bringing Together Indigenous and Mainstream Knowledges and Ways of Knowing," *Journal of Environmental Studies and Sciences* 2, no. 4 (November 2012): 331–40 at 335, https://doi.org/10.1007/s13412-012-0086-8.

65. Andrea J. Reid et al., "'Two-Eyed Seeing': An Indigenous Framework to Transform Fisheries Research and Management," *Fish and Fisheries* 22, no. 2 (March 2021): 243–61, https://doi.org/10.1111/faf.12516.

66. Samuel Pironon et al., "Potential Adaptive Strategies for 29 Sub-Saharan Crops Under Future Climate Change," *Nature Climate Change* 9, no. 10 (October 2019): 758–63, https://doi.org/10.1038/s41558-019-0585-7.

67. Aoife Cantwell-Jones et al., "Global Plant Diversity as a Reservoir of Micronutrients for Humanity," *Nature Plants* 8, no. 3 (February 24, 2022): 225–32, https://doi.org/10.1038/s41477-022-01100-6.

68. Alexandre Antonelli, "Indigenous Knowledge Is Key to Sustainable Food Systems," *Nature* 613, no. 7943 (January 12, 2023): 239–42, https://doi.org/10.1038/d41586-023-00021-4.

69. Antonelli.

70. "Teff Patents Declared Invalid, 'Great News' for Ethiopia," *Kluwer Patent Blog*, February 11, 2019, https://patentblog.kluweriplaw.com/2019/02/12/teff-patents-declared-invalid-great-news-for-ethiopia/.

71. Matson et al.

72. Antonelli.

73. "African BioGenome Project—Genomics in the Service of Conservation and Improvement of African Biological Diversity."

74. ThankGod Echezona Ebenezer et al., "Africa: Sequence 100,000 Species to Safeguard Biodiversity," *Nature* 603, no. 7901 (March 17, 2022): 388–92, https://doi.org/10.1038/d41586-022-00712-4.

75. Dhanya Vijayan et al., "Indigenous Knowledge in Food System Transformations," *Communications Earth & Environment* 3, no. 1 (September 17, 2022): 213, https://doi.org/10.1038/s43247-022-00543-1.

76. "White House Releases First-of-a-Kind Indigenous Knowledge Guidance for Federal Agencies," The White House, December 1, 2022, https://www.whitehouse.gov/ceq/news-updates/2022/12/01/white-house-releases-first-of-a-kind-indigenous-knowledge-guidance-for-federal-agencies/.

77. Sandra Díaz et al., "The IPBES Conceptual Framework—Connecting Nature and People," *Current Opinion in Environmental Sustainability* 14 (June 2015): 1–16, https://doi.org/10.1016/j.cosust.2014.11.002.

78. Kaitlin Almack et al., "Building Trust Through the Two-Eyed Seeing Approach to Joint Fisheries Research," *Journal of Great Lakes Research* 49 (June 2023): S46–57, https://doi.org/10.1016/j.jglr.2022.11.005.

79. Reid et al.

80. Almack et al., S49.

81. Almack et al., S55.

82. Almack et al., S49.

83. Almack et al., S55.

84. William Nikolakis and Ngaio Hotte, "Implementing 'Ethical Space': An Exploratory Study of Indigenous-Conservation Partnerships," *Conservation Science and Practice* 4, no. 1 (January 2022): e580, https://doi.org/10.1111/csp2.580.

85. Almack et al., S52.

86. Almack et al., S54.

87. Elise Hugus, "The Cape Cod Ark," *Edible Cape Cod*, December 16, 2014, https://ediblecapecod.ediblecommunities.com/food-thought/cape-cod-ark.

88. Steve Rose, "The New Alchemists: Could the Past Hold the Key to Sustainable Living?," *The Guardian*, September 29, 2019, http://www.theguardian.com

/lifeandstyle/ng-interactive/2019/sep/29/the-new-alchemists-could-the-past-hold
-the-key-to-sustainable-living.

89. Rose.

90. Wade Greene, "The New Alchemists," *The New York Times*, August 8, 1976, https://www.nytimes.com/1976/08/08/archives/the-new-alchemists-seeking-a
-soft-technology-to-heal-an-unhealthy.html.

91. Elise Hugus, "The Cape Cod Ark," *Edible Cape Cod*, December 16, 2014, https://ediblecapecod.ediblecommunities.com/food-thought/cape-cod-ark.

92. "What Is Biomimicry?," *Biomimicry Institute* (blog), accessed July 6, 2024, https://biomimicry.org/what-is-biomimicry/.

93. "What is Biomimicry?"

94. "AskNature," *AskNature*, accessed June 24, 2024, https://asknature.org
/about/.

95. "Perennial Grain Crop Development," The Land Institute, accessed July 6, 2024, https://landinstitute.org/our-work/perennial-crops/.

96. Emily Monaco, "GMO Technology's Role in the Future of Food Is Not What You Might Expect," *Organic Authority*, August 14, 2018, https://www.organicauthority
.com/buzz-news/genetic-technology-has-a-role-in-the-future-of-our-food-just-not-in
-the-way-you-might-expect.

97. "The State of Kernza®," Kernza®, accessed July 6, 2024, https://kernza
.org/the-state-of-kernza/.

98. Monaco.

99. "AskNature," *AskNature*, accessed June 24, 2024, https://asknature.org
/about/.

100. *Rick Haney—Soil Testing for Soil Health*, 2021, https://www.youtube.com
/watch?v=zaz2hPvirdY.

101. Soil & Climate Alliance Meeting, Arkansas, June 2022 (attended by author).

102. Andrea Vetter, "The Matrix of Convivial Technology—Assessing Technologies for Degrowth," *Journal of Cleaner Production* 197 (October 2018): 1778–86, https://doi.org/10.1016/j.jclepro.2017.02.195.

103. "farmOS," farmOS, accessed July 6, 2024, https://farmos.org/.

104. Kelly Bronson, "Looking Through a Responsible Innovation Lens at Uneven Engagements with Digital Farming," *NJAS: Wageningen Journal of Life Sciences* 90–91, no. 1 (December 1, 2019): 1–6, https://doi.org/10.1016/j.njas.2019.03
.001, 4.

105. Bronson, 4.

106. "OpenTEAM—Open Technology Ecosystem for Agricultural Management," OpenTEAM, accessed June 28, 2024, https://openteam.community/.

107. "Access Tools and Support—OpenTEAM," OpenTEAM, accessed July 6, 2024, https://openteam.community/access-tools-and-support/.

108. "Tool Library," Farm Hack, accessed June 27, 2024, https://farmhack.org/tools.

109. Chris Gaillard, "L'Atelier Paysan," L'Atelier Paysan, accessed July 6, 2024, https://www.latelierpaysan.org.

110. "Qui Sommes-Nous?," L'Atelier Paysan, accessed July 6, 2024, https://www.latelierpaysan.org/Qui-sommes-nous.

111. "The Bionutrient Meter," Bionutrient Food Association, accessed July 6, 2024, https://www.bionutrient.org/bionutrientmeter.

112. "GOAT—Gathering for Open Agricultural Technology," GOAT, accessed July 6, 2024, https://goatech.org/.

CHAPTER 7: A REGENERATIVE MINDSCAPE

1. Mateo Mier Y. Terán Giménez Cacho et al., "Bringing Agroecology to Scale: Key Drivers and Emblematic Cases," *Agroecology and Sustainable Food Systems* 42, no. 6 (July 3, 2018): 637–69, https://doi.org/10.1080/21683565.2018.1443313.

2. Michele-Lee Moore, Darcy Riddell, and Dana Vocisano, "Scaling Out, Scaling Up, Scaling Deep: Strategies of Non-Profits in Advancing Systemic Social Innovation," *Journal of Corporate Citizenship* 2015, no. 58 (June 1, 2015): 67–84, https://doi.org/10.9774/GLEAF.4700.2015.ju.00009.

3. Christine Wamsler et al., "Theoretical Foundations Report: Research and Evidence for the Potential of Consciousness Approaches and Practices to Unlock Sustainability and Systems Transformation," UNDP Conscious Food Systems Alliance (CoFSA), United Nations Development Programme, 2022, 18, https://www.contemplative-sustainable-futures.com/_files/ugd/4cc31e_143f3bc24f2c43ad94316cd50fbb8e4a.pdf.

4. United Nations Development Programme (UNDP), "Human Development Report 2020: The Next Frontier: Human Development and the Anthropocene," 2020, 398, https://hdr.undp.org/content/human-development-report-2020.

5. Hannah Gosnell, Nicholas Gill, and Michelle Voyer, "Transformational Adaptation on the Farm: Processes of Change and Persistence in Transitions to 'Climate-Smart' Regenerative Agriculture," *Global Environmental Change* 59 (November 2019): 101965, https://doi.org/10.1016/j.gloenvcha.2019.101965.

6. Mitchell Hora, "With Ray Archuleta, It's All About the Soil," *Field Work*, accessed July 12, 2024, https://www.fieldworktalk.org/episode/2022/06/29/with-ray-archuleta-its-all-about-the-soil.

7. Sheryl Karas, "Regenerative Agriculture Is a State of Mind," California State University Chico, Regenerative Agriculture and Resilient Systems, accessed March 3, 2024, https://www.csuchico.edu/regenerativeagriculture/blog/regen-ag-state-of-mind.shtml.

8. Jamie Bristow et al., "The System Within: Addressing the Inner Dimensions of Sustainability and Systems Change," The Club of Rome, 2024, 4, https://www

.clubofrome.org/wp-content/uploads/2024/05/Earth4All_Deep_Dive_Jamie
_Bristow.pdf.

9. Sara Grenni, Katriina Soini, and Lummina Geertruida Horlings, "The Inner Dimension of Sustainability Transformation: How Sense of Place and Values Can Support Sustainable Place-Shaping," *Sustainability Science* 15, no. 2 (March 2020): 411–22, https://doi.org/10.1007/s11625-019-00743-3.

10. Christoph Woiwode et al., "Inner Transformation to Sustainability as a Deep Leverage Point: Fostering New Avenues for Change through Dialogue and Reflection," *Sustainability Science* 16, no. 3 (May 2021): 841–58, https://doi.org/10.1007/s11625-020-00882-y.

11. Ethan Gordon, Federico Davila, and Chris Riedy, "Transforming Landscapes and Mindscapes Through Regenerative Agriculture," *Agriculture and Human Values* 39, no. 2 (June 2022): 809–26, https://doi.org/10.1007/s10460-021-10276-0.

12. Bristow et al., 3.

13. Bristow et al., 3.

14. Annie L. Booth, Harvey L. Jacobs, and Center for Environmental Philosophy, The University of North Texas, "Ties That Bind: Native American Beliefs as a Foundation for Environmental Consciousness," *Environmental Ethics* 12, no. 1 (1990): 27–43, https://doi.org/10.5840/enviroethics199012114.

15. Michael T. Schmitt et al., "'Indigenous' Nature Connection? A Response to Kurth, Narvaez, Kohn, and Bae (2020)," *Ecopsychology* 13, no. 1 (March 1, 2021): 64–67, https://doi.org/10.1089/eco.2020.0066.

16. Karen O'Brien and Linda Sygna, "Responding to Climate Change: The Three Spheres of Transformation," in *Proceedings of Transformation in a Changing Climate* (Oslo, Norway: University of Oslo, 2013), 16–23, http://www.cchange.no/.

17. David J. Abson et al., "Leverage Points for Sustainability Transformation," *Ambio* 46, no. 1 (February 2017): 30–39, https://doi.org/10.1007/s13280-016-0800-y.

18. Philipe Bujold and Madhuri Karak, "To Scale Behavior Change: Target Early Adopters, Then Leverage Social Proof and Social Pressure," *Behavioral Scientist*, March 1, 2021, https://behavioralscientist.org/to-scale-behavior-change-target-early-adopters-then-leverage-social-proof-and-social-pressure/.

19. Richard M. Ryan and Edward L. Deci, "Self-Determination Theory and the Facilitation of Intrinsic Motivation, Social Development, and Well-Being," *American Psychologist* 55, no. 1 (2000): 68–78, https://doi.org/10.1037/0003-066X.55.1.68.

20. Jose Luis Vivero-Pol et al., "Food as Commons: Towards a New Relationship Between the Public, the Civic and the Private," in *Routledge Handbook of Food as a Commons* (London and New York: Earthscan from Routledge, 2019), 279–80.

21. Hannah Gosnell, "Regenerative Agriculture: Putting the Heart and Soul Back in Farming," *Scientia*, 2021, https://doi.org/10.33548/SCIENTIA606.

22. Gosnell.

23. Gosnell, Gill, and Voyer, 11.

24. Mier Y. Terán Giménez Cacho et al., 645.

25. Gosnell, Gill, and Voyer, 6.

26. "Adam Chappell," *Farmer's Footprint*, accessed July 8, 2024, https://farmersfootprint.us/adam-chappell/.

27. Susan Chenery and Vanessa Gorman, "How the Regenerative Farming Movement Transformed Charles Massy's Sheep Station," *ABC News*, September 27, 2020, https://www.abc.net.au/news/2020-09-28/charlie-massy-regenerative-farming-movement/12438352.

28. Julie Snorek, Susanne Freidberg, and Geneva Smith, "Relationships of Regeneration in Great Plains Commodity Agriculture," *Agriculture and Human Values*, April 11, 2024, https://doi.org/10.1007/s10460-024-10558-3.

29. Snorek, Freidberg, and Smith, 6.

30. Snorek, Freidberg, and Smith, 6.

31. Chris Clayton, "High Input Costs Might Turn More Farmers to Regenerative Practices," *Investigate Midwest*, February 2, 2023, http://investigatemidwest.org/2023/02/02/high-input-costs-might-turn-more-farmers-to-regenerative-practices/.

32. Miller-Klugesherz and Sanderson, 6.

33. Snorek, Freidberg, and Smith, 6.

34. Miller-Klugesherz and Sanderson, 5.

35. Lee Allen, "Organic Almond Acreage Is Increasing," *Farm Progress*, June 2, 2021, https://www.farmprogress.com/tree-nuts/organic-almond-acreage-is-increasing.

36. Eve Andrews, "Where Soil Is Holy, and Climate Change Is Seldom Mentioned," *Ambrook*, October 18, 2023, https://ambrook.com/research/sustainability/climate-change-regenerative-farming-soil-Ohio.

37. Hannah Gosnell, "Regenerating Soil, Regenerating Soul: An Integral Approach to Understanding Agricultural Transformation," *Sustainability Science* 17, no. 2 (March 2022): 603–20, https://doi.org/10.1007/s11625-021-00993-0, 616.

38. Miller-Klugesherz and Sanderson, 4.

39. Mitchell Hora, "With Ray Archuleta, It's All About the Soil," *Field Work*, accessed July 12, 2024, https://www.fieldworktalk.org/episode/2022/06/29/with-ray-archuleta-its-all-about-the-soil.

40. Hora.

41. Gosnell, 605.

42. Steve Gliessman, "Balancing Nature and Agriculture," *Agroecology and Sustainable Food Systems* 46, no. 2 (February 7, 2022): 163–64, https://doi.org/10.1080/21683565.2022.2009171.

43. Miller-Klugesherz and Sanderson, 8.

44. Miller-Klugesherz and Sanderson, 7.

45. Melissa Schnyder, "Examining Value-Based Framing of Agroecology by Experts in Training Centers in Belgium, France, and Spain," *Agroecology and*

Sustainable Food Systems 46, no. 1 (January 2, 2022): 82–107 at 94, https://doi.org/10.1080/21683565.2021.1935395.

46. Madison Seymour and Sean Connelly, "Regenerative Agriculture and a More-than-Human Ethic of Care: A Relational Approach to Understanding Transformation," *Agriculture and Human Values* 40, no. 1 (March 2023): 231–44, https://doi.org/10.1007/s10460-022-10350-1.

47. Miller-Klugesherz and Sanderson, 8.

48. Snorek, Freidberg, and Smith, 10.

49. Gordon, Davila, and Riedy.

50. Miller-Klugesherz and Sanderson, 5.

51. Kelly Donati, "Going Against the Grain in the West Australian Wheatbelt," in *Beyond Global Food Supply Chains*, eds. Victoria Stead and Melinda Hinkson (Singapore: Springer Nature Singapore, 2022), 55–67 at 63, https://doi.org/10.1007/978-981-19-3155-0_5.

52. Randall Hyman, "The Power of Perception: Culture Change in Farming," *Great Lakes Protection Fund* (blog), February 22, 2019, https://glpf.org/blog/power-of-perception-culture-change-in-farming/.

53. Rob J. F. Burton et al., *The Good Farmer: Culture and Identity in Food and Agriculture*, 1st ed., Earthscan Food and Agriculture Series (New York, NY: Routledge, 2020), 1, https://doi.org/10.4324/9781315190655.

54. Burton et al.

55. Miller-Klugesherz and Sanderson, 5.

56. Liz Carlisle, "Factors Influencing Farmer Adoption of Soil Health Practices in the United States: A Narrative Review," *Agroecology and Sustainable Food Systems* 40, no. 6 (July 2, 2016): 583–613 at 604, https://doi.org/10.1080/21683565.2016.1156596.

57. Michael S. Carolan, "Social Change and the Adoption and Adaptation of Knowledge Claims: Whose Truth Do You Trust in Regard to Sustainable Agriculture?," *Agriculture and Human Values* 23, no. 3 (October 30, 2006): 325–39, https://doi.org/10.1007/s10460-006-9006-4.

58. Hannah Gosnell, "Regenerating Soil, Regenerating Soul: An Integral Approach to Understanding Agricultural Transformation," *Sustainability Science* 17, no. 2 (March 2022): 603–20 at 612, https://doi.org/10.1007/s11625-021-00993-0.

59. Miller-Klugesherz and Sanderson, 5.

60. "2022 Census of Agriculture: Share of Farmland Rented Holds Steady at 39 Percent," USDA ERS, accessed July 11, 2024, http://www.ers.usda.gov/data-products/chart-gallery/gallery/chart-detail/?chartId=109182.

61. Pranay Ranjan et al., "Understanding Barriers and Opportunities for Adoption of Conservation Practices on Rented Farmland in the US," *Land Use Policy* 80 (January 2019): 214–23, https://doi.org/10.1016/j.landusepol.2018.09.039.

62. Charles John Massy, "Transforming the Earth: A Study in the Change of Agricultural Mindscapes," The Australian National University, 2013, 231, https://doi.org/10.25911/5D74E3A700128.

63. "Diffusion of Innovation Theory," Behavioral Change Models, accessed July 19, 2024, https://sphweb.bumc.bu.edu/otlt/MPH-Modules/SB/Behavioral ChangeTheories/BehavioralChangeTheories4.html.

64. Talia Smith et al., "Accelerating the 10 Critical Transitions: Positive Tipping Points for Food and Land Use Systems Transformation," The Food and Land Use Co-alition, 2021, https:// www.foodandlandusecoalition.org/ accelerating-the-10-critical-transitionspositive-tipping-points-for-food-andland-use-systems-transformation/.

65. Christina Prell, Social Network Analysis (Los Angeles: SAGE, 2012).

66. Alina M. Udall et al., "How Do I See Myself? A Systematic Review of Identities in Pro-Environmental Behaviour Research," Journal of Consumer Behaviour 19, no. 2 (March 2020): 108–41, https://doi.org/10.1002/cb.1798.

67. Eija Soini Coe and Richard Coe, "Agroecological Transitions in the Mind," Elementa: Science of the Anthropocene 11, no. 1 (February 20, 2023): 00026, 6, https://doi.org/10.1525/elementa.2022.00026.

68. Markus Barth et al., "Collective Responses to Global Challenges: The Social Psychology of Pro-Environmental Action," Journal of Environmental Psychology 74 (April 2021): 101562, https://doi.org/10.1016/j.jenvp.2021.101562.

69. Mark Reed et al., "What Is Social Learning?," Ecology and Society 15, no. 4 (October 19, 2010), https://doi.org/10.5751/ES-03564-150401.

70. Carlisle, 599.

71. Mark Lubell, Meredith Niles, and Matthew Hoffman, "Extension 3.0: Managing Agricultural Knowledge Systems in the Network Age," Society & Natural Resources 27, no. 10 (October 3, 2014): 1089–1103, https://doi.org/10.1080/08941920.2014.933496.

72. IPES-Food, "Breaking Away from Industrial Food and Farming Systems: Seven Case Studies of Agroecological Transition," 2018, https://ipes-food.org/report/breaking-away-from-industrial-food-and-farming-systems/.

73. Gosnell, Gill, and Voyer, 11.

74. "Dr. Elaine's Soil Food Web School," Soil Food Web School, accessed July 20, 2024, https://www.soilfoodweb.com/.

75. "Regen.Ag Academy," Regen.Ag Academy, accessed July 13, 2024, https:// academyregenag.thinkific.com/.

76. "Rodale Institute Virtual Campus," Rodale Institute Virtual Campus, accessed July 14, 2024, https://courses.rodaleinstitute.org/courses/category/courses-for-farmers.

77. Clara I. Nicholls and Miguel A. Altieri, "Pathways for the Amplification of Agroecology," Agroecology and Sustainable Food Systems 42, no. 10 (November 26, 2018): 1170–93 at 1172, https://doi.org/10.1080/21683565.2018.1499578.

78. Jennifer Blesh and Steven A. Wolf, "Transitions to Agroecological Farming Systems in the Mississippi River Basin: Toward an Integrated Socioecological Analysis," *Agriculture and Human Values* 31, no. 4 (December 2014): 621–35 at 629, https://doi.org/10.1007/s10460-014-9517-3.

79. Blesh and Wolf, 629.

80. Rebecca Cross and Peter Ampt, "Exploring Agroecological Sustainability: Unearthing Innovators and Documenting a Community of Practice in Southeast Australia," *Society & Natural Resources* 30, no. 5 (May 4, 2017): 585–600, https://doi.org/10.1080/08941920.2016.1230915.

81. Snorek, Freidberg, and Smith, 2.

82. Mitchell Hora and Zach Johnson, "The Bleeding Edge: Families That Led the Conservation Charge in Washington County," *Field Work*, accessed July 12, 2024, https://www.fieldworktalk.org/episode/2021/03/31/the-bleeding-edge.

83. "Mitchell Hora," *Farmer's Footprint*, accessed July 12, 2024, https://farmersfootprint.us/mitchell-hora/.

84. "Beyond the Yield," *Soil Regen*, accessed July 20, 2024, https://www.agsoilregen.com/beyondtheyield.

85. "Fuller Field School," Fuller Field School | Circle 7 by Fuller Farms, accessed July 20, 2024, https://www.fullerfieldschool.com.

86. "Fuller Field School."

87. "Fuller Field School."

88. "No-till on the Plains | Agriculture Production Systems Modeling Nature," No-till on the Plains, accessed July 13, 2024, https://www.notill.org/.

89. "Upcoming Events," No-till on the Plains | Agriculture Production Systems Modeling Nature, accessed July 20, 2024, https://www.notill.org/upcoming-events.

90. Snorek, Freidberg, and Smith, 9.

91. "Overview," Soil & Climate Initiative, accessed July 16, 2024, https://www.soilclimateinitiative.org/about.

92. "Women, Food and Agriculture Network," Women, Food and Agriculture Network, accessed July 18, 2024, https://wfan.org.

93. Snorek, Freidberg, and Smith, 9.

94. Nicholls and Altieri, 1171.

95. "Cuba's Journey to Becoming a Global Leader in Organic Agriculture," *Global Alliance for the Future of Food* (blog), May 20, 2021, https://futureoffood.org/insights/farmer-to-farmer-agroecology-movement-cuba/.

96. Shreehari Paliath, "In Andhra Pradesh, Women Are Taking a Lead Role in the Transition to Natural, Sustainable Farming," Scroll.in, October 9, 2022, https://scroll.in/article/1034434/in-andhra-pradesh-women-are-taking-a-lead-role-in-the-transition-to-natural-sustainable-farming.

97. Samara Brock et al., "Knowledge Democratization Approaches for Food Systems Transformation," *Nature Food* 5, no. 5 (May 9, 2024): 342–45 at 344, https://doi.org/10.1038/s43016-024-00966-3.

98. Ashlesha Khadse et al., "Taking Agroecology to Scale: The Zero Budget Natural Farming Peasant Movement in Karnataka, India," *The Journal of Peasant Studies* 45, no. 1 (January 2, 2018): 192–219, https://doi.org/10.1080/03066150.2016.1276450.

99. Brock et al., 344.

100. *Andhra Pradesh to Africa: Taking Agroecology to Scale through Farmer-to-Farmer Online Exchanges*, 2023, https://www.youtube.com/watch?v=jQLLGgGAqfM.

101. *Farmers Experience the Climate Crisis First-Hand*, 2019, https://www.youtube.com/watch?v=EXBOkRwJXXw.

102. Randall Hyman, "The Power of Perception: Culture Change in Farming," *Great Lakes Protection Fund* (blog), February 22, 2019, https://glpf.org/blog/power-of-perception-culture-change-in-farming/.

103. Hyman.

104. "Grow More," National Wildlife Federation—Growing Outreach, July 30, 2024, https://growingoutreach.nwf.org/grow-more/.

105. "Conservation Champions," National Wildlife Federation—Growing Outreach, accessed July 18, 2024, https://growingoutreach.nwf.org/conservation-champions/.

106. "Conservation Champions."

107. "Lands for Life," *Rare*, accessed July 10, 2024, https://rare.org/program/lands-for-life/.

108. Bujold and Karak.

109. Philipe Bujold and Madhuri Karak, "To Scale Behavior Change: Target Early Adopters, Then Leverage Social Proof and Social Pressure," *Behavioral Scientist*, March 1, 2021, https://behavioralscientist.org/to-scale-behavior-change-target-early-adopters-then-leverage-social-proof-and-social-pressure/.

110. "Reviving Ancestral Farming Practices through Behavioral Science: How the Center for Behavior & the Environment at Rare Equipped Local Changemakers in Chiapas, Mexico," Center for Behavior & the Environment, 2020, 3, https://behavior.rare.org/wp-content/uploads/2020/10/BE.Center-Impact-Story.-Chiapas-Mexico.-9.14.pdf.

111. "Reviving Ancestral Farming Practices through Behavioral Science," 3.

112. "Reviving Ancestral Farming Practices through Behavioral Science," 3.

113. "Reviving Ancestral Farming Practices through Behavioral Science," 3.

114. "Reviving Ancestral Farming Practices through Behavioral Science," 6.

115. Nicholls and Altieri, 1189.

CHAPTER 8: REGENERATION RISING

1. Donella Meadows, "Leverage Points: Places to Intervene in a System," *The Academy for Systems Change* (blog), accessed May 23, 2024, https://donellameadows.org/archives/leverage-points-places-to-intervene-in-a-system/.

2. Daniel Christian Wahl, "Constructing a Regenerative Future," *RSA Journal*, November 19, 2021, https://www.thersa.org/rsa-journal/2021/issue-4/feature/constructing-a-regenerative-future.

3. Jenny Odell, "Notes of a Bioregional Interloper," *Open Space*, October 9, 2017, https://openspace.sfmoma.org/2017/10/notes-of-a-bioregional-interloper/.

4. "Core Principles of Bioregionalism," Cascadia Department of Bioregion, accessed August 9, 2024, https://cascadiabioregion.org/bioregionalism-core-principles.

5. "Our Theory of Change," Department of Bioregions, accessed July 24, 2024, https://deptofbioregion.org/principles/.

6. Thich Nhất Hạnh, *Interbeing*, 4th ed. (Berkeley, CA: Parallax Press, 2020).

7. Arne Naess, "The Shallow and the Deep, Long-Range Ecology Movement: A Summary," *OpenAirPhilosophy.Org*, 2005, https://openairphilosophy.org/wp-content/uploads/2018/11/OAP_Naess_Shallow_and_the_Deep.pdf.

8. "Our Bioregion," Cascadia Department of Bioregion, accessed August 9, 2024, https://cascadiamovement.org/our-bioregion/.

9. "About Cascadia," Cascadia Department of Bioregion, accessed August 7, 2024, https://cascadiabioregion.org/a-cascadia-primer.

10. "The Cascadia Movement," Cascadia Department of Bioregion, accessed August 7, 2024, https://cascadiabioregion.org/the-cascadia-movement.

11. "Basic Goals," Cascadia Department of Bioregion, accessed November 26, 2024, https://cascadiabioregion.org/basic-goals.

12. TP4D, "Territorial Approaches for Sustainable Development. White Paper for Policy Formulation and Project Implementation," 2023, https://api.ecoagriculture.org/uploads/Plaq_TP_4_D_ENG_HD_0b266a1f01.pdf.

13. Leonard Charles et al., "Where You At? A Bioregional Quiz," *Coevolution Quarterly* 32 (Winter 1981).

14. Gary Snyder, "Reinhabitation," *Manoa* 25, no. 1 (2013): 44–48, https://doi.org/10.1353/man.2013.0010.

15. Robin Wall Kimmerer, *Braiding Sweetgrass: Indigenous Wisdom, Scientific Knowledge, and the Teachings of Plants* (Minneapolis, MN: Milkweed Editions, 2020), 9.

16. Patrick Sisson, "In Age of Climate Change, Can Our Lawns Be More than Landscaping?," *Curbed*, May 1, 2019, https://archive.curbed.com/2019/5/1/18524512/landscaping-gardening-lawns-front-yards.

17. Susannah B. Lerman et al., "Humanity for Habitat: Residential Yards as an Opportunity for Biodiversity Conservation," *BioScience* 73, no. 9 (October 11, 2023): 671–89, https://doi.org/10.1093/biosci/biad085.

18. Eric Holthaus, "Lawns Are the No. 1 Irrigated 'Crop' in America. They Need to Die," *Grist*, May 2, 2019, https://grist.org/article/lawns-are-the-no-1-agricultural -crop-in-america-they-need-to-die/.

19. "Doug Tallamy," Homegrown National Park, accessed July 26, 2024, https:// homegrownnationalpark.org/doug-tallamy/.

20. "Founders," Homegrown National Park, accessed July 26, 2024, https:// homegrownnationalpark.org/founders/.

21. "About Us," Homegrown National Park, accessed July 26, 2024, https:// homegrownnationalpark.org/about-us/.

22. Dave Goulson et al., "Bee Declines Driven by Combined Stress from Parasites, Pesticides, and Lack of Flowers," *Science* 347, no. 6229 (March 27, 2015): 1255957, https://doi.org/10.1126/science.1255957.

23. "History of Food Not Lawns," August 17, 2020, https://www.foodnotlawns .com/about/.

24. Gabriel R. Valle, "The Past in the Present: What Our Ancestors Taught Us About Surviving Pandemics," *Food Ethics* 6, no. 2 (October 2021): 7, https://doi.org /10.1007/s41055-021-00088-7.

25. Dimosthenis Vasiloudis, "'Chinampas': The Ancient Aztec Floating Gardens That Hold Promise for Future Urban Agriculture," *The Archaeologist*, August 22, 2021, https://www.thearchaeologist.org/blog/chinampas-the-ancient-aztec-floating -gardens-that-hold-promise-for-future-urban-agriculture.

26. Kayo Tajima, "The Marketing of Urban Human Waste in the Early Modern Edo/Tokyo Metropolitan Area," *Environnement Urbain / Urban Environment*, 1 (September 9, 2007), https://journals.openedition.org/eue/1039.

27. Steven R. McGreevy et al., "Sustainable Agrifood Systems for a Post-Growth World," *Nature Sustainability* 5, no. 12 (August 4, 2022): 1011–17 at 1013, https://doi .org/10.1038/s41893-022-00933-5.

28. M. A. Altieri and C. I. Nicholls, "Agroecología Urbana: Diseño de Granjas Urbanas Ricas En Biodiversidad, Productivas y Resilientes," *Agro Sur* 46, no. 2 (2018): 49–60, https://doi.org/10.4206/agrosur.2018.v46n2-07.

29. Florian Thomas Payen et al., "How Much Food Can We Grow in Urban Areas? Food Production and Crop Yields of Urban Agriculture: A Meta-Analysis," *Earth's Future* 10, no. 8 (August 2022): e2022EF002748, https://doi.org/10.1029/2022EF002748.

30. "Climate Victory Gardens," Green America, accessed August 14, 2024, https://www.greenamerica.org/climate-victory-gardens.

31. Megan E. Springate, "Victory Gardens on the World War II Home Front," US National Park Service, accessed July 24, 2024, https://www.nps.gov/articles /000/victory-gardens-on-the-world-war-ii-home-front.htm.

32. "Climate Victory Gardens."

33. Valle, "The Past in the Present: What Our Ancestors Taught Us about Surviving Pandemics," 10.

34. McGreevy et al., 1013.

35. "About Organic Land Care," NOFA Organic Land Care, accessed July 30, 2024, https://nofa.organiclandcare.net/about-organic-land-care/.

36. Rachel Carson, *Silent Spring*, 40th anniversary ed., 1st Mariner Books ed. (Boston: Houghton Mifflin, 2002), 129.

37. "Our History," Slow Food, accessed August 2, 2024, https://www.slowfood.com/our-history/.

38. Hugh Campbell, "Breaking New Ground in Food Regime Theory: Corporate Environmentalism, Ecological Feedbacks and the 'Food from Somewhere' Regime?," *Agriculture and Human Values* 26, no. 4 (December 2009): 309–19, https://doi.org/10.1007/s10460-009-9215-8.

39. Hannah Ritchie, Veronika Samborska, and Max Roser, "Urbanization," *Our World in Data*, February 23, 2024, https://ourworldindata.org/urbanization.

40. "2020 Census Urban Areas Facts," United States Census Bureau, June 2023, https://www.census.gov/programs-surveys/geography/guidance/geo-areas/urban-rural/2020-ua-facts.html.

41. John H. Vandermeer, *The Ecology of Agroecosystems* (Sudbury, MA: Jones and Bartlett, 2011), 17.

42. Fritjof Capra, "The New Facts of Life," ecoliteracy.org, June 29, 2009, https://www.ecoliteracy.org/article/new-facts-life.

43. Capra.

44. Joe Fassler, "Regenerative Agriculture Needs a Reckoning," *The Counter*, May 3, 2021, https://thecounter.org/regenerative-agriculture-racial-equity-climate-change-carbon-farming-environmental-issues/.

45. Liz Carlisle et al., "Transitioning to Sustainable Agriculture Requires Growing and Sustaining an Ecologically Skilled Workforce," *Frontiers in Sustainable Food Systems* 3 (November 1, 2019): 96, https://doi.org/10.3389/fsufs.2019.00096.

46. Fassler.

47. David W. Orr, "The Effective Shape of Our Future," *Conservation Biology* 8, no. 3 (1994): 622–24.

48. Jason Bradford, "The Future Is Rural: Food Systems Adaptations to the Great Simplification" (Post Carbon Institute, 2019), https://www.postcarbon.org/publications/the-future-is-rural/.

49. Bradford, 1.

50. Bradford, 1.

51. Capra.

52. Zlatina Tsvetkova, "Scaling in the Context of a Landscape Partnership—5 Lessons from Mother Nature," *The 4 Returns Community Platform*, March 24, 2023, https://4returns.commonland.com/stories/scaling-in-the-context-of-a-landscape-partnership/.

53. Sandra Waddock, *Catalyzing Transformation: Making System Change Happen* (New York, NY: Business Expert Press, 2024), 63.

54. Michael Quinn Patton, *Blue Marble Evaluation: Premises and Principles* (New York: The Guilford Press, 2020), 157.

55. Patton, 157.

56. Margaret J. Wheatley and Deborah Frieze, "Using Emergence to Take Social Innovations to Scale," 2006, https://www.margaretwheatley.com/articles /emergence.html.

57. Vanessa Machado de Oliveira, *Hospicing Modernity: Facing Humanity's Wrongs and the Implications for Social Activism* (Berkeley, CA: North Atlantic Books, 2021).

58. Per Olsson and Michele-Lee Moore, "Transformations, Agency and Positive Tipping Points: A Resilience-Based Approach," in *Positive Tipping Points Towards Sustainability*, ed. J. David Tàbara et al. (Cham: Springer International Publishing, 2024), 59–77, https://doi.org/10.1007/978-3-031-50762-5_4.

59. Waddock, 63.

60. Adrienne M. Brown, *Emergent Strategy* (Chico, CA: AK Press, 2017).

61. Wheatley and Frieze.

62. IPES-Food, "Breaking Away from Industrial Food and Farming Systems: Seven Case Studies of Agroecological Transition," 2018, 5, https://ipes-food.org /report/breaking-away-from-industrial-food-and-farming-systems/.

63. Waddock, 63.

64. Emily Kawano, "Imaginal Cells of the Solidarity Economy," *Nonprofit Quarterly*, September 8, 2021, https://nonprofitquarterly.org/imaginal-cells-of-the -solidarity-economy/.

65. Michael Quinn Patton, "The Global Alliance Formally Adopts a Theory of Transformation," *Global Alliance for the Future of Food* (blog), accessed July 23, 2024, https://futureoffood.org/insights/theory-of-transformation/.

66. Ethan Roland Soloviev and Gregory Landua, "Levels of Regenerative Agriculture," Terra Genesis International, September 2016, 13, https://www .ethansoloviev.com/wp-content/uploads/2019/02/Levels-of-Regenerative -Agriculture.pdf.

67. Jamie Bristow et al., "The System Within: Addressing the Inner Dimensions of Sustainability and Systems Change," The Club of Rome, 2024, Earth4All: deep-dive paper 17, 11.

68. "What Is Transition? | Circular Model & Reconomy," *Transition Network*, July 28, 2016, https://transitionnetwork.org/about-the-movement/what-is -transition/.

69. "What Is a Practice of Change and Why Are They Important?," *Practising Transition*, June 7, 2024, https://practise.transitionmovement.org/what-is-a-transition -practice/.

70. Jorge Garcia-Arias, Carlos Tornel, and María Flores Gutiérrez, "Weaving a Rhizomatic Pluriverse: Allin Kawsay, the Crianza Mutua Networks, and the Global Tapestry of Alternatives," *Globalizations*, May 21, 2024, 1–21 at 2, https://doi.org/10 .1080/14747731.2024.2352942.

71. "An Introduction to the GTA," GTA, February 26, 2020, https:// globaltapestryofalternatives.org/introduction.

72. Garcia-Arias, Tornel, and Flores Gutiérrez, 2.

73. Gideon Kossoff, "Cosmopolitan Localism: The Planetary Networking of Everyday Life in Place," *Cuadernos Del Centro de Estudios de Diseño y Comunicación*, no. 73 (September 20, 2019), https://doi.org/10.18682/cdc.vi73.1037.

74. Waddock, 48.

75. "Camp Altiplano, Spain," Ecosystem Restoration Communities, accessed August 11, 2024, https://www.ecosystemrestorationcommunities.org/community /camp-altiplano-spain/.

76. Liora Adler, "Restoring the Earth, One Camp at a Time," *Resilience*, August 12, 2019, https://www.resilience.org/stories/2019-08-12/restoring-the-earth -one-camp-at-a-time/.

77. Constance Neely, Delia Catacutan, and Rob Youl, "Globalizing Local Actions: An Introduction to the Ever-Expanding Story," in D. Catacutan, C. Neely, M. Johnson, H. Poussard, & R. Youl (eds.), *Landcare: Local Action—Global Progress* (Nairobi, Kenya: World Agroforestry Centre, 2009), 5–1 2 at 7, https://apps.worldagroforestry .org/downloads/Publications/PDFS/B16017.pdf.

78. Andrew Campbell, "The Essence of Landcare-Flourishing in the Philippines," *Australian Centre for Intenational Agriucltural Research* (blog), August 28, 2018, https:// www.aciar.gov.au/media-search/blogs/essence-landcare-flourishing-philippines.

79. Ezio Manzini, "Small, Local, Open and Connected, Resilient Systems and Sustainable Qualities," *Design Observer*, February 6, 2013, http://designobserver .com/feature/small-local-open-and-connected-resilient-systems-and-sustainable -qualities/37670.

80. Sara J. Scherr et al., "Public Policy to Support Landscape and Seascape Partnerships: Meeting Sustainable Development Goals through Collaborative Territorial Action (White Paper)," EcoAgriculture Partners, GALLOP initiative, Cornell University and Columbia University, 2022, 28, https://api.ecoagriculture.org /uploads/Public_Policy_White_Paper_db85218dd1.pdf.

81. "International Model Forest Network » About," International Model Forest Network, accessed August 7, 2024, https://imfn.net/about/.

82. "Model Forests," International Model Forest Network, accessed September 2, 2024, https://imfn.net/model-forest/.

83. "Latin-American Model Forest Network," International Model Forest Network, accessed August 5, 2024, https://imfn.net/about/regional-networks/ibero -american-model-forest-network/.

84. Scherr et al., 10.

85. Brianna Van Matre, "Look Beyond the Farm to Solve Global Food System Challenges," *Agrilinks*, May 16, 2023, https://agrilinks.org/post/look-beyond-farm -solve-global-food-system-challenges.

86. L. Buck et al., "Using Integrated Landscape Management to Scale Agroforestry: Examples from Ecuador," *Sustainability Science* 15, no. 5 (September 2020): 1401–15, https://doi.org/10.1007/s11625-020-00839-1.

87. Van Matre.

88. 1000 Landscapes for 1 Billion People, "A Practical Guide to Integrated Landscape Management," EcoAgriculture Partners, on behalf of 1000 Landscapes for 1 Billion People, 2022, https://landscapes.global/wp-content/uploads/2022/12 /ILM_Practical_Guide_DEC22.pdf.

89. "UN Decade on Restoration," UN Decade on Restoration, accessed August 12, 2024, http://www.decadeonrestoration.org/node.

90. "About the UN Decade," UN Decade on Restoration, accessed August 12, 2024, http://www.decadeonrestoration.org/about-un-decade.

91. *Landscapes for Life: Approaches to Landscape Management for Sustainable Food and Agriculture* (Rome, Italy: Food and Agriculture Organization of the United Nations, 2017), https://openknowledge.fao.org/server/api/core/bitstreams/ff1e6259 -a55d-4a79-b239-81d18e243538/content.

92. "Landscapes for Life," 3.

93. Koen van Seijen, "Sara Scherr on How to Work on Landscape Scale Regeneration on 1000 Landscapes for 1 Billion People—Investing in Regenerative Agriculture," *Investing in Regenerative Agriculture*, accessed August 4, 2024, https:// investinginregenerativeagriculture.com/2022/03/15/sara-scherr-2/.

94. "Willem Ferwerda," Erasmus Research Institute of Management—ERIM, accessed July 23, 2024, https://www.erim.eur.nl/eco-transformation/about/team /willem-ferwerda/.

95. "4 Returns Framework," Commonland, accessed June 18, 2024, https:// commonland.com/4-returns-framework/.

96. Claire Patterson, "Commonland Annual Report 2023," Commonland, June 27, 2024, https://commonland.com/commonland-annual-report-2023/.

97. FAO, Cirad, and Rythu Sadhikara Samstha, "Re-Thinking Food Systems in Andhra Pradesh, India: How Natural Farming Could Feed the Future," 2023, https:// www.fao.org/family-farming/detail/en/c/1679556/.

98. "Restoring the Landscape by Building a Movement," *The 4 Returns Community Platform*, accessed August 5, 2024, https://4returns.commonland.com /landscapes/revitalizing-land-and-community-in-the-altiplano/.

99. "Manifesto for a Regenerative Territory," July 2022, https://territorioregen- erativo.com/Manifiesto-Territorio-Regenerativo-Julio-22-EN.pdf.

100. "The Commonland Annual Report 2023," Commonland, 2023, https://commonland.com/wp-content/uploads/2024/06/COM-AnnualReport-2023.pdf.

101. "Bioregional Weaving Labs," *Ashoka*, accessed July 22, 2024, https://www.ashoka.org/en/program/bioregional-weaving-labs-collective.

102. "Who We Are," 1000 Landscapes for 1 Billion People, accessed August 4, 2024, https://landscapes.global/who-we-are/.

103. "EcoAgriculture | Thousand Landscapes for One Billion People," EcoAgriculture Partners, accessed August 4, 2024, https://ecoagriculture.org/our-work/1000L.

104. Rene Kuppe, "Protected Areas and Indigenous Territories: A Problematic Coexistence," International Work Group for Indigenous Affairs (IWGIA), October 3, 2023, https://www.iwgia.org/en/news/5270-protected-areas-and-indigenous-territories-a-problematic-coexistence.html?utm_source=substack.

105. Jake M. Robinson et al., "Traditional Ecological Knowledge in Restoration Ecology: A Call to Listen Deeply, to Engage with, and Respect Indigenous Voices," *Restoration Ecology* 29, no. 4 (May 2021): e13381, 4, https://doi.org/10.1111/rec.13381.

106. FAO, IUCN CEM, and SER, "Principles for Ecosystem Restoration to Guide the United Nations Decade 2021–2030" (Rome, Italy: FAO, 2021), 10, https://cdn.ymaws.com/www.ser.org/resource/resmgr/publications/principles_for_ecosystem_res.pdf.

107. "UN Declaration on the Rights of Indigenous Peoples," A/RES/61/295 (2007), https://www.ohchr.org/en/indigenous-peoples/un-declaration-rights-indigenous-peoples.

108. Robinson et al., 4.

109. Tom Lovett, "Caring for Country in Western Australia," The 4 Returns Community Platform, January 6, 2022, https://4returns.commonland.com/landscapes/caring-for-country-in-western-australia/.

110. "Regenerating a Global Biodiversity Hotspot," Commonland, accessed July 23, 2024, https://commonland.com/landscapes/regenerating-a-global-biodiversity-hotspot/.

111. Kelly Donati, "Going Against the Grain in the West Australian Wheatbelt," in *Beyond Global Food Supply Chains*, eds. Victoria Stead and Melinda Hinkson (Singapore: Springer Nature Singapore, 2022), 55–67 at 56, https://doi.org/10.1007/978-981-19-3155-0_5.

112. C. J. A. Bradshaw, "Little Left to Lose: Deforestation and Forest Degradation in Australia Since European Colonization," *Journal of Plant Ecology* 5, no. 1 (March 1, 2012): 109–20, https://doi.org/10.1093/jpe/rtr038.

113. Donati, 56.

114. "About Wheatbelt Integrity Group," Wheatbelt Integrity Group, accessed August 5, 2024, https://www.wig.farm/about.php.

115. Lovett.

116. "Walking Together," Danjoo Koorliny, accessed August 11, 2024, https://danjookoorliny.com.au/pages/walking-together.

117. "Regenerating a Global Biodiversity Hotspot."

118. "Grocery Delivery Perth | Regenerative Fresh Produce," Dirty Clean Food, accessed August 5, 2024, https://www.dirtycleanfood.com.au/.

119. Ben Cole, "WOA's Infinity-Loop Business Model," The 4 Returns Community Platform, October 21, 2020, https://4returns.commonland.com/stories/woas-infinity-loop-business-model/.

120. "Regenerating a Global Biodiversity Hotspot."

121. James (Sa'ke'j) Youngblood Henderson, "Wild Buffalo Recovery and Ecological Restoration of the Grasslands," Centre for International Governance Innovation, June 27, 2019, https://www.cigionline.org/articles/wild-buffalo-recovery-and-ecological-restoration-grasslands/.

122. Lina Tran, "The Return of the American Bison Is an Environmental Boon—and a Logistical Mess," *Grist*, November 9, 2022, https://grist.org/indigenous/return-of-american-bison-environmental-boon-yellowstone-buffalo/.

123. Henderson.

124. Henderson.

125. "Buffalo Project," Prairie Island Indian Community, accessed August 7, 2024, https://prairieisland.org/who-we-are/our-culture/buffalo-project.

126. "Bison Bellows: The 'Buffalo' Treaty," US National Park Service, November 6, 2017, https://www.nps.gov/articles/bison-bellows-1-21-16.htm.

127. Henderson.

128. "Treaty," Buffalo Treaty, accessed August 7, 2024, https://www.buffalotreaty.com/treaty.

129. Henderson.

130. "Efforts to Restore Buffalo and Re-Establish Indigenous Lifeways Expands in Unprecedented Alliance Between Indigenous Leaders, Environmental Nonprofits," The Nature Conservancy, July 17, 2024, https://www.nature.org/en-us/newsroom/historic-alliance-supporting-buffalo-rematriation/. "Tribal Buffalo Lifeways Collaboration," accessed August 7, 2024, https://tribes.nativephilanthropy.org/tribal-buffalo.

131. "Efforts to Restore Buffalo and Re-Establish Indigenous Lifeways Expands in Unprecedented Alliance Between Indigenous Leaders, Environmental Nonprofits."

132. Tran.

133. US Department of the Interior, "Grasslands Keystone Initiative: Protecting an Integral Ecosystem," 2023, https://www.doi.gov/sites/doi.gov/files/grasslands-keystone-initiative.pdf.

134. "Central Grasslands Roadmap," Central Grasslands Roadmap, accessed August 13, 2024, https://www.grasslandsroadmap.org.

135. "Central Grasslands Roadmap."

136. "Central Grasslands Roadmap."

137. J. Boone Kauffman et al., "Livestock Use on Public Lands in the Western USA Exacerbates Climate Change: Implications for Climate Change Mitigation and Adaptation," *Environmental Management* 69, no. 6 (June 2022): 1137–52, https://doi .org/10.1007/s00267-022-01633-8.

138. Jake M. Robinson et al., "Traditional Ecological Knowledge in Restoration Ecology: A Call to Listen Deeply, to Engage with, and Respect Indigenous Voices," *Restoration Ecology* 29, no. 4 (May 2021): e13381, https://doi.org/10.1111/rec.13381.

139. Thomas Berry, *The Great Work: Our Way into the Future*, 1st ed. (New York: Bell Tower, 1999).

140. "Earth Law Center | Champions for the Rights of Nature," Earth Law Center, accessed May 2, 2024, https://www.earthlawcenter.org.

141. "Earth Jurisprudence," Wild Law Institute, accessed May 2, 2024, https:// www.wildlaw.net/earth-jurisprudence.

142. "Global Alliance for the Rights of Nature (GARN)," October 12, 2020, https://www.garn.org/.

143. "Buen Vivir: The Rights of Nature in Bolivia and Ecuador," December 2, 2018, https://rapidtransition.org/stories/the-rights-of-nature-in-bolivia-and-ecuador/.

144. Gary Brierley et al., "A Geomorphic Perspective on the Rights of the River in Aotearoa New Zealand," *River Research and Applications* 35, no. 10 (December 2019): 1640–51, https://doi.org/10.1002/rra.3343.

145. Bioneers, "Legalizing Nature's Rights: How Tribal Nations Are Leading the Fastest Growing Environmental Movement in History," *Bioneers* (blog), February 14, 2023, https://bioneers.org/legalizing-natures-rights-tribal-nations-leading-fastest -growing-environmental-movement-history/.

146. "Rights of Manoomin," Center for Democratic and Environmental Rights, accessed August 14, 2024, https://www.centerforenvironmentalrights.org/rights-of -manoomin.

147. Jon E. Hess et al., "Robust Recolonization of Pacific Lamprey Following Dam Removals," *Transactions of the American Fisheries Society* 150, no. 1 (January 2021): 56–74, https://doi.org/10.1002/tafs.10273.

148. Sandi Schwartz, Beth Styler Barry, and ELC Team, "Dam Removal 101: Why Earth Law Supports Free-Flowing Rivers," *Earth Law Center* (blog), February 28, 2024, https://www.earthlawcenter.org/blog-entries/2024/2/dam-removal-101-why -earth-law-supports-free-flowing-rivers.

149. Yurok Tribal Council, "Resolution Establishing the Rights of the Klamath River," Pub. L. No. 19–40 (2019), https://ecojurisprudence.org/wp-content/uploads /2022/02/US_Klamath_Resolution-Establishing-Rights-of-the-Klamath-River _168.pdf.

150. "Lake Erie Bill of Rights," Lake Erie Advocates, accessed August 14, 2024, https://www.lakeerieadvocates.org/lake-erie-bill-of-rights.

151. "Buen Vivir."

152. Inspired by an exercise designed by Rob Hopkins, founder of the Transition Towns Movement. Rob Hopkins, "When a Resilient Future Calls by to See If You Want to Come out to Play," *Rob Hopkins* (blog), April 21, 2020, https://www .robhopkins.net/2020/04/21/when-a-resilient-future-calls-by-to-see-if-you-want-to -come-out-to-play/.

Selected Bibliography

1000 Landscapes for 1 Billion People. "A Practical Guide to Integrated Landscape Management." Washington, DC, USA: EcoAgriculture Partners, on behalf of 1000 Landscapes for 1 Billion People, 2022. https://landscapes.global/wp-content/uploads/2022/12/ILM_Practical_Guide_DEC22.pdf.

Almack, Kaitlin, Erin S. Dunlop, Ryan Lauzon, Sidney Nadjiwon, and Alexander T. Duncan. "Building Trust Through the Two-Eyed Seeing Approach to Joint Fisheries Research." *Journal of Great Lakes Research* 49 (June 2023): S46–57. https://doi.org/10.1016/j.jglr.2022.11.005.

Alonso-Fradejas, Alberto, Lyda Fernanda Forero, Delphine Ortega-Espès, Martín Drago, and Kirtana Chandrasekaran. "'Junk Agroecology': The Corporate Capture of Agroecology for a Partial Ecological Transition Without Social Justice." Friends of the Earth International, Transnational Institute, and Crocevia, 2020. https://www.foei.org/publication/junk-agroecology/.

Altieri, Miguel A., and Clara I. Nicholls. "Agroecology and the Reconstruction of a Post-COVID-19 Agriculture." *The Journal of Peasant Studies* 47, no. 5 (July 28, 2020): 881–98. https://doi.org/10.1080/03066150.2020.1782891.

An Ecomodernist Manifesto. "An Ecomodernist Manifesto." Accessed April 8, 2024. http://www.ecomodernism.org/manifesto-english.

Anderson, Colin R., and Chris Maughan. "'The Innovation Imperative': The Struggle over Agroecology in the International Food Policy Arena." *Frontiers in Sustainable Food Systems* 5 (February 18, 2021): 619185. https://doi.org/10.3389/fsufs.2021.619185.

Anderson, Colin Ray, Janneke Bruil, M. Jahi Chappell, Csilla Kiss, and Michel Patrick Pimbert. "Conceptualizing Processes of Agroecological Transformations: From Scaling to Transition to Transformation." In *Agroecology Now!*, edited by Colin Ray Anderson, Janneke Bruil, M. Jahi Chappell, Csilla Kiss, and Michel Patrick Pimbert, 29–46. Cham: Springer International Publishing, 2021. https://doi.org/10.1007/978-3-030-61315-0_3.

Anthony, Mark A., S. Franz Bender, and Marcel G. A. Van Der Heijden. "Enumerating Soil Biodiversity." *Proceedings of the National Academy of Sciences* 120, no. 33 (August 15, 2023): e2304663120. https://doi.org/10.1073/pnas.2304663120.

Antonelli, Alexandre. "Indigenous Knowledge Is Key to Sustainable Food Systems." *Nature* 613, no. 7943 (January 12, 2023): 239–42. https://doi.org/10.1038/d41586-023-00021-4.

Bailey, Rob, and Laura Wellesley. *Chokepoints and Vulnerabilities in Global Food Trade.* Toronto, ON, CA: Chatham House, 2017.

Bajželj, Bojana, Thomas E. Quested, Elin Röös, and Richard P. J. Swannell. "The Role of Reducing Food Waste for Resilient Food Systems." *Ecosystem Services* 45 (October 2020): 101140. https://doi.org/10.1016/j.ecoser.2020.101140.

Bakker, Lieneke, Wopke Van Der Werf, Pablo Tittonell, Kris A. G. Wyckhuys, and Felix J. J. A. Bianchi. "Neonicotinoids in Global Agriculture: Evidence for a New Pesticide Treadmill?" *Ecology and Society* 25, no. 3 (2020): art26. https://doi.org/10.5751/ES-11814-250326.

Baldridge, Abigail S., Mark D. Huffman, Fraser Taylor, Dagan Xavier, Brooke Bright, Linda V. Van Horn, Bruce Neal, and Elizabeth Dunford. "The Healthfulness of the US Packaged Food and Beverage Supply: A Cross-Sectional Study." *Nutrients* 11, no. 8 (July 24, 2019): 1704. https://doi.org/10.3390/nu11081704.

Barrett, Hannah, and David Christian Rose. "Perceptions of the Fourth Agricultural Revolution: What's In, What's Out, and What Consequences Are Anticipated?" *Sociologia Ruralis* 62, no. 2 (April 2022): 162–89. https://doi.org/10.1111/soru.12324.

Barrios, Edmundo, Barbara Gemmill-Herren, Abram Bicksler, Emma Siliprandi, Ronnie Brathwaite, Soren Moller, Caterina Batello, and Pablo Tittonell. "The 10 Elements of Agroecology: Enabling Transitions Towards Sustainable Agriculture and Food Systems Through Visual Narratives." *Ecosystems and People* 16, no. 1 (January 2020): 230–47. https://doi.org/10.1080/26395916.2020.1808705.

Bartlett, Cheryl, Murdena Marshall, and Albert Marshall. "Two-Eyed Seeing and Other Lessons Learned Within a Co-Learning Journey of Bringing Together Indigenous and Mainstream Knowledges and Ways of Knowing." *Journal of Environmental Studies and Sciences* 2, no. 4 (November 2012): 331–40. https://doi.org/10.1007/s13412-012-0086-8.

Basso, Carla Giovana, Anderson Tadeu De Araújo-Ramos, and Anderson Joel Martino-Andrade. "Exposure to Phthalates and Female Reproductive Health: A Literature Review." *Reproductive Toxicology* 109 (April 2022): 61–79. https://doi.org/10.1016/j.reprotox.2022.02.006.

Bastian, Brock, and Steve Loughnan. "Resolving the Meat-Paradox: A Motivational Account of Morally Troublesome Behavior and Its Maintenance." *Personality and Social Psychology Review* 21, no. 3 (August 2017): 278–99. https://doi.org/10.1177/1088868316647562.

Baur, Patrick, and Alastair Iles. "Replacing Humans with Machines: A Historical Look at Technology Politics in California Agriculture." *Agriculture and Human Values* 40, no. 1 (March 2023): 113–40. https://doi.org/10.1007/s10460-022-10341-2.

Bellon-Maurel, Véronique, Evelyne Lutton, Pierre Bisquert, Ludovic Brossard, Stéphanie Chambaron-Ginhac, Pierre Labarthe, Philippe Lagacherie et al. "Digital Revolution for the Agroecological Transition of Food Systems: A Responsible Research and

Innovation Perspective." *Agricultural Systems* 203 (December 2022): 103524. https://doi.org/10.1016/j.agsy.2022.103524.

Benbrook, Charles M. "Trends in Glyphosate Herbicide Use in the United States and Globally." *Environmental Sciences Europe* 28, no. 1 (December 2016): 3. https://doi.org/10.1186/s12302-016-0070-0.

Benton, Tim G., and Rob Bailey. "The Paradox of Productivity: Agricultural Productivity Promotes Food System Inefficiency." *Global Sustainability* 2 (2019): e6. https://doi.org/10.1017/sus.2019.3.

Berry, Wendell. "The Agrarian Standard." *Orion Magazine.* Accessed June 4, 2024. https://orionmagazine.org/article/the-agrarian-standard/.

Biohabitats. "Lessons from Indigenous Traditions and Innovation / Expert Q&A: Dr. Enrique Salmón," 2018. https://www.biohabitats.com/newsletter/ecology-culture-and-economy-lessons-from-indigenous-traditions-and-innovation/expert-qa-enrique-salmon/.

Biovision Foundation for Ecological Development and Global Alliance for the Future of Food. "Beacons of Hope: Accelerating Transformations to Sustainable Food Systems." Global Alliance for the Future of Food, 2019. https://futureoffood.org/wp-content/uploads/2021/02/BeaconsOfHope_Report_082019.pdf.

Blesh, Jennifer, and Steven A. Wolf. "Transitions to Agroecological Farming Systems in the Mississippi River Basin: Toward an Integrated Socioecological Analysis." *Agriculture and Human Values* 31, no. 4 (December 2014): 621–35. https://doi.org/10.1007/s10460-014-9517-3.

Blum, Winfried E. H., Sophie Zechmeister-Boltenstern, and Katharina M. Keiblinger. "Does Soil Contribute to the Human Gut Microbiome?" *Microorganisms* 7, no. 9 (August 23, 2019): 287. https://doi.org/10.3390/microorganisms7090287.

Borrelli, Pasquale, David A. Robinson, Larissa R. Fleischer, Emanuele Lugato, Cristiano Ballabio, Christine Alewell, Katrin Meusburger, et al. "An Assessment of the Global Impact of 21st Century Land Use Change on Soil Erosion." *Nature Communications* 8, no. 1 (December 8, 2017): 2013. https://doi.org/10.1038/s41467-017-02142-7.

Bovarnick, Andrew, and Thomas Legrand. "Beyond Sustainability." UNDP, June 10, 2022. https://www.undp.org/facs/blog/beyond-sustainability.

Bradford, Jason. "The Future Is Rural: Food Systems Adaptations to the Great Simplification." Post Carbon Institute, 2019. https://www.postcarbon.org/publications/the-future-is-rural/.

Breier, Jannes, Luana Schwarz, Jonathan F. Donges, Dieter Gerten, and Johan Rockström. "Regenerative Agriculture for Food Security and Ecological Resilience: Illustrating Global Biophysical and Social Spreading Potentials." Potsdam Institute for Climate Impact Research, 2023. https://doi.org/10.48485/PIK.2023.001.

Bristow, Jamie, Rose Bell, Christine Wamsler, Tomas Björkman, Phoebe Tickell, Julia Kim, and Otto Scharmer. "The System Within: Addressing the Inner Dimensions of Sustainability and Systems Change." The Club of Rome, 2024. https://www

.clubofrome.org/wp-content/uploads/2024/05/Earth4All_Deep_Dive_Jamie
_Bristow.pdf.

Bronson, Kelly. "Looking Through a Responsible Innovation Lens at Uneven Engagements with Digital Farming." *NJAS: Wageningen Journal of Life Sciences* 90–91, no. 1 (December 1, 2019): 1–6. https://doi.org/10.1016/j.njas.2019.03.001.

Brooks, Sally. "Configuring the Digital Farmer: A Nudge World in the Making?" *Economy and Society* 50, no. 3 (July 3, 2021): 374–96. https://doi.org/10.1080/03085147.2021 .1876984.

Buckton, Sam J., Ioan Fazey, Bill Sharpe, Eugyen Suzanne Om, Bob Doherty, Peter Ball, Katherine Denby, et al. "The Regenerative Lens: A Conceptual Framework for Regenerative Social-Ecological Systems." *One Earth* 6, no. 7 (July 2023): 824–42. https://doi.org/10.1016/j.oneear.2023.06.006.

Buffalo Treaty. "Treaty." Accessed August 7, 2024. https://www.buffalotreaty.com/treaty.

Burton, Rob J. F. "Seeing Through the 'Good Farmer's' Eyes: Towards Developing an Understanding of the Social Symbolic Value of 'Productivist' Behaviour." *Sociologia Ruralis* 44, no. 2 (April 2004): 195–215. https://doi.org/10.1111/j.1467-9523.2004 .00270.x.

Calafat, Antonia M., Xiaoyun Ye, Lee-Yang Wong, John A. Reidy, and Larry L. Needham. "Exposure of the U.S. Population to Bisphenol A and 4- *Tertiary* -Octylphenol: 2003–2004." *Environmental Health Perspectives* 116, no. 1 (January 2008): 39–44. https://doi.org/10.1289/ehp.10753.

Campanale, Claudia, Carmine Massarelli, Ilaria Savino, Vito Locaputo, and Vito Felice Uricchio. "A Detailed Review Study on Potential Effects of Microplastics and Additives of Concern on Human Health." *International Journal of Environmental Research and Public Health* 17, no. 4 (January 2020): 1212. https://doi.org/10.3390/ ijerph17041212.

Cantwell-Jones, Aoife, Jenny Ball, David Collar, Mauricio Diazgranados, Ruben Douglas, Félix Forest, Julie Hawkins, et al. "Global Plant Diversity as a Reservoir of Micronutrients for Humanity." *Nature Plants* 8, no. 3 (February 24, 2022): 225–32. https://doi.org/10.1038/s41477-022-01100-6.

Capra, Fritjof. "The New Facts of Life." ecoliteracy.org, June 29, 2009. https://www .ecoliteracy.org/article/new-facts-life.

Capra, Fritjof. *The Web of Life: A New Scientific Understanding of Living Systems.* New York: Anchor Books, 1997.

Carbonell, Isabelle M. "The Ethics of Big Data in Big Agriculture." *Internet Policy Review* 5, no. 1 (March 31, 2016). https://doi.org/10.14763/2016.1.405.

Carlisle, Liz. "Factors Influencing Farmer Adoption of Soil Health Practices in the United States: A Narrative Review." *Agroecology and Sustainable Food Systems* 40, no. 6 (July 2, 2016): 583–613. https://doi.org/10.1080/21683565.2016.1156596.

Carlisle, Liz, Maywa Montenegro De Wit, Marcia S. DeLonge, Alastair Iles, Adam Calo, Christy Getz, Joanna Ory, et al. "Transitioning to Sustainable Agriculture Requires

Growing and Sustaining an Ecologically Skilled Workforce." *Frontiers in Sustainable Food Systems* 3 (November 1, 2019): 96. https://doi.org/10.3389/fsufs.2019.00096.

Carolan, Michael. "Acting like an Algorithm: Digital Farming Platforms and the Trajectories They (Need Not) Lock-In." *Agriculture and Human Values* 37, no. 4 (December 2020): 1041–53. https://doi.org/10.1007/s10460-020-10032-w.\

Carrington, Damian. "Earth's Sixth Mass Extinction Event Under Way, Scientists Warn." *The Guardian*, July 10, 2017. https://www.theguardian.com/environment/2017/jul/10/earths-sixth-mass-extinction-event-already-underway-scientists-warn.

Cascadia Department of Bioregion. "Core Principles of Bioregionalism." Accessed August 9, 2024. https://cascadiabioregion.org/bioregionalism-core-principles.

Cassidy, Emily S., Paul C. West, James S. Gerber, and Jonathan A. Foley. "Redefining Agricultural Yields: From Tonnes to People Nourished per Hectare." *Environmental Research Letters* 8, no. 3 (September 1, 2013): 034015. https://doi.org/10.1088/1748-9326/8/3/034015.

Cho, Renee. "How Climate Change Will Affect Plants." *State of the Planet* (blog), January 27, 2022. https://news.climate.columbia.edu/2022/01/27/how-climate-change-will-affect-plants/.

Clapp, Jennifer, and Gyorgy Scrinis. "'Big Food, Nutritionism and Corporate Power.'" *Globalizations* 14, no. 4 (2017): 578–95. http://www.tandfonline.com/doi/abs/10.1080/14747731.2016.1239806.

Committee on the Guidance on PFAS Testing and Health Outcomes, Board on Environmental Studies and Toxicology, Board on Population Health and Public Health Practice, Division on Earth and Life Studies, Health and Medicine Division, and National Academies of Sciences, Engineering, and Medicine. *Guidance on PFAS Exposure, Testing, and Clinical Follow-Up*. Washington, DC: National Academies Press, 2022. https://doi.org/10.17226/26156.

Commonland. "4 Returns Framework." Accessed June 18, 2024. https://commonland.com/4-returns-framework/.

Crippa, M., E. Solazzo, D. Guizzardi, F. Monforti-Ferrario, F. N. Tubiello, and A. Leip. "Food Systems Are Responsible for a Third of Global Anthropogenic GHG Emissions." *Nature Food* 2, no. 3 (March 8, 2021): 198–209. https://doi.org/10.1038/s43016-021-00225-9.

Day, Cathy, and Sarah Cramer. "Transforming to a Regenerative U.S. Agriculture: The Role of Policy, Process, and Education." *Sustainability Science* 17, no. 2 (March 2022): 585–601. https://doi.org/10.1007/s11625-021-01041-7.

Duke, Stephen O. "Perspectives on Transgenic, Herbicide- Resistant Crops in the United States Almost 20 Years After Introduction." *Pest Management Science* 71, no. 5 (May 2015): 652–57. https://doi.org/10.1002/ps.3863.

Duncan, Jessica, J. S. C. Wiskerke, and Michael S. Carolan, eds. *Routledge Handbook of Sustainable and Regenerative Food Systems*. New York: Routledge, Taylor & Francis Group, 2021.

Ellis, Erle C., Nicolas Gauthier, Kees Klein Goldewijk, Rebecca Bliege Bird, Nicole Boivin, Sandra Díaz, Dorian Q. Fuller, et al. "People Have Shaped Most of Terrestrial Nature for at Least 12,000 Years." *Proceedings of the National Academy of Sciences* 118, no. 17 (April 27, 2021): e2023483118. https://doi.org/10.1073/pnas .2023483118.

Evenson, R. E., and D. Gollin. "Assessing the Impact of the Green Revolution, 1960 to 2000." *Science* 300, no. 5620 (May 2, 2003): 758–62. https://doi.org/10.1126 /science.1078710.

Eyhorn, Frank, Adrian Muller, John P. Reganold, Emile Frison, Hans R. Herren, Louise Luttikholt, Alexander Mueller, et al. "Sustainability in Global Agriculture Driven by Organic Farming." *Nature Sustainability* 2, no. 4 (April 9, 2019): 253–55. https:// doi.org/10.1038/s41893-019-0266-6.

Farley, Joshua. "The Foundations for an Ecological Economy: An Overview." In J. Farley & D. Malghan (eds.), *Beyond Uneconomic Growth, A Festschrift in Honor of Herman Daly*, 2: 1–17. Burlington, VT: University of Vermont, 2016.

Food and Agriculture Organization of the United Nations (FAO). "Discards in the World's Marine Fisheries. An Update." Accessed March 23, 2024. https://www.fao .org/3/y5936e/y5936e09.htm.

Food and Agriculture Organization of the United Nations (FAO). "The 10 Elements of Agroecology: Guiding the Transition to Sustainable Food and Agricultural Systems." Rome, Italy, 2017. https://openknowledge.fao.org/handle/20.500.14283 /i9037en.

Food and Agriculture Organization of the United Nations (FAO), Cirad, and Rythu Sadhikara Samstha. "Re-Thinking Food Systems in Andhra Pradesh, India: How Natural Farming Could Feed the Future." 2023. https://www.fao.org/family -farming/detail/en/c/1679556/.

Food and Agriculture Organization of the United Nations (FAO) and Intergovernmental Technical Panel on Soils (ITPS). "Status of the World's Soil Resources (SWSR)— Main Report."2015. https://www.fao.org/3/bc590e/bc590e.pdf.

Food and Agriculture Organization of the United Nations (FAO), International Fund for Agricultural Development (IFAD), United Nations Children's Fund (UNICEF), World Food Programme (WFP), and World Health Organization (WHO). "The State of Food Security and Nutrition in the World 2024." Rome, Italy, July 23, 2024. https://doi.org/10.4060/cd1254en.

Francis, C., G. Lieblein, S. Gliessman, T. A. Breland, N. Creamer, R. Harwood, L. Salomonsson, et al. "Agroecology: The Ecology of Food Systems." *Journal of Sustainable Agriculture* 22, no. 3 (July 17, 2003): 99–118. https://doi.org/10.1300 /J064v22n03_10.

Gearhardt, Ashley N., and Erica M. Schulte. "Is Food Addictive? A Review of the Science." *Annual Review of Nutrition* 41, no. 1 (October 11, 2021): 387–410. https://doi.org/10 .1146/annurev-nutr-110420-111710.

Gliessman, Steve. "Balancing Nature and Agriculture." *Agroecology and Sustainable Food Systems* 46, no. 2 (February 7, 2022): 163–64. https://doi.org/10.1080/21683565 .2022.2009171.

Gliessman, Steve. "Defining Agroecology." *Agroecology and Sustainable Food Systems* 42, no. 6 (July 3, 2018): 599–600. https://doi.org/10.1080/21683565.2018.1432329.

Gliessman, Steve. "Transforming Food Systems with Agroecology." *Agroecology and Sustainable Food Systems* 40, no. 3 (March 15, 2016): 187–89. https://doi.org/10.1080 /21683565.2015.1130765.

Global Alliance for the Future of Food. "Community Managed Natural Farming: An Agroecology Movement Takes Root in India's Andhra Pradesh State." Accessed April 26, 2024. https://futureoffood.org/insights/community-managed-natural -farming-india/.

Gordon, Ethan, Federico Davila, and Chris Riedy. "Transforming Landscapes and Mindscapes Through Regenerative Agriculture." *Agriculture and Human Values* 39, no. 2 (June 2022): 809–26. https://doi.org/10.1007/s10460-021-10276-0.

Gosnell, Hannah. "Regenerating Soil, Regenerating Soul: An Integral Approach to Understanding Agricultural Transformation." *Sustainability Science* 17, no. 2 (March 2022): 603–20. https://doi.org/10.1007/s11625-021-00993-0.

Gosnell, Hannah, Nicholas Gill, and Michelle Voyer. "Transformational Adaptation on the Farm: Processes of Change and Persistence in Transitions to 'Climate-Smart' Regenerative Agriculture." *Global Environmental Change* 59 (November 2019): 101965. https://doi.org/10.1016/j.gloenvcha.2019.101965.

Hamant, Olivier. "Plant Scientists Can't Ignore Jevons Paradox Anymore." *Nature Plants* 6, no. 7 (July 14, 2020): 720–22. https://doi.org/10.1038/s41477-020-0722-3.

Herrero, Mario, Philip K. Thornton, Daniel Mason-D'Croz, Jeda Palmer, Tim G. Benton, Benjamin L. Bodirsky, Jessica R. Bogard, et al. "Innovation Can Accelerate the Transition Towards a Sustainable Food System." *Nature Food* 1, no. 5 (May 19, 2020): 266–72. https://doi.org/10.1038/s43016-020-0074-1.

Herrero, Mario, Philip K. Thornton, Daniel Mason-D'Croz, Jeda Palmer, Benjamin L. Bodirsky, Prajal Pradhan, Christopher B. Barrett, et al. "Articulating the Effect of Food Systems Innovation on the Sustainable Development Goals." *The Lancet Planetary Health* 5, no. 1 (January 2021): e50–62. https://doi.org/10.1016/S2542 -5196(20)30277-1.

High Level Panel of Experts on Food Security and Nutrition (HLPE). "Agroecological and Other Innovative Approaches for Sustainable Agriculture and Food Systems That Enhance Food Security and Nutrition." Rome, Italy: High Level Panel of Experts on Food Security and Nutrition of the Committee on World Food Security, 2019. https://www.globalagriculture.org/fileadmin/files/weltagrarbericht/IAASTD -Buch/01Reports/11FAOAgroecology/HLPEAgroecologyReport.pdf.

Hirt, Heribert. "Healthy Soils for Healthy Plants for Healthy Humans: How Beneficial Microbes in the Soil, Food and Gut Are Interconnected and How Agriculture Can

Contribute to Human Health." *EMBO Reports* 21, no. 8 (August 5, 2020): e51069. https://doi.org/10.15252/embr.202051069.

Holt-Giménez, Eric, and Miguel A. Altieri. "Agroecology, Food Sovereignty and the New Green Revolution." *Journal of Sustainable Agriculture*, September 4, 2012, 120904081412003. https://doi.org/10.1080/10440046.2012.716388.

Howard, Philip H. *Concentration and Power in the Food System: Who Controls What We Eat?* Revised edition. Contemporary Food Studies: Economy, Culture and Politics. London: Bloomsbury Academic, 2021.

International Panel of Experts on Sustainable Food Systems (IPES-Food). "Breaking Away from Industrial Food and Farming Systems: Seven Case Studies of Agroecological Transition. " 2018. https://ipes-food.org/report/breaking-away-from -industrial-food-and-farming-systems/.

International Panel of Experts on Sustainable Food Systems (IPES-Food). "From Uniformity to Diversity: A Paradigm Shift from Industrial Agriculture to Diversified Agroecological Systems." 2016. https://ipes-food.org/report/from-uniformity-to -diversity/.

International Panel of Experts on Sustainable Food Systems (IPES-Food). "Smoke and Mirrors: Examining Competing Framings of Food System Sustainability: Agroecology, Regenerative Agriculture, and Nature Based Solutions." 2022. https://ipes -food.org/wp-content/uploads/2024/03/SmokeAndMirrors.pdf.

International Panel of Experts on Sustainable Food Systems (IPES-Food). "The Politics of Protein: Examining Claims A bout Livestock, Fish, 'Alternative Proteins' and Sustainability. " 2022. https://ipes-food.org/report/the-politics-of-protein/.

International Panel of Experts on Sustainable Food Systems (IPES-Food). "Too Big to Feed: Exploring the Impacts of Mega-Mergers, Consolidation, and Concentration of Power in the Agri-Food Sector. " 2017. https://www.ipes-food.org/_img/upload /files/Concentration_FullReport.pdf.

International Panel of Experts on Sustainable Food Systems (IPES-Food). "Unravelling the Food–Health Nexus: Addressing Practices, Political Economy, and Power Relations to Build Healthier Food Systems." The Global Alliance for the Future of Food and IPES-Food, 2017. https://futureoffood.org/wp-content/uploads/2021/01 /FoodHealthNexus_Full-Report_FINAL.pdf.

Jacobs, Donald Trent, and Darcia Narvaez. *Restoring the Kinship Worldview: Indigenous Voices Introduce 28 Precepts for Rebalancing Life on Planet Earth.* Berkeley, CA: North Atlantic Books, 2022.

Jaffee, Daniel, and Philip H. Howard. "Corporate Cooptation of Organic and Fair Trade Standards." *Agriculture and Human Values* 27, no. 4 (December 2010): 387–99. https://doi.org/10.1007/s10460-009-9231-8.

Johnston, Lyla June. "Architects of Abundance: Indigenous Regenerative Food and Land Management Systems and the Excavation of Hidden History." University of Alaska Fairbanks, 2022. https://scholarworks.alaska.edu/handle/11122/13122.

Kassam, Amir, and Leila Kassam, eds. *Rethinking Food and Agriculture: New Ways Forward*. Woodhead Publishing Series in Food Science, Technology and Nutrition. Cambridge, MA, United States: Woodhead Publishing, an imprint of Elsevier, 2021.

Kawano, Emily, and Julie Matthaei. "System Change: A Basic Primer to the Solidarity Economy." *Non Profit News | Nonprofit Quarterly*, July 8, 2020. https://nonprofitquarterly.org/system-change-a-basic-primer-to-the-solidarity-economy/.

Kelly, Marjorie. "Breaking Up With Capitalism." *YES! Magazine*, May 7, 2024. https://www.yesmagazine.org/opinion/2024/05/07/money-wealth-democracy-capitalism.

Kimmerer, Robin Wall. *Braiding Sweetgrass: Indigenous Wisdom, Scientific Knowledge, and the Teachings of Plants*. Minneapolis, MN: Milkweed Editions, 2020.

Kimmerer, Robin Wall. "Corn Tastes Better on the Honor System." *Emergence Magazine*. Accessed June 24, 2024. https://emergencemagazine.org/feature/corn-tastes-better/.

Klerkx, Laurens, and David Rose. "Dealing with the Game-Changing Technologies of Agriculture 4.0: How Do We Manage Diversity and Responsibility in Food System Transition Pathways?" *Global Food Security* 24 (March 2020): 100347. https://doi.org/10.1016/j.gfs.2019.100347.

Klerkx, Laurens, Emma Jakku, and Pierre Labarthe. "A Review of Social Science on Digital Agriculture, Smart Farming and Agriculture 4.0: New Contributions and a Future Research Agenda." *NJAS: Wageningen Journal of Life Sciences* 90–91, no. 1 (December 1, 2019): 1–16. https://doi.org/10.1016/j.njas.2019.100315.

Klerkx, Laurens, and Pablo Villalobos. "Are AgriFoodTech Start-Ups the New Drivers of Food Systems Transformation? An Overview of the State of the Art and a Research Agenda." *Global Food Security* 40 (March 2024): 100726. https://doi.org/10.1016/j.gfs.2023.100726.

La Via Campesina—EN. "Food Sovereignty, a Manifesto for the Future of Our Planet," October 13, 2021. https://viacampesina.org/en/food-sovereignty-a-manifesto-for-the-future-of-our-planet-la-via-campesina/.

LaDuke, Winona. "The Political Economy of Wild Rice Indigenous Heritage and University Research." *The Multinational Monitor* 25, no. 4 (April 2004). https://www.multinationalmonitor.org/mm2004/04012004/april04corp4.html.

Loring, Philip, Brent Loken, Mariella Meyer, Sharelle Polack, Alayna Paolini, Richard Nyiawung, and Murli Dhar. "Solving the Great Food Puzzle: Right Innovation, Right Impact, Right Place." WWF, 2023. https://wwfint.awsassets.panda.org/downloads/solving-the-great-food-puzzle-right-innovation--right-impact--right-place.pdf.

Macdiarmid, Jennie I., Flora Douglas, and Jonina Campbell. "Eating like There's No Tomorrow: Public Awareness of the Environmental Impact of Food and Reluctance to Eat Less Meat as Part of a Sustainable Diet." *Appetite* 96 (January 2016): 487–93. https://doi.org/10.1016/j.appet.2015.10.011.

Mann, Charles C. *The Wizard and the Prophet: Two Remarkable Scientists and Their Dueling Visions to Shape Tomorrow's World*. New York: Alfred A. Knopf, 2018.

Manzini, Ezio. "Small, Local, Open and Connected. Resilient Systems and Sustainable Qualities." *Design Observer*, February 6, 2013. http://designobserver.com/feature /small-local-open-and-connected-resilient-systems-and-sustainable-qualities /37670.

Massy, Charles John. "Transforming the Earth: A Study in the Change of Agricultural Mindscapes." The Australian National University, 2013. https://openresearch -repository.anu.edu.au/handle/1885/115203.

Matson, Laura, G.-H. Crystal Ng, Michael Dockry, Madeline Nyblade, Hannah Jo King, Mark Bellcourt, Jeremy Bloomquist, et al. "Transforming Research and Relationships Through Collaborative Tribal-University Partnerships on Manoomin (Wild Rice)." *Environmental Science & Policy* 115 (January 2021): 108–15. https://doi .org/10.1016/j.envsci.2020.10.010.

McDonald, Garry W., and Murray G. Patterson. "Bridging the Divide in Urban Sustainability: From Human Exceptionalism to the New Ecological Paradigm." *Urban Ecosystems* 10, no. 2 (June 1, 2007): 169–92. https://doi.org/10.1007/s11252 -006-0017-0.

McGreevy, Steven R., Christoph D. D. Rupprecht, Daniel Niles, Arnim Wiek, Michael Carolan, Giorgos Kallis, Kanang Kantamaturapoj, et al. "Sustainable Agrifood Systems for a Post-Growth World." *Nature Sustainability* 5, no. 12 (August 4, 2022): 1011–17. https://doi.org/10.1038/s41893-022-00933-5.

Meadows, Donella. "Leverage Points: Places to Intervene in a System." *The Academy for Systems Change* (blog). Accessed May 23, 2024. https://donellameadows.org /archives/leverage-points-places-to-intervene-in-a-system/.

Meadows, Donella H., Club of Rome, and Potomac Associates, eds. *The Limits to Growth: A Report for the Club of Rome's Project on the Predicament of Mankind*. 2nd ed. A Potomac Associates Book. New York: Universe Books, 1974.

Mier Y. Terán Giménez Cacho, Mateo, Omar Felipe Giraldo, Miriam Aldasoro, Helda Morales, Bruce G. Ferguson, Peter Rosset, Ashlesha Khadse, and Carmen Campos. "Bringing Agroecology to Scale: Key Drivers and Emblematic Cases." *Agroecology and Sustainable Food Systems* 42, no. 6 (July 3, 2018): 637–65. https://doi.org/10 .1080/21683565.2018.1443313.

Milgroom, Jessica. "Wild Rice and the Ojibwe." *MNopedia*, July 20, 2020. https://www .mnopedia.org/thing/wild-rice-and-ojibwe.

Miller-Klugesherz, Jacob A., and Matthew R. Sanderson. "Good for the Soil, but Good for the Farmer? Addiction and Recovery in Transitions to Regenerative Agriculture." *Journal of Rural Studies* 103 (October 2023): 103123. https://doi.org/10.1016/j.jrurstud .2023.103123.

Monbiot, George. *Regenesis: Feeding the World Without Devouring the Planet*. Great Britain: Penguin Books, 2022.

Monteiro, Carlos Augusto, Renata Bertazzi Levy, Rafael Moreira Claro, Inês Rugani Ribeiro De Castro, and Geoffrey Cannon. "Increasing Consumption of Ultra-Processed Foods and Likely Impact on Human Health: Evidence from Brazil." *Public Health Nutrition* 14, no. 1 (December 20, 2010): 5–13. https://doi.org/10.1017/S1368980010003241.

Montenegro De Wit, Maywa. "Can Agroecology and CRISPR Mix? The Politics of Complementarity and Moving Toward Technology Sovereignty." *Agriculture and Human Values* 39, no. 2 (June 2022): 733–55. https://doi.org/10.1007/s10460-021 -10284-0.

Montenegro De Wit, Maywa. "Democratizing CRISPR? Stories, Practices, and Politics of Science and Governance on the Agricultural Gene Editing Frontier." Edited by Anne R. Kapuscinski and Elizabeth Fitting. *Elementa: Science of the Anthropocene* 8 (January 1, 2020): 9. https://doi.org/10.1525/elementa.405.

Montgomery, David R. *Dirt: The Erosion of Civilizations.* Berkeley, CA: University of California Press, 2012.

Montgomery, David R. "Soil Health and the Revolutionary Potential of Conservation Agriculture." In *Rethinking Food and Agriculture.* Elsiever, 2021.

Montgomery, David R., and Anne Biklé. *The Hidden Half of Nature: The Microbial Roots of Life and Health.* New York: W.W. Norton & Company, 2016.

Montgomery, David R., Anne Biklé, Ray Archuleta, Paul Brown, and Jazmin Jordan. "Soil Health and Nutrient Density: Preliminary Comparison of Regenerative and Conventional Farming." *PeerJ* 10 (January 27, 2022): e12848. https://doi.org/10 .7717/peerj.12848.

Moore, Michele-Lee, Darcy Riddell, and Dana Vocisano. "Scaling Out, Scaling Up, Scaling Deep: Strategies of Non-Profits in Advancing Systemic Social Innovation." *Journal of Corporate Citizenship* 2015, no. 58 (June 1, 2015): 67–84. https://doi.org /10.9774/GLEAF.4700.2015.ju.00009.

Nicholls, Clara I., and Miguel A. Altieri. "Pathways for the Amplification of Agroecology." *Agroecology and Sustainable Food Systems* 42, no. 10 (November 26, 2018): 1170–93. https://doi.org/10.1080/21683565.2018.1499578.

Nyéléni: International Movement for Food Sovereignty. "Declaration of Nyéléni," February 27, 2007. https://nyeleni.org/en/declaration-of-nyeleni/.

O'Brien, Karen, and Linda Sygna. "Responding to Climate Change: The Three Spheres of Transformation." In *Proceedings of Transformation in a Changing Climate*, 16–23. Oslo, Norway: University of Oslo, 2013. http://www.cchange.no/.

Olsson, Per, and Michele-Lee Moore. "Transformations, Agency and Positive Tipping Points: A Resilience-Based Approach." In *Positive Tipping Points Towards Sustainability*, edited by J. David Tàbara, Alexandros Flamos, Diana Mangalagiu, and Serafeim Michas, 59–77. Springer Climate. Cham: Springer International Publishing, 2024. https://doi.org/10.1007/978-3-031-50762-5_4.

Ortiz-Bobea, Ariel, Toby R. Ault, Carlos M. Carrillo, Robert G. Chambers, and David B. Lobell. "Anthropogenic Climate Change Has Slowed Global Agricultural

Productivity Growth." *Nature Climate Change* 11, no. 4 (April 2021): 306–12. https://doi.org/10.1038/s41558-021-01000-1.

Pew. "Pew Commission Says Industrial Scale Farm Animal Production Poses 'Unacceptable' Risks to Public Health, Environment." April 29, 2008. http://pew.org/1Q6ZCrz.

Pigford, Ashlee-Ann E., Gordon M. Hickey, and Laurens Klerkx. "Beyond Agricultural Innovation Systems? Exploring an Agricultural Innovation Ecosystems Approach for Niche Design and Development in Sustainability Transitions." *Agricultural Systems* 164 (July 2018): 116–21. https://doi.org/10.1016/j.agsy.2018.04.007.

"Politics of Technology." A Growing Culture & ETC Group, 2023. https://www.etcgroup.org/sites/www.etcgroup.org/files/files/politics_of_technology_en_v2.pdf.

Reid, Andrea J., Lauren E. Eckert, John-Francis Lane, Nathan Young, Scott G. Hinch, Chris T. Darimont, Steven J. Cooke, Natalie C. Ban, and Albert Marshall. "'Two-Eyed Seeing': An Indigenous Framework to Transform Fisheries Research and Management." *Fish and Fisheries* 22, no. 2 (March 2021): 243–61. https://doi.org/10.1111/faf.12516.

Rimanoczy, Isabel, and Ana Maria Llamazares. "Twelve Principles to Guide a Long-Overdue Paradigm Shift." *Journal of Management, Spirituality & Religion* 18, no. 6 (December 1, 2021): 54–76. https://doi.org/10.51327/JKKI4753.

Ripple, William J., Christopher Wolf, Thomas M. Newsome, Phoebe Barnard, and William R. Moomaw. "Corrigendum: World Scientists' Warning of a Climate Emergency." *BioScience* 70, no. 1 (January 1, 2020): 100–100. https://doi.org/10.1093/biosci/biz152.

Rockström, Johan, Will Steffen, Kevin Noone, Åsa Persson, Stuart Chapin, Eric F. Lambin, Timothy M. Lenton, et al. "Planetary Boundaries: Exploring the Safe Operating Space for Humanity." *Ecology and Society* 14, no. 2 (2009). http://www.jstor.org/stable/26268316.

Ronald, Pamela C., and Raoul W. Adamchak. *Tomorrow's Table: Organic Farming, Genetics, and the Future of Food.* New York, NY: Oxford University Press, 2008.

Rose, David Christian, Anna Barkemeyer, Auvikki De Boon, Catherine Price, and Dannielle Roche. "The Old, the New, or the Old Made New? Everyday Counter-Narratives of the So-Called Fourth Agricultural Revolution." *Agriculture and Human Values* 40, no. 2 (June 2023): 423–39. https://doi.org/10.1007/s10460-022-10374-7.

Rose, David Christian, and Jason Chilvers. "Agriculture 4.0: Broadening Responsible Innovation in an Era of Smart Farming." *Frontiers in Sustainable Food Systems* 2 (December 21, 2018): 87. https://doi.org/10.3389/fsufs.2018.00087.

Rose, Steve. "The New Alchemists: Could the Past Hold the Key to Sustainable Living?" *The Guardian*, September 29, 2019. http://www.theguardian.com/lifeandstyle/ng-interactive/2019/sep/29/the-new-alchemists-could-the-past-hold-the-key-to-sustainable-living.

Rushton, Jonathan, Barry J. McMahon, Mary E. Wilson, Jonna A. K. Mazet, and Bhavani Shankar. "A Food System Paradigm Shift: From Cheap Food at Any Cost to Food Within a One Health Framework." *NAM Perspectives* 11 (November 22, 2021). https://doi.org/10.31478/202111b.

Sands, Bryony, Mario Reinaldo Machado, Alissa White, Egleé Zent, and Rachelle Gould. "Moving Towards an Anti-Colonial Definition for Regenerative Agriculture." *Agriculture and Human Values* 40, no. 4 (December 2023): 1697–716. https://doi .org/10.1007/s10460-023-10429-3.

Sanford, Carol. *The Regenerative Life: Transform Any Organization, Our Society, and Your Destiny.* Boston: Nicholas Brealey Publishing, 2020.

Scherr, Sara J., Louise E. Buck, Renilde Becqué, Carolina Moeller, Carlos Agnes, and Mike Keller. "Landscape Collaboration for Regenerative Food Systems: Towards an Action Agenda." Meridian Institute, on behalf of Regen10 and EcoAgriculture Partners, on behalf of 1000 Landscapes for 1 Billion, n.d. https://api.ecoagriculture .org/uploads/Landscape_collaboration_for_regenerative_food_systems_FINAL _cb369d2342.pdf.

Scrinis, Gyorgy. *Nutritionism: The Science and Politics of Dietary Advice.* New York, NY: Columbia University Press, 2013.

Seneff, Stephanie. *Toxic Legacy: How the Weedkiller Glyphosate Is Destroying Our Health and the Environment.* White River Junction, VT : Chelsea Green Publishing, 2021.

Seymour, Madison, and Sean Connelly. "Regenerative Agriculture and a More-than-Human Ethic of Care: A Relational Approach to Understanding Transformation." *Agriculture and Human Values* 40, no. 1 (March 2023): 231–44. https://doi.org/10 .1007/s10460-022-10350-1.

Shroff, Ruchi, and Carla Ramos Cortés. "The Biodiversity Paradigm: Building Resilience for Human and Environmental Health." *Development* 63, no. 2–4 (December 2020): 172–80. https://doi.org/10.1057/s41301-020-00260-2.

Snorek, Julie, Susanne Freidberg, and Geneva Smith. "Relationships of Regeneration in Great Plains Commodity Agriculture." *Agriculture and Human Values*, April 11, 2024. https://doi.org/10.1007/s10460-024-10558-3.

Soini Coe, Eija, and Richard Coe. "Agroecological Transitions in the Mind." *Elementa: Science of the Anthropocene* 11, no. 1 (February 20, 2023): 00026. https://doi.org/10 .1525/elementa.2022.00026.

Stilgoe, Jack, Richard Owen, and Phil Macnaghten. "Developing a Framework for Responsible Innovation." *Research Policy* 42, no. 9 (November 2013): 1568–80. https://doi.org/10.1016/j.respol.2013.05.008.

Stockholm Resilience Center. "Planetary Boundaries." Accessed August 15, 2024. https://www.stockholmresilience.org/research/planetary-boundaries.html.

Stone, Glenn Davis. "Commentary: New Histories of the Indian Green Revolution." *The Geographical Journal* 185, no. 2 (June 2019): 243–50. https://doi.org/10.1111 /geoj.12297.

Taub, Daniel R., Brian Miller, and Holly Allen. "Effects of Elevated CO_2 on the Protein Concentration of Food Crops: A Meta-analysis." *Global Change Biology* 14, no. 3 (March 2008): 565–75. https://doi.org/10.1111/j.1365-2486.2007.01511.x.

Thackara, John. "Bioregioning: Pathways to Urban-Rural Reconnection." *She Ji: The Journal of Design, Economics, and Innovation* 5, no. 1 (2019): 15–28. https://doi.org /10.1016/j.sheji.2019.01.002.

The Consilience Project. "Technology Is Not Values Neutral: Ending the Reign of Nihilistic Design." Civilization Research Institute, June 26, 2022. https:// consilienceproject.org/technology-is-not-values-neutral-ending-the-reign-of -nihilistic-design-2/.

The Economics of Ecosystems and Biodiversity (TEEB). "Agriculture & Food." Accessed March 25, 2024. https://teebweb.org/our-work/agrifood/.

The Lancet. "Addressing the Vulnerability of the Global Food System." *The Lancet* 390, no. 10090 (July 2017): 95. https://doi.org/10.1016/S0140-6736(17)31803-2.

Tittonell, Pablo, Veronica El Mujtar, Georges Felix, Yodit Kebede, Luciana Laborda, Raquel Luján Soto, and Joris De Vente. "Regenerative Agriculture—Agroecology without Politics?" *Frontiers in Sustainable Food Systems* 6 (August 2, 2022): 844261. https://doi.org/10.3389/fsufs.2022.844261.

Tittonell, Pablo, Gervasio Piñeiro, Lucas A. Garibaldi, Santiago Dogliotti, Han Olff, and Esteban G. Jobbagy. "Agroecology in Large Scale Farming—A Research Agenda." *Frontiers in Sustainable Food Systems* 4 (December 18, 2020): 584605. https://doi .org/10.3389/fsufs.2020.584605.

Van Der Burg, Simone, Marc-Jeroen Bogaardt, and Sjaak Wolfert. "Ethics of Smart Farming: Current Questions and Directions for Responsible Innovation Towards the Future." *NJAS: Wageningen Journal of Life Sciences* 90–91, no. 1 (December 1, 2019): 1–10. https://doi.org/10.1016/j.njas.2019.01.001.

Waddock, Sandra. *Catalyzing Transformation: Making System Change Happen.* eBook. New York, NY: Business Expert Press, 2024.

Waddock, Sandra. *Transforming Towards Life-Centered Economies: How Business, Government, and Civil Society Can Build a Better World.* eBook. Business Expert Press, 2020.

Wahl, Daniel Christian. "Constructing a Regenerative Future." *RSA Journal*, November 19, 2021. https://www.thersa.org/rsa-journal/2021/issue-4/feature /constructing-a-regenerative-future.

Wamsler, Christine, Jamie Bristow, Kira Cooper, Gretchen Steidle, Sara Taggart, Lena Søvold, Tom H. Oliver, Jessica Brockler, and Thomas Legrand. "Theoretical Foundations Report: Research and Evidence for the Potential of Consciousness Approaches and Practices to Unlock Sustainability and Systems Transformation." UNDP Conscious Food Systems Alliance (CoFSA), United Nations Development Programme (UNDP), 2022. https://www.contemplative-sustainable-futures .com/_files/ugd/4cc31e_143f3bc24f2c43ad94316cd5ofbb8e4a.pdf.

Webb, Patrick, Tim G. Benton, John Beddington, Derek Flynn, Niamh M. Kelly, and Sandy M. Thomas. "The Urgency of Food System Transformation Is Now Irrefutable." *Nature Food* 1, no. 10 (September 28, 2020): 584–85. https://doi.org /10.1038/s43016-020-00161-0.

Wezel, A., S. Bellon, T. Doré, C. Francis, D. Vallod, and C. David. "Agroecology as a Science, a Movement and a Practice. A Review." *Agronomy for Sustainable Development* 29, no. 4 (December 2009): 503–15. https://doi.org/10.1051/agro /2009004.

Willett, Walter, Johan Rockström, Brent Loken, Marco Springmann, Tim Lang, Sonja Vermeulen, Tara Garnett, et al. "Food in the Anthropocene: The EAT–Lancet Commission on Healthy Diets from Sustainable Food Systems." *The Lancet* 393, no. 10170 (February 2019): 447–92. https://doi.org/10.1016/S0140-6736(18)31788-4.

Woiwode, Christoph, Niko Schäpke, Olivia Bina, Stella Veciana, Iris Kunze, Oliver Parodi, Petra Schweizer-Ries, and Christine Wamsler. "Inner Transformation to Sustainability as a Deep Leverage Point: Fostering New Avenues for Change Through Dialogue and Reflection." *Sustainability Science* 16, no. 3 (May 2021): 841–58. https://doi.org/10.1007/s11625-020-00882-y.

Index

About the Author

Nicole Negowetti is an expert on the future of food—an internationally recognized food law and policy scholar, advocate, and attorney. Since 2011, she has developed, led, and implemented a broad range of federal, state, and local policy initiatives addressing the health, environmental, and economic impacts of the food system and promoting sustainable, equitable, and healthy food and agricultural production. She has vast experience working with innovative food entrepreneurs, companies, and advocacy organizations, particularly in her previous role of vice president of policy and food systems at the Plant Based Foods Association, the first and only trade association for the plant-based foods industry in the United States, where she was also managing director of the Plant Based Foods Institute, a nonprofit organization that is catalyzing a regenerative plant-based food movement. Negowetti also serves on the United Nations Development Programme's Conscious Food Systems Alliance.

Currently a visiting lecturer at the Tufts Friedman School of Nutrition Science and Policy, Negowetti teaches food law and regulation. She is also the founder of Food for Us, a growing movement for food democracy rooted in care and connection. Negowetti has previously held several academic positions at Harvard Law School, and her interdisciplinary scholarship has been published by the Brookings Institution and in journals. She earned her JD from the Franklin Pierce Law Center (University of New Hampshire School of Law); her MA in peace and development studies from the University of Limerick, Ireland; and her BA in political science and philosophy from the University of Scranton.